Annals of Mathematics Studies

Number 73

LIE EQUATIONS
VOLUME I: GENERAL THEORY

BY

ANTONIO KUMPERA AND
DONALD SPENCER

PRINCETON UNIVERSITY PRESS
AND
UNIVERSITY OF TOKYO PRESS

PRINCETON, NEW JERSEY
1972

LC Card: 77-39055

ISBN: 0-691-08111-5

AMS 1970: 58H05

Published in Japan exclusively by
University of Tokyo Press;
in other parts of the world by
Princeton University Press

Printed in the United States of America

FOREWORD

In his papers [13(a), (b)], Spencer developed a general mechanism for the local deformation of structures on manifolds defined by transitive continuous pseudogroups. A new version of this theory, based on the differential calculus in the analytic spaces of Grothendieck, has been given by Malgrange [9(c)] in his proof of the integrability (existence of local coordinates) of almost-structures defined by elliptic transitive continuous pseudogroups (or elliptic Lie equations), under a certain integrability condition.

The authors here redevelop the theory by two different approaches. The starting point of one approach is based on the idea of B. Malgrange in which the jet sheaves and the operators on them are defined by factoring sheaves and operators defined on the product of the manifold with itself modulo powers of the ideal defining the diagonal. This idea is exploited in a completely systematic way, which brings computational simplicity. The second approach is developed in the context of derivations where the theory finds its most natural expression. The equivalence of the two approaches is shown.

In Chapter I we define the various jet sheaves and the linear complexes and give a brief outline of formal integrability for partial differential equations which is essentially borrowed from Malgrange [9(a), (b)]. The theory is described on the sheaf level and the only essential innovation (better retrogression) is a direct proof of the exactness of the first linear complex without using δ. We show that D is equivalent to the sheaf map (on the level of germs) associated to the bundle map $J_k(\Lambda^p T^*) \to J_{k-1}(\Lambda^{p+1} T^*)$ defined by $j_k\omega(x) \mapsto j_{k-1} d\omega(x)$. The exactness of the ensuing vector bundle complex is then a fibrewise problem which transcribed in coordinates is simply the Poincaré lemma for exterior differential forms with polynomial coefficients.

In Chapter II we define linear Lie equations. The compensated bracket $[\![\ ,\ ]\!]$ in $\underline{J_k T}$ or $\underline{\tilde{J}_k T}$ is defined following Malgrange [9(c)] and we prove that Lie equations are invariant by prolongation.

In Chapter III we introduce the compensated bracket $[\![\ ,\ ]\!]$ for vector valued differential forms (jet forms) and we prove the main identities which relate this bracket with the linear operator $\tilde{D}$. The definition of $[\![\ ,\ ]\!]$ follows a pattern similar to the original definition given by Spencer in [13(a)].

In Chapter IV we describe the non-linear complexes which are finite forms of (the initial portions of) the linear ones and we prove the main non-linear identities. The non-linear operator $\tilde{\mathfrak{D}}$ is defined following Spencer's idea [13(c)] of *twisting* d which, in the present context, appears as the twisting of the vector 1-form χ by the representation $\mathcal{A}d$.

In Chapter V the theory is transformed into the context of derivations of sheaves of jet forms, a technique extensively used in the original papers [13(a), (b)]. Many *ad hoc* definitions and constructions of earlier chapters appear naturally in this setting. For example, the twisting of χ appears in this context as the twisting of the exterior differential operator d. This chapter is, in fact, the best expression of our work and provides additional insight.

The appendix gives an introduction to Lie groupoids. We hope this will benefit the readers who are acquainted with the work of Ehresmann [2] and the pointwise jet theory. Some constructions of Chapter IV are presented in this context.

Let us finally make a few comments on the analogies and differences between Malgrange's approach [9(c)] and the present one. To start with, the linear complexes as well as the compensated bracket in $\underline{\check{J}_k T}$ are obviously the same. The difference begins in the definition of the compensated bracket in $\Lambda \underline{T}^* \otimes_{\mathcal{O}} \check{\mathcal{J}}_k T$ $(\Lambda(\underline{J^0 T})^* \otimes (\underline{J^k T} + \underline{\tilde{J}^k T})$ in Malgrange). The two approaches are entirely different. It is possible to relate the two previously mentioned spaces by a "transposition" procedure which, roughly speaking, consists in transposing horizontal forms into vertical ones on X^2. However, one bracket does not transpose into the other. In fact, our representation $\mathcal{ad}$ does not agree with the corresponding transposed operation, a certain Lie derivative, in Malgrange, since $\mathcal{ad}$ is not left $\mathcal{O}_X$-linear on $\wedge\underline{T}^* \otimes_{\mathcal{O}} \check{\mathcal{J}}_k T$ but only on $\wedge\underline{T}^* \otimes_{\mathcal{O}} \tilde{\mathcal{J}}_k T$. It is not easy to pinpoint the precise difference in the two definitions. Our approach somehow discards the higher order terms, in the covariant part, which are present in Malgrange's treatment due to a differential d'. However, the two brackets agree, by transposition, on a sufficiently large domain, namely $(\wedge\underline{T}^* \otimes_{\mathcal{O}} \tilde{\mathcal{J}}_k T) \oplus R\mathcal{X}_k$. This accounts for the fact that the two approaches will eventually meet at the Buttin formula (22.12), hence the non-linear operators $\tilde{\mathfrak{D}}$ and $\tilde{\mathfrak{D}}_1$ will be the same. Another substantial difference is the following: Malgrange claims that his bracket is invariant by $\tilde{F}^{-1}$ which means, after transposition, that our bracket should be invariant by $\mathfrak{A}dF$. This however is not the case. In our approach, the invariance only holds for elements in a certain subsheaf which, however, is large enough to meet all the requirements of the theory(*).

Finally, we do not claim to give complete references for all the known results in the text. The reader can find many references and an extensive bibliography in Spencer's survey article [13 (d)].

The authors are greatly indebted to B. Malgrange whose ideas provide the foundation for part of this monograph. The second-named author is especially indebted to him for many suggestions communicated in a correspondence extending over two years; in fact, the starting point of this work was a manuscript which was originally intended as a collaboration of B. Malgrange and the second-named author. The second author is also grateful to C. Buttin for many helpful discussions and clarifications.

(*) During the course of proof-reading this manuscript, a definitive version of Malgrange's work [9 (c)] appeared, namely "Equations de Lie," Université de Grenoble (to appear in the *Journal of Differential Geometry*), which is listed as [9(d)] in the references.

The authors(**) also wish to express their acknowledgements to C. M. de Barros and T. Klemola for many helpful discussions on the subject of this monograph.

(**) The first-named author was partially supported by the National Research Council of Canada, grant A-5604, and the second-named author by National Science Foundation Grant GP-31917X.

GLOSSARY OF SYMBOLS

$E, \underline{E}, \mathcal{E}, \mathcal{E}(a), \mathcal{O}, \mathcal{O}_X$	49
$\underline{\phi}: \underline{E} \to \underline{F}$	49
I_a, I_a^{k+1}	50
$\mathcal{O}_\Delta, \mathcal{I}, \mathcal{I}^{k+1}, \mathcal{I}_k$	53
Δ	52
$J_kE, \underline{J_kE}$	51
$J_kX, \underline{J_kX}$	136
$\mathcal{J}_kE, \mathcal{G}_kE$	56, 91
$\mathcal{J}'_kE$	57
$\mathcal{J}_kE = \mathcal{J}_k \otimes_{\mathcal{O}} \mathcal{E}$	59
$\mathcal{J}'_kE = \mathcal{E} \otimes_{\mathcal{O}} \mathcal{J}_k$	60
$\mathcal{E} \otimes_{\mathcal{O}} \mathcal{J}_k \otimes_{\mathcal{O}} \mathcal{F}$	61
$\mathcal{C}_k^{p+1}\mathcal{E}$	75
$\mathcal{C}_\ell^{p+1}\mathcal{R}_k$	86
$\tilde{\mathcal{C}}_k^p T$	202
$\tilde{\mathcal{C}}_{k+1}^\circ T$	203, 242
$\check{\mathcal{J}}_kT, \tilde{\mathcal{J}}_kT, \mathcal{J}_kT$	91
$\tilde{\mathcal{J}}_\infty T$	252
$\mathcal{R}_k$	79, 87, 211
$\mathcal{R}_{k+\ell}$	80, 87
$\mathcal{S}$	82
$\mathcal{G}_\ell, \mathcal{G}$	83
g_ℓ, g	84
$H^q_m\mathcal{G}$	83
$H^q_m g$	84
Π_1, Π_2	54
$\Pi_kX, \underline{\Pi_kX}$	136, 141
Γ_kX	137, 141, 144, 156
F_k, F'_k	57
$\tilde{\mathcal{C}}^\infty X^2$	140
$\mathcal{C}^\infty X$	136
$\mathcal{C}^\infty_V X^2$	138
$\mathcal{A}ut\ X$	136
$\mathcal{A}ut_V X^2$	139
$\widetilde{\mathcal{A}ut}\ X^2, \widetilde{\mathcal{A}ut_V} X^2, \overline{\mathcal{A}ut_V} X^2$	141
$\mathcal{F}(X), \mathcal{X}(X), A^r(X), B^r(X)$	104
$\check{\Phi}^r, \tilde{\Phi}^r, \check{\Phi}, \tilde{\Phi}$	109
$\Phi^r_V, \Phi^r_{PH}, \Phi_{PH}$	110
Ψ^ℓ, Ψ	116
$\mathcal{X}(X^2), \check{\mathcal{X}}(X^2)$	89, 99
$\tilde{\mathcal{X}}(X^2), \mathcal{X}(X^2)_{PH}, \mathcal{X}(X^2)_{PV}$	90
$\mathcal{X}(X^2)_V, \mathcal{X}(X^2)_H$	89
$\check{\Theta}, \tilde{\Theta}, \Theta_V, \Theta_{PH}, \Theta_{PV}$	90

$\Gamma(X, \tilde{\mathcal{J}}_k T)$, $\Gamma_c(X, \tilde{\mathcal{J}}_k T)$, $\Gamma_a(X, \Gamma_k X)$	144
$\Gamma(X, \mathfrak{g})$, $\Gamma_c(X, \mathfrak{g})$	264
$F: (x,y) \mapsto (f(x), \varphi(x,y))$, $f(x) = \varphi(x,x)$	139
$[F]_{(x,y)} \bullet [\xi]_{(x,y)}$	157
$F\bullet$, $F^*\bullet$	160, 229, 230
${}_uF$	165, 185, 211
Der $\mathcal{B}_k(X)$	221
Der A(X), Der^pA(X), Der A(U)	105
$\mathcal{D}er\ \Lambda\underline{T}^*$	107, 246
$\mathcal{D}er\ \Psi$	116
$\mathcal{D}er_{\mathbf{R},a}\mathcal{J}_k$	217
$\mathcal{D}er_a\mathcal{B}_k(X)$	216, 222
$\mathcal{D}er_\sigma\mathcal{B}_k(X)$	237
$\mathcal{D}er_\Sigma\mathcal{B}_k(X)$	240, 245
$\mathcal{D}er^r_{\Sigma,k}$	242, 245
$\mathcal{D}er_{\Sigma,\infty}$	252
Der $\Gamma(X, \mathcal{B}_k X)$	215
$\mathrm{Der}_a\Gamma(X, \mathcal{B}_k X)$	216
$\mathcal{A}^r_k(X)$, $\mathcal{B}^r_k(X)$	213
$\mathcal{A}lt^r_{\mathcal{O}}(\tilde{\mathcal{J}}_k T, \mathcal{J}_k)$	212
$\Lambda(\tilde{\mathcal{J}}_k T)^* \otimes_{\mathcal{O}} \mathcal{J}_k$	213
$\mathcal{O}^r_k(X) = \mathcal{A}lt^r_{\mathcal{O}}(\tilde{\mathcal{J}}_k T, \tilde{\mathcal{J}}_k T) \simeq \Lambda^r(\tilde{\mathcal{J}}_k T)^* \otimes_{\mathcal{O}} \tilde{\mathcal{J}}_k T$	214
i_k, j_k	51, 53
$\tilde{j}_k$	91
j_k	136, 141, 142
$\tilde{j}_k$	142
$\nu_{k+\ell}$, $J_k\phi$	52
ρ_k	52, 69
$\tilde{\rho}_h$, $\breve{\rho}_h$	92
α_k	136
β_k	54, 136
a_k, b_k	137
r, r_k	54
ψ_k, ψ'_k	60
ε_k, ζ_k	70, 92
$\iota_{k+\ell}$, $\mathrm{Id} \otimes \phi$	77
$p_\ell\phi$	79
δ	74
$\underline{\delta}$	71, 72, 74, 189
$u \barwedge v$	104
i(u), i	105, 223
L(u), L	118
$\mathcal{L}(u)$, $\mathcal{L}$	106, 116, 117, 226
$\mathcal{K}(u)$, $\mathcal{K}$	231
$ad(u)$, ad	117,123,135,173,239,241
ad_∞	253
$\mathcal{A}d$ F, $\mathcal{A}d$	170,172,178,182,209,248
Exp	144,147,151,154,189,261
$\sim_k$	138
Ω	126
χ	126

χ_∞	256
Ξ_∞	254
Σ_∞	255
V, H, T_V, T_H	64
d	105, 220
d_H	65, 125, 126, 162
d_V	69, 125
d_∞	256
D	66, 67, 68, 71, 75, 128
D′	76
$\tilde{D}$	114, 128, 134
$\tilde{D}_\infty$	253
$\tilde{D}'$	207
$\hat{D}$	233, 245
$\hat{D}_\infty$	254
$\mathfrak{D}$	159, 167, 169
$\mathfrak{D}_1$	162
$\tilde{\mathfrak{D}}$	174, 176, 183
$\tilde{\mathfrak{D}}_1$	183, 186, 187
$\tilde{\mathfrak{D}}'$	204, 210, 211
$\mathfrak{D}'_1$	206, 211
$\hat{\mathfrak{D}}$	247, 250, 252
$\hat{\mathfrak{D}}_1$	247, 252
$\tilde{\mathfrak{D}}_\infty$, $\tilde{\mathfrak{D}}_{1,\infty}$, $\hat{\mathfrak{D}}_\infty$, $\hat{\mathfrak{D}}_{1,\infty}$	256
{ , }	120
[,]	88, 93, 102, 128
[,]	94, 95, 100, 102, 114, 117, 128, 230, 260
⟦ , ⟧	95, 100, 102
$[\ ,\ ]_\infty$, $\mathbf{[\ ,\ ]}_\infty$	253

TABLE OF CONTENTS

LIE EQUATIONS

INTRODUCTION

As the title indicates, the content of these notes is a lengthy construction of techniques devised to study specific differential geometric problems some of which we expect to treat in Part II. In this introduction we state our main objectives and illustrate by examples some of their geometric implications. A fairly detailed summary of the present monograph is given in [43]. The references will provide further motivation.

The main goal of these notes is the construction of the non-linear complexes $(23.9)_{k+1}$, $(25.3)_{k+1}$ and $(30.1)_{k+1}$ as well as their linearizations $(6.2)_{k+1}$, $(7.1)_{k+1}$ and the first two lines of $(29.6)_k$ (more precisely $(23.18)_{k+1}$, $(25.7)_{k+1}$ and the first five terms of $(29.6)_k$). To each non-linear complex are attached some fundamental identities and a structure equation. Similarly, to each linear complex are attached some fundamental identities which are, in most cases, infinitesimal counterparts of the non-linear ones and can be obtained from these by differentiation. The construction process is however reversed. It is quite clear what the first linear complex should be. The operators involved should measure the deviation of a section in a jet bundle from being holonomic. It turns out that these operators are extensions of the ordinary exterior differentiation d. The second linear complex is a quotient of the first one. Restricted to an elliptic Lie equation it gives a resolution of the sheaf of solutions of the given equation and generalizes the classical resolution of de Rham. The theory of harmonic differential forms extends to general elliptic systems (satisfying some regularity conditions: see Spencer [13(a), (b)]). In particular, the resolution associated to the equation $\frac{\partial f}{\partial x_i} = 0$, f a real valued function (i.e., $df = 0$), identifies with the classical resolution of de Rham. On the contrary, there are many non-linear complexes which are finite forms of the initial portion of the above mentioned complexes. For example, Guillemin and Sternberg [6] and more recently Quê [10(b)] have approached deformation theory using the first non-linear complex $(21.8)_{k+1}$. Malgrange [9(c), (d)] based his proof of integrability for elliptic almost-structures on his version of the third non-linear complex $(25.3)_{k+1}$. The first and second non-linear complexes are finite forms of the first linear complex. The third non-linear complex, which is essentially obtained by quotienting the second non-linear complex, is a finite form of the second linear complex. Another finite form of the latter is a non-linear complex obtained by quotienting the first non-linear complex (which is not considered in the present notes). The $\hat{\mathfrak{D}}$-complex discussed in Chapter V is a finite form of the restricted $\hat{D}$-complex. The former is canonically isomorphic to the third non-linear complex and the latter is isomorphic to the second linear complex.

A basic idea in the construction of the non-linear complexes consists first in expressing the linear operator D as the *adjoint* of a certain element with respect to a bracket and then in defining a finite form for this adjoint representation. Here is a trivial example. Let ξ be a fixed vector field and consider the first order linear operstor $D\colon \theta \in \mathfrak{X}(X) \mapsto \mathfrak{L}(\theta)\xi \in \mathfrak{X}(X)$. Then $D\theta = [\theta, \xi] = \mathrm{ad}\theta(\xi)$, a finite form of $\theta \mapsto \mathrm{ad}\theta$ is $f \mapsto f_*^{-1}$ and a finite form of D is $\mathfrak{D}\colon f \mapsto f_*^{-1}\xi - \xi$. In fact $D\theta = \frac{d}{dt}\mathfrak{D}(\exp t\theta)|_{t=0}$. In Theorem 2 of Section 16 we show that the operators D of the first linear complex satisfy the equation $Du = [\chi_k, u]$ with respect to the Nijenhuis bracket of vector forms (or rather, a straightforward extension to *jet forms*), hence D is the adjoint of χ_k and, correspondingly, the non-linear operator $\mathfrak{D}$ is defined by $F \mapsto \chi_k - F^* \bullet \chi_k$. The same theorem shows that $\tilde{D}$, a simple shifting of D to an isomorphic sheaf, satisfies the equation $\tilde{D}u = [\![\chi_k, u]\!]$ with respect to the compensated bracket obtained by adding two extra terms to the Nijenhuis bracket. The corresponding non-linear operator $\tilde{\mathfrak{D}}$ is then defined by $F \mapsto \chi_k - \mathfrak{A}dF \bullet \chi_k$ where

$F \mapsto \mathfrak{A}dF$ is a finite form of $\Xi \mapsto ad\Xi$, $\Xi \in \tilde{\mathfrak{J}}_{k+1}T$, $F \in \Gamma_{k+1}X$ and $\tilde{D}\Xi = -[\![\Xi, \chi_k]\!] = -ad\Xi(\chi_k)$. Similarly, the operator $\hat{\mathfrak{D}}$ is defined by the twisting of d, namely $F \mapsto d - \mathfrak{A}dF \bullet d$. The non-linear operators $\mathfrak{D}_1$, $\tilde{\mathfrak{D}}_1$ and $\hat{\mathfrak{D}}_1$ are defined in such a way that $\mathfrak{D}_1 \circ \mathfrak{D} = 0$ etc., are restatements of the structure equations.

Let us now say a few words about linear Lie equations. These are linear differential equations whose formal solutions (of order k) are endowed with a Lie bracket induced by a canonically defined bracket on the sheaf $\underline{J_kT}$, where T is the tangent bundle of the manifold X. The classical construction goes like this. Let P_k be the principal bundle of k-th order (holonomic) frames and J_kT the vector bundle of k-jets of vector fields. Libermann [46(c)] pointed out that the standard prolongation, to P_k, of vector fields θ on X yields a canonical isomorphism of the fibre $(J_kT)_y$ with the tangent space T_AP_k where $y = \beta(A) =$ target of A. From this we infer that the R-vector space $\Gamma(X, J_kT)$ of global sections of J_kT is canonically isomorphic to the vector space of all right-invariant vector fields on P_k (with respect to the right action of the structure group G_k). The bracket of invariant vector fields transports to a bracket on $\Gamma(X, J_kT)$ and defines a Lie algebra structure on the sheaf $\underline{J_kT}$. In this monograph we indicate the bracket on $\underline{J_kT}$ by $[\![\ ,\ ]\!]$. If θ and θ' are vector fields on X then $[\![j_k\theta, j_k\theta']\!] = j_k[\theta, \theta']$. More details on the previous classical construction can be found in Section 35, example b, of this monograph. In Chapter II the bracket $[\![\ ,\ ]\!]$ is introduced by a completely different approach. It is worthwhile to observe that this bracket has been extensively used by Sophus Lie. Let H_k be a G-structure of order k on X where G is a Lie subgroup of G_k (for simplicity, the reader can assume that $k = 1$, i.e., H is a G-structure in the usual sense with $G \subset GL(n, R)$). The aforementioned isomorphisms associate to H_k a vector sub-bundle $R_k \subset J_kT$, i.e., a homogeneous linear differential equation of order k, where $(R_k)_y$ corresponds to the subspace T_AH_k, hence $R_k \simeq TH_k/G$. Sections of R_k

correspond to right-invariant vector fields on P_k which are tangent to H_k, hence correspond to all the right-invariant vector fields on H_k. We infer that $\underline{R_k}$ is a Lie subalgebra of $\underline{J_k T}$, i.e., a Lie equation. Any G'-structure H'_k similar to H_k (obtained by right translation on P_k, the group G' being a conjugate of G) has the same Lie equation. Moreover, the Lie algebra of all right-invariant vector fields on P_k which are tangent to H_k induces an integrable field of contact elements. If G and X are connected then the maximal integral leaves are precisely the G'-structures similar to H_k (otherwise, these are unions of leaves). The converse statement requires some regularity conditions on R_k. We conclude, therefore, that a Lie equation has a more intrinsic nature than any of the corresponding similar G-structures. The solutions of R_k, i.e., the vector fields θ such that $j_k\theta \subset R_k$, are the infinitesimal automorphisms of H_k. They are characterized by the property that the standard prolongation $\mathfrak{p}_k\theta$ to P_k is a right-invariant vector field which is tangent to the submanifold H_k. If H_k is integrable (has local coordinates) then the vector fields associated to these coordinates are solutions of R_k. In many instances the Lie equation can be constructed as follows. Assume that $k = 1$ and let H be a G-structure associated to a tensor field τ on X. H is the set of all frames (admissible frames) which transform τ into a typical tensor τ_0 on the vector space R^n (see examples below). The corresponding Lie equation R is of order 1 and is given by

$$R = \{j_1\theta(x) \in J_1T \mid \mathfrak{L}(\theta)\tau(x) = 0\} .$$

Other examples of linear Lie equations are the defining equations for the infinitesimal transformations of a Lie pseudogroup Γ. Non-linear Lie equations are finite forms of linear Lie equations. Examples of these are the Lie groupoids associated to G-structures and the non-linear differential equations defining the finite transformations of a pseudogroup. In general, non-linear Lie equations arise by exponentiating linear Lie equations.

In this monograph we consider Lie equations on the sheaf level $\underline{R_k}$ and denote them by $\mathcal{R}_k$. Similarly, the non-linear Lie equations are considered as sub-sheaves of the sheaf of germs of admissible local α-sections of $\Pi_k X$, the groupoid of invertible k-jets.

We now examine two geometric problems for which these techniques have been devised.

A. INTEGRABILITY OF LIE STRUCTURES

Let $\mathcal{R}_k$ be a linear Lie equation on the manifold X and denote by $\Gamma(\mathcal{R}_k)$ the corresponding non-linear Lie equation. The integrability problem for a $\Gamma(\mathcal{R}_k)$-structure on a manifold Y (for the definition see [13(a)] and [9(c)] Def. (6.17) or [9(d)] Def. (8.1)) can be formulated in such a way that its solution involves essentially two steps. The first one consists in proving formal (power series) integrability of the structure and is provided by the integrability condition which involves the operator $\tilde{\mathfrak{D}}'$ (cf. [9(c)] Lemma (6.18) or [9(d)] Prop. (8.2) where $\hat{\mathfrak{D}}$ stands for $\tilde{\mathfrak{D}}'$). The second and usually difficult step consists in proving (actual) integrability (cf. [9(c)] Def. (6.20) or [9(d)] Def. (8.4)). It amounts to integrating the non-linear equation defining the Lie structure. The integrability of this equation is shown to be equivalent to the exactness of the third non-linear complex restricted to the non-linear Lie equation $\Gamma(\mathcal{R}_k)$. The latter is an integrability problem which is *elliptic* when the equation $\Gamma(\mathcal{R}_k)$ (or $\mathcal{R}_k$) is elliptic. A special case of Lie structures is the almost Γ-structure where Γ is a continuous pseudogroup of local transformations on X (cf. [13(a), (b)]). In this case, the linear Lie equation $\mathcal{R}_k$ (k the order of Γ) is the linear partial differential equation defining the infinitesimal transformations of Γ. The non-linear Lie equation $\Gamma(\mathcal{R}_k)$ is the k-th order groupoid (or rather the sheaf of germs of admissible local α-sections of this groupoid) defining the finite transformations of Γ. An integrable almost Γ-structure is known in literature as a Γ-structure. For example, let X be given an almost-complex structure (see example 7 below). Let Γ be the pseudogroup of local biholomorphic transformations of $C^n = R^{2n}$, denote by G the real form of the complex group GL(n, C) and by $\mathfrak{g}$ its Lie algebra (the real form of $\mathfrak{gl}(n, C)$). Γ is defined as the set of local transformations $f: R^{2n} \to R^{2n}$ such that $f_*(x) \in G$ ($f_*(x)$ preserves the complex structure of the tangent spaces), i.e., the set of local transformations which are solutions of the non-linear Lie equation $\Gamma(R_1) = \{\lambda \in \Pi_1(R^{2n}) \mid \lambda \in G\}$ where $\Pi_1(R^{2n})$ is the manifold of invertible 1-jets of R^{2n} and these identify with the invertible linear transformations $T_x R^{2n} = R^{2n} \to T_y R^{2n} = R^{2n}$. $\Gamma(R_1)$ is simply the Cauchy-Riemann equation together with a non-vanishing condition on the jacobian matrix. The associated linear Lie equation R_1 is again the Cauchy-Riemann equation for vector fields on R^{2n} namely $\xi_*(x) \in \mathfrak{g}$. Its solutions are the infinitesimal automorphisms of the complex structure of C^n or, equivalently, the holomorphic vector fields. Let Λ be the set of all invertible 1-jets $\mu \in \Pi_1(R^{2n}, X)$ which are complex linear isomorphisms of $T_x R^{2n} = C^n$ onto $T_y X$ where x is the source and y is the target of λ. The groupoid $\Gamma(R_1)$ operates to the right on Λ and the action is of principal type. The almost-complex structure is entirely determined by the equation Λ together with the action of $\Gamma(R_1)$, i.e., by the almost Γ-structure Λ. Other examples of elliptic almost-structures (ellipticity assumption on the linear Lie equation) are those associated to any complex transitive continuous pseudogroup. We find in this class the G-structures where G is any (real form of a) complex Lie subgroup of $GL(n, C)$. Almost-structures associated to transitive continuous pseudogroups of finite type are also elliptic. In particular we find the G-structures of finite type such as the Riemannian or Lorentz structures and the conformal structures $(n > 2)$. A G-structure is elliptic if and only if the Lie algebra of G does not contain any real subalgebra generated by elements of rank 1.

We shall now discuss several examples of integrability. Some of these do not fit properly into the scope of the present general mechanism. Nevertheless, they have been included in order to give a wider view which is by no means exhaustive.

Generally speaking, a structure defined on a differentiable manifold X is said to be integrable if for any point $x \in X$ there is a coordinate neighborhood such that the structure, expressed in terms of these coordinates, becomes a *standard* structure on the numerical space R^n, $n = \dim X$. If H_k is a G-structure of order k then the standard structure is the flat G-structure on R^n and the integrability is equivalent to the existence of local coordinates whose associated fields of k-frames are local sections of H_k. If the structure is defined by a tensor field then integrability is equivalent to the existence of local coordinates such that the tensor field, expressed in these coordinates, becomes at each point a fixed tensor on the vector space R^n. In the general case of an almost Γ-structure, integrability is equivalent to the existence of a Γ-structure, i.e., an atlas on X with the property that the change of overlapping charts is given by elements of Γ, which induces the given almost-structure. If Γ operates (as usual) in Euclidean space, the standard almost Γ-structure is given by the k-th order groupoid of Γ where k is greater or equal to the order of Γ. For a structure to be integrable it must satisfy some necessary conditions referred to as *integrability* conditions. The problem of integrability of structures consists in proving that these conditions are also sufficient.

EXAMPLE (1): *The Frobenius theorem*

Let Σ be a field of contact elements of dimension p on the manifold X, i.e., for each $x \in X$ we are given a p-dimensional subspace Σ_x of T_xX (all the data are always assumed to be differentiable). The standard structure in R^n is given by the field of p-planes which are parallel to the plane of the first p coordinates. The equation of this field is $dx_{p+1} = \dots = dx_n = 0$. Σ defines a G-structure H of order 1 whose *adapted* frames at the point x are the frames whose first p vectors are contained in Σ_x and G is the group of all $g \in GL(n, R)$ that preserve the plane of the first p coordinates, i.e., $g = (g^i_j)$ with $g^i_j = 0$ for $p + 1 \leq i \leq n$ and $1 \leq j \leq p$. The integrability of Σ means that for any $x \in X$ there exist local coordinates (U, x_i) such that Σ_y is spanned by $\{\frac{\partial}{\partial x_1}, \dots, \frac{\partial}{\partial x_p}\}$ for any $y \in U$. Equivalently, it means that for any $x \in X$ there exists a p-dimensional integral submanifold Y of Σ containing x, i.e., $T_yY = \Sigma_y$ for any $y \in Y$. There are many equivalent ways of stating the integrability condition. We state a few:

a) For any vector fields θ, θ' belonging to Σ (i.e., $\theta_x \in \Sigma_x$ for any x) the Lie bracket $[\theta, \theta'] = \mathcal{L}(\theta)\theta'$ also belongs to Σ.

b) For any θ belonging to Σ the local transformation $\exp t\theta$ is a local automorphism of Σ in the sense that $(\exp t\theta)_*\Sigma = \Sigma$ (this is a finite form of (a)).

Dual conditions are obtained by replacing Σ with $\Sigma^\perp$ where $\Sigma_x^\perp$ is the annihilator of Σ_x in T_x^*X:

a′) For any vector field θ belonging to Σ and any 1-form ω belonging to $\Sigma^\perp$ the Lie derivative $\mathcal{L}(\theta)\omega$ also belongs to $\Sigma^\perp$.

b′) For any θ belonging to Σ, $\exp t\theta$ is a local automorphism of $\Sigma^\perp$.

c) For any 1-form ω which vanishes on Σ (i.e., a section of $\Sigma^\perp$) the 2-form $d\omega$ also vanishes on Σ.

d) The pre-sheaf $\Gamma_{loc}\Sigma^\perp$ generates an ideal closed under d.

The sufficiency of these integrability conditions is the classical Frobenius theorem. The associated Lie equation R_1 is equal to the set of 1-jets $j_1\theta(x)$ such that $\mathcal{L}(\theta)\xi(x) \in \Sigma_x$ for any ξ belonging to Σ.

EXAMPLE (2): *Fields of p-frames, parallelism*

Consider the structure defined on X by p vector fields $\theta_1, \ldots, \theta_p$ which are assumed to be linearly independent at each point. The corresponding G-structure H is composed of the frames whose first p vectors are the vectors induced by the θ_i and G is composed of the matrices $g = (g_j^i)$ with $g_j^i = \delta_j^i$ for $1 \leq i \leq n$ and $1 \leq j \leq p$. Integrability means that there are local coordinates (U, x_i) such that $\theta_i|_U = \dfrac{\partial}{\partial x_i}$, $1 \leq i \leq p$. The integrability condition is given by $[\theta_i, \theta_j] = 0$, $1 \leq i, j \leq p$, and the sufficiency of this condition is essentially again the Frobenius theorem. The Lie equation is given by $R_1 = \{j_1\theta(x) \mid \mathcal{L}(\theta)\theta_i(x) = 0, 1 \leq i \leq p\}$. If $p = 1$, the integrability condition is void (always satisfied). In fact, any non-singular vector field θ can always be written locally as $\theta = \dfrac{\partial}{\partial x_1}$. If $p = n$, the structure is known as a parallelism on X.

EXAMPLE (3): *Riemannian structures*

Let X be given a Riemannian structure defined by a metric tensor g. The corresponding G-structure H is composed of the orthonormal frames and $G = 0(n)$. The standard structure in R^n is defined by the Euclidean scalar product. Integrability means there are local coordinates (U, x_i) such that $g|_U = \Sigma dx_i^2$, or equivalently, such that $\{\frac{\partial}{\partial x_i}\}$ is an orthonormal frame. The integrability condition is given by the vanishing of the curvature tensor

$$R(\xi, \eta) = [\nabla_\xi, \nabla_\eta] - \nabla_{[\xi, \eta]}$$

where ∇_ξ is the covariant derivative along the vector field ξ (with respect to the Levi-Civita connection). The sufficiency of this condition is again essentially the Frobenius theorem. The Lie equation is defined by $R_1 = \{j_1\theta(x) \mid \mathfrak{L}(\theta)g(x) = 0\}$.

EXAMPLE (4): *Linear connections*

Let X be given a linear connection, i.e., a law of derivation $\xi \mapsto \nabla_\xi$. The standard structure on R^n is the flat (Euclidean) connection where $\nabla_\xi \eta = (\mathfrak{L}(\xi)\eta_i)\frac{\partial}{\partial x_i}$ and $\eta = \eta_i \frac{\partial}{\partial x_i}$. The structure is integrable if there are local coordinates (U, x_i) such that $\nabla_\xi|_U$ has the above *flat* expression. In terms of the frame bundle P and the linear connection given by a field of horizontal contact elements Σ, it means that the fields of frames associated to these coordinate systems are integral sections of Σ. The integrability condition is expressed by the vanishing of the curvature tensor and the torsion tensor

$$T(\xi, \eta) = \nabla_\xi \eta - \nabla_\eta \xi - [\xi, \eta] .$$

The sufficiency of the above conditions results by a double application of Frobenius' theorem.

EXAMPLE (5): *Canonical differential forms*

We recall some classical theorems on the reduction of exterior differential forms to canonical expressions.

a) Let Ω be a volume element on X. The standard volume element in R^n is $dx_1 \wedge \ldots \wedge dx_n$. The structure (X, Ω) is always integrable since for any $x \in X$ there is a coordinate system (U, x_i) such that $\Omega|_U$ has the desired expression. R_1 is equal to $\{j_1\theta(x) \mid div_\Omega \theta = 0\}$.

b) Let Ω be an exterior $(n-1)$-form which is everywhere non-zero. In R^n the standard $(n-1)$-form is, for example, $dx_2 \wedge \ldots \wedge dx_n$. The integrability condition is $d\Omega = 0$. Observe that if $(d\Omega)_x \neq 0$ then, in the neighborhood of x, Ω has the canonical form $x_1 dx_2 \wedge \ldots \wedge dx_n$.

c) Let Ω be an exterior 2-form of constant rank $2p$ (the rank being always even). The canonical form is $\sum_{1 \leq i \leq p} dx_i \wedge dx_{p+i}$ and the integrability condition is $d\Omega = 0$. If *rank* Ω = *dim* X, the structure is called almost-symplectic. In this case the G-structure H is composed of the symplectic frames (with respect to Ω) and G is the symplectic subgroup $Sp(2p)$ of $GL(2p, R)$. If Ω is integrable then the structure is called symplectic or Hamiltonian.

d) Let Ω be a differential 1-form on X and assume that Ω is everywhere non-zero. Consider the sequence of forms $\Omega, d\Omega, \Omega \wedge d\Omega, \ldots, (d\Omega)^p, \Omega \wedge (d\Omega)^p, \ldots$. The form Ω is said to be of class $2p$ (in the sense of Darboux) in a neighborhood U of a point x if $(d\Omega)^p \neq 0$ at every point of U and $\Omega \wedge (d\Omega)^p = 0$ on U; Ω is of class $2p+1$ if $\Omega \wedge (d\Omega)^p \neq 0$ and $(d\Omega)^{p+1} = 0$ at every point of U. The standard form in R^n of class $2p$ is $\sum_{1 \leq i \leq p} x_i dx_{p+i}$ and of class $2p+1$ is $dx_1 + \sum_{1 \leq i \leq p} x_{i+1} dx_{i+p+1}$. The integrability condition is given by the Darboux class of Ω. If Ω is everywhere of class $2p$ (or $2p+1$) then in the neighborhood of each point it can be expressed as the standard form with respect to some coordinate system. If *class* Ω = *dim* X then (X, Ω) is called a contact structure (even or odd according to the parity of *class* Ω). The standard odd contact structure on R^{2p+1} is given by $dx_1 + \sum_{1 \leq i \leq p} x_{i+1} dx_{i+p+1}$ and the standard even contact structure on R^{2p} is given by $\sum_{1 \leq i \leq p} e^{x_i} dx_{i+p}$. More generally, a contact structure on X is given by a covering $(U_\alpha, \Omega_\alpha)$ of X such that *class* Ω_α = *dim* X for all α (i.e., each Ω_α defines a local contact structure on the open set U_α) and $\Omega_\alpha = f_{\alpha\beta}\Omega_\beta$ on the overlap. Let $\Sigma = \ker \Omega_\alpha$. Σ is a field of hyperplanes. It can be proved ([29], [42]) that a field of hyperplanes Σ defines an even (resp. odd) contact structure if and only if the dimension of the Cauchy characteristics of Σ is 1 (resp. 0). This dimension at the point x is given by $\dim \ker (d\Omega|\Sigma_x) = \dim \{v \in \Sigma_x | i(v)d\Omega_x \equiv 0 \mod \Omega_x\}$ where Ω is a local basis of $\Sigma^\perp$. If the structure is defined by a global form Ω then X is clearly orientable. An odd contact structure is defined by a global form Ω if and only if X is orientable ([29], [47]). The same does not hold for even contact structures.

The previous results are due to Darboux. (a) and (b) are trivial requiring only quadratures whereas (c) and (d) are rather clever applications of Frobenius' theorem. A few things in the way of canonical expressions can also be stated for degrees other than those considered ([15], [27], [28], [60]).

EXAMPLE (6): *Almost-product structure*

Let $X = Y \times Z$ be a product of two manifolds. TX is canonically isomorphic to $TY \times TZ$ and each tangent space has a splitting in two subspaces, namely the null spaces of the projections. X is locally isomorphic to $R^p \times R^q$, $p = \dim Y$ and $q = \dim Z$. Conversely, let X be given a field (Σ, Ξ) of supplementary contact elements, i.e., $T_xX = \Sigma_x \oplus \Xi_x$ and assume that each field Σ and Ξ is of constant dimension, $\dim \Sigma = p$ and $\dim \Xi = q$ with $p + q = n$. This structure, introduced by Schouten [57] and Nickerson-Spencer [52] (see also [31]) is called an almost-product structure. The standard structure in R^n is given by the field of supplementary planes, the first being parallel to the plane of the first p coordinates and the second being parallel to the plane of the last q coordinates. (Σ, Ξ) defines a G-structure H of order 1 whose *adapted* frames at x are those whose first p vectors belong to Σ_x and whose last q vectors belong to Ξ_x and G is the subgroup of matrices (g^i_j) such that $g^i_j = 0$ for $p + 1 \leq i \leq n$, $1 \leq j \leq p$ and

$1 \leq i \leq p$, $p+1 \leq j \leq n$. In other words $G \simeq GL(p, R) \times GL(q, R)$. Integrability means the existence of local coordinates (U, x_i) such that Σ is spanned by $\{\frac{\partial}{\partial x_1}, \ldots, \frac{\partial}{\partial x_p}\}$ and Ξ is spanned by $\{\frac{\partial}{\partial x_{p+1}}, \ldots, \frac{\partial}{\partial x_n}\}$. Equivalently, it means that for each $x \in X$ there exist integral submanifolds of Σ and Ξ of maximal dimension and containing x, hence X is locally a product $R^p \times R^q$. The integrability condition is a Frobenius condition on Σ and Ξ separately. The condition can be written simultaneously as follows. The structure (Σ, Ξ) is equivalent to a field π of projections, i.e., π_x is a projection in $T_x X$ $(\pi_x^2 = \pi_x)$ and *rank* π_x is constant. The integrability condition is the vanishing of the *torsion* tensor $[\pi, \pi]$ defined in terms of the Nijenhuis bracket of the vector 1-form π (Section 13). The Lie equation is equal to $R_1 = \{j_1\theta(x) \mid \mathcal{L}(\theta)\pi(x) = 0\}$. It can also be defined by extending the definition in example (1) in an obvious way. Given an almost-product structure (Σ, Ξ) one might be interested only in partial integrability, i.e., integrability of Σ or Ξ. Let π be the field of projections such that $im\ \pi = \Sigma$, $ker\ \pi = \Xi$ and let $\pi' = Id - \pi$ be the complementary projection. The integrability condition for Σ can also be given by $\pi' \circ [\pi, \pi] = 0$ and for Ξ by $\pi \circ [\pi', \pi'] = 0$ or $\pi \circ [\pi, \pi] = 0$ since $[\pi, \pi] = [\pi', \pi']$ (see [3], p. 358). If X is paracompact, the integrability condition for any field of contact elements Σ on X can always be expressed as above. In fact, let g be a Riemannian metric on X. The field $(\Sigma, \Sigma^{\perp})$ is an almost-product structure (orthogonality with respect to g). Let (Σ, Ξ) be an almost-product structure with associated projection π. This structure can also be defined by a field I of endomorphisms (a vector 1-form) satisfying $I^2 = Id$, the binding relation being $I = 2\pi - Id = \pi - \pi'$. $ker\ \pi$ (resp. $im\ \pi$) is the eigenspace of I corresponding to -1 (resp. 1). The integrability condition is again $[I, I] = 0$. If $dim\ \Sigma = dim\ \Xi$ (hence $dim\ X$ is even) the structure is sometimes called almost-paracomplex. Examples are the connections on TX.

The integrability theorem for almost-product structures has a natural generalization due to Haantjes [34]. Let h be a vector 1-form on X, or equivalently, a field $x \to h_x$ where h_x is an endomorphism of T_xX. Assume that each h_x has a complete real spectrum (i.e., reduces to diagonal form with real eigenvalues) the multiplicity of each eigenvalue being constant. If Σ_λ denotes the field of eigenspaces belonging to λ then every sum $\Sigma = \Sigma_{\lambda_1} + \dots + \Sigma_{\lambda_k}$ is integrable if and only if the tensor $\mathcal{H}$ vanishes where $\mathcal{H}(u,v) = h^2H(u,v) + H(hu,hv) - hH(hu,v) - hH(u,hv)$ and $H = [h,h]$. A simple proof of this statement can be found in [3]. Observe that a projection has the complete real spectrum $\{0,1\}$. Other generalizations will be mentioned in example 10. We shall see there that the vanishing of $\mathcal{H}$ is actually equivalent to the vanishing of H.

Let π be one of the projections associated to an almost-product structure and let d_π ($\mathcal{L}(\pi)$ in the notation of Section 13) be the derivation of type d on $\Lambda\mathfrak{X}(X)^*$ associated to the vector 1-form π. The integrability condition can also be expressed by $d_\pi^2 = 0$ (cf. [31], [52]), i.e., d_π defines a complex. Integrability (existence of local coordinates) is shown to imply the exactness, on the level of germs of a restricted d_π-complex (local d_π-Poincaré lemma) the description of which is given in example 10 (in the special case of an integrable almost-product structure one can actually prove stronger results than those described in example 10 – cf. [31] and [52]). This restricted d_π-complex yields a de Rham type resolution of the sheaf of germs of functions which are constant on the leaves of the integrable field $im\ \pi$.

EXAMPLE (7): *Almost-complex structure*

Let X be a complex manifold, i.e., X is a real manifold covered by an atlas with values in $\mathbb{C}^n \equiv \mathbb{R}^{2n}$ (we consider the identification $(x_1+iy_1,\dots) \mapsto (x_1,\dots,x_n,y_1,\dots,y_n)$) and the change of charts is holomorphic. Since the differential of a holomorphic map is $\mathbb{C}$-linear, we infer that each real tangent space T_xX has a canonical $\mathbb{C}$-vector space structure and consequently a natural orientation. Conversely, let X be a real manifold and assume that each tangent space carries a structure of $\mathbb{C}$-vector space compatible with its $\mathbb{R}$-linear structure and varying smoothly with the point x. The real dimension of X is *a fortiori* even. This structure, introduced by Ehresmann [24 (a)], is called almost-complex. It can be defined in several equivalent ways:

a) The C-linear structure of T_xX is entirely determined when scalar multiplication by $i = \sqrt{-1}$ is known hence an almost-complex structure is also defined by a (real) vector 1-form J on X such that $J^2 = -\mathrm{Id}$ where J(v) stands for iv. It can also be defined by a complex vector 1-form as will be seen below.

b) Let $T_{\mathbf{C}}X$ be the complexified tangent bundle (or equivalently, the bundle of complex tangent vectors, i.e., derivations of C-valued functions on X). The real 1-form J extends to a complex vector 1-form $J_{\mathbf{C}}$ on $T_{\mathbf{C}}X$ which satisfies the equations $(J_{\mathbf{C}})^2 = -\mathrm{Id}$ and $J_{\mathbf{C}}\overline{v} = \overline{J_{\mathbf{C}}v}$ (conjugation under the complexified structure). It follows that $T_{\mathbf{C}}X$ carries two C-vector space structures, one defined by the complexification (hence for any real manifold X) and the other by $J_{\mathbf{C}}$. The two structures permute, i.e., $J_{\mathbf{C}} \circ i = i \circ J_{\mathbf{C}}$. Furthermore, there are two C-linear supplementary and conjugate (under the complexification) subspaces $T_{\mathbf{C}}^{(1,0)}$ and $T_{\mathbf{C}}^{(0,1)}$ characterized by the property that $T_{\mathbf{C}}^{(1,0)}$ is the largest subspace on which both structures coincide, i.e., $T_{\mathbf{C}}^{(1,0)} = \{v \in T_{\mathbf{C}}X \mid iv = J_{\mathbf{C}}v\}$ and $T_{\mathbf{C}}^{(0,1)}$ is the largest subspace on which the two structures are conjugate, i.e., $i = -J_{\mathbf{C}}$. They are in fact the eigenspaces of $J_{\mathbf{C}}$ relative to the only eigenvalues i and $-i$ (with respect to the complexified structure). The corresponding supplementary projections $P = \frac{1}{2}(\mathrm{Id} - iJ_{\mathbf{C}})$ and $Q = \frac{1}{2}(\mathrm{Id} + iJ_{\mathbf{C}})$ are C-linear for both structures and conjugate $(\overline{P(\overline{v})} = Q(v))$. If TX is the bundle of real vectors (which identifies with the sub-bundle of complex vectors v for which $\overline{v} = v$) then $P: TX \to T_{\mathbf{C}}^{(1,0)}$ is a bijection which transforms the J-complex structure of TX onto the i-complex structure of $T_{\mathbf{C}}^{(1,0)}$ (i.e., the one induced by $T_{\mathbf{C}}X$). The inverse map is given by $v \to \mathit{Re}\, v = \frac{1}{2}(v + \overline{v})$. Similarly, $Q: TX \to T_{\mathbf{C}}^{(0,1)}$ is a bijection which transforms J into $-i$ and the inverse is the map *Re*. We infer that an almost-complex structure defines a complex almost-product structure on X with the additional property that the splitting $\mathbb{C}$-linear subspaces are conjugate. $J_{\mathbb{C}}$, P and Q are bound by the relations $J_{\mathbb{C}} = i(2P - \mathrm{Id}) = i(\mathrm{Id} - 2Q) = i(P - Q)$. Conversely, any such complex conjugate almost-product structure determines an almost-complex structure on X obtained by transporting the induced $\mathbb{C}$-linear structure of any factor to TX via the restricted projection. In fact, we obtain two structures J and J′, one corresponding to each factor, such that $J = -J'$.

c) Let $T_{\mathbb{C}}X^*$ denote the $\mathbb{C}$-dual of $T_{\mathbb{C}}X$. Then, by duality, an almost-complex structure on X is defined by a complex conjugate almost-product structure in $T_{\mathbb{C}}X^*$. The corresponding factors are the subspaces of complex 1-forms of type $(1,0)$ and $(0,1)$. Observing that $T_{\mathbb{C}}X^*$ is canonically isomorphic to the bundle of $\mathbb{C}$-valued real 1-forms on TX, we infer that an almost-complex structure is also defined in terms of a complex conjugate almost-product structure given by a splitting of $\mathbb{C}$-valued real 1-forms.

d) Let X carry an almost-complex structure. Each T_xX is a $\mathbb{C}$-vector space of (complex) dimension n. Let $\mathrm{Lis}_{\mathbb{C}}(T_xX, \mathbb{C}^n)$ be the set of all $\mathbb{C}$-linear isomorphisms $T_xX \to \mathbb{C}^n$. $\mathrm{Lis}_{\mathbb{C}}(T_xX, \mathbb{C}^n)$ is a simply transitive homogeneous space under the left action of $GL(n, \mathbb{C})$. Taking the canonical basis of $\mathbb{C}^n$, each $u \in \mathrm{Lis}_{\mathbb{C}}(T_xX, \mathbb{C}^n)$ is uniquely represented by a set $(\omega_1, \dots, \omega_n)$ of $\mathbb{C}$-linear scalar 1-forms on T_xX which satisfy the property $\omega_1 \wedge \dots \wedge \omega_n \neq 0$ or equivalently $(\frac{i}{2})^n\, \omega_1 \wedge \dots \wedge \omega_n \wedge \bar{\omega}_1 \wedge \dots \wedge \bar{\omega}_n > 0$ (i.e., this real volume element is non-zero and defines the orientation of T_xX induced by the $\mathbb{C}$-linear structure). The non-vanishing of the real volume element means that the real linear map $T_xX \to \mathbb{R}^{2n} \equiv \mathbb{C}^n$ associated to $(Re\,\omega_1, \dots, Re\,\omega_n, Im\,\omega_1, \dots, Im\,\omega_n)$ is an isomorphism. Conversely, let X be an even dimensional manifold and assume that at each point $x \in X$ we associate a subset of $\mathrm{Lis}_{\mathbb{R}}(T_xX, \mathbb{C}^n)$ which is a simply transitive homogeneous space under the left action of $GL(n,\mathbb{C})$. Then on each T_xX there is canonically defined a $\mathbb{C}$-linear structure. The smoothness condition is assured, for example, by assuming the existence of local sections. In terms of scalar forms it means that to each x is associated a class of co-frames $(\omega_1, \ldots, \omega_n)$ where each ω_i is a $\mathbb{C}$-valued real 1-form on T_xX, such that $(\frac{i}{2})^n \omega_1 \wedge \ldots \wedge \omega_n \wedge \bar{\omega}_1 \wedge \ldots \wedge \bar{\omega}_n \neq 0$ and any two co-frames are linear combinations of each other, the matrix of coefficients belonging to $GL(n, \mathbb{C})$. If an *a priori* orientation is assigned to X then the previously defined almost-complex structure induces the given orientation if and only if the chosen elements in $\mathrm{Lis}_{\mathbb{R}}(T_xX, \mathbb{C}^n)$ are orientation preserving or, equivalently, if the volume element is > 0. Let $GL_{\mathbb{R}}(n, \mathbb{C}) \subset GL(2n, \mathbb{R})$ be the real form of $GL(n, \mathbb{C})$, i.e., a complex matrix $A + iB$ is represented by

$$\begin{pmatrix} A & -B \\ B & A \end{pmatrix}.$$

If in the previous definition $\mathbb{C}^n$ is replaced by $\mathbb{R}^{2n}$ and ω_i by $\alpha_i = Re\, \omega_i$ and $\beta_i = Im\, \omega_i$ then $GL(n, \mathbb{C})$ should be replaced by $GL_{\mathbb{R}}(n, \mathbb{C})$ and the volume element is then equal to $\alpha_1 \wedge \ldots \wedge \alpha_n \wedge \beta_1 \wedge \ldots \wedge \beta_n$.

e) Let X be given an almost-complex structure. Denote by H the set of all frames in P_1 of the form $(e_1, \ldots, e_n, Je_1, \ldots, Je_n)$ where $(e_1, \ldots, e_n)$ is a complex basis of T_xX. H is also the set of real frames such that the matrix of J with respect to these frames reduces to

$$\begin{pmatrix} 0 & -\mathrm{Id} \\ \mathrm{Id} & 0 \end{pmatrix}.$$

H is a $GL_{\mathbb{R}}(n, \mathbb{C})$ structure. It corresponds to a reduction of the structure group $GL(2n, \mathbb{R})$ of TX to $GL_{\mathbb{R}}(n, \mathbb{C})$. Conversely, such a structure H defines an almost-complex structure where J is defined, with respect to any adapted frame, by the previous matrix.

f) Ehresmann [24 (a)] observed that the existence of an almost-complex structure on X implies the existence of an almost-symplectic structure and conversely. In other terms, to each almost-complex structure is associated (non canonically) an almost-symplectic structure (hence many almost-symplectic structures) and vice-versa. We can thus define and study the properties of almost-complex structures via the associated almost-symplectic structure. Moreover, this procedure has been useful in proving the existence of almost-complex structures on given manifolds. The standard method is the following. Let J be an almost-complex structure on X. There exists a Riemannian metric g which is invariant under J, i.e., $g(Jv, Jw) = g(v, w)$, or equivalently, $g(v, Jv) = 0$ (assuming X paracompact). To construct g take any g' and define $g(v, w) = g'(v, w) + g'(Jv, Jw)$. Next define the almost-symplectic 2-form Ω by $\Omega(v, w) = g(v, Jw)$. We observe that Ω is compatible with g in the sense that $\hat{\Omega} \circ (\hat{g})^{-1} \circ \hat{\Omega} + \hat{g} = 0$ where the $\hat{}$ denotes the duality $TX \to TX^*$ associated to a quadratic form, i.e., $\hat{g}(v)(w) = g(v, w)$ and similarly for $\hat{\Omega}$. Conversely, given Ω we choose a compatible Riemannian metric g (this is not so trivial) and define $J = -(\hat{g})^{-1} \circ \hat{\Omega}$. A proof of the latter, in a more general context, can be found in [16]. We remark that de Barros [17(a), (b), (c)] has studied corresponding properties for a much wider class of structures.

An almost-complex structure on X is integrable if it emanates from a complex manifold structure on X. This means that X can be covered by real coordinate systems (x_i, y_i), $x_i + iy_i = z_i$, such that in each set of coordinates the almost-complex structure becomes the standard structure on $\mathbf{R}^{2n} \equiv \mathbf{C}^n$ or, in a more suggestive way, it means that in each coordinate patch the *almost-complex* calculus on X becomes the standard complex variables calculus of $\mathbf{C}^n$. In terms of the $GL_{\mathbf{R}}(n, \mathbf{C})$-structure H it means that the field of frames $\left(\frac{\partial}{\partial x_1}, \dots, \frac{\partial}{\partial x_n}, \frac{\partial}{\partial y_1}, \dots, \frac{\partial}{\partial y_n}\right)$ is a local section of H or, equivalently, that $\left(\frac{\partial}{\partial z_1}, \dots, \frac{\partial}{\partial z_n}\right)$ is a complex local basis of $T_{\mathbf{C}}^{(1,0)}$ or $\left(\frac{\partial}{\partial \bar{z}_1}, \dots, \frac{\partial}{\partial \bar{z}_n}\right)$ is a complex local basis of $T_{\mathbf{C}}^{(0,1)}$. By duality, $(dz_1, \dots, dz_n)$ is a complex basis of the space of forms of type $(1, 0)$ and $(d\bar{z}_1, \dots, d\bar{z}_n)$ a basis of forms of type $(0, 1)$. The integrability condition can be stated in several equivalent ways of which we list a few.

1 – (de Rham). For any local field of coframes $(\omega_1, \ldots, \omega_n)$ as in (d) there exist C-valued real 1-forms ω_{ij} such that

$$d\omega_i = \sum_{j=1}^{n} \omega_{ij} \wedge \omega_j$$

for all i, or equivalently, $d\omega_i \equiv 0 \; mod \; (\omega_1, \ldots, \omega_n)$ which is a special case of the complex Frobenius theorem. The necessity of the condition is trivial. In fact, assuming X integrable, let $(z_1, \ldots, z_n)$ be the complex valued functions which are the components of a chart for the complex manifold structure. The corresponding field of coframes $(dz_1, \ldots, dz_n)$ defines locally the almost-complex structure hence $dz_i = z_i^j \omega_j$ where (z_i^j) is a function with values in $GL(n, \mathbb{C})$. Differentiation gives $0 = (dz_i^j) \wedge \omega_j + z_i^j d\omega_j$ and this equation can be solved for the $d\omega_j$.

2 – (Ehresmann [24 (a)]). Given any field of coframes $(\omega_1, \ldots, \omega_n)$ defining the almost-complex structure, we always have

$$d\omega_i = \Sigma\omega_{ij} \wedge \omega_j + \Sigma a_{ijk}\overline{\omega}_j \wedge \overline{\omega}_k .$$

Let $(\xi_1, \ldots, \xi_n)$ be the dual field of frames. The expression

$$\Sigma a_{ijk}\overline{\omega}_j \wedge \overline{\omega}_k \otimes \xi_i$$

is the local expression of a tensor field on X, namely the Hermitian torsion tensor in the terminology of Ehresmann [24 (a)]. Its vanishing is obviously equivalent to de Rham's condition. This torsion tensor can be redefined in terms of the Nijenhuis bracket $[J, J]$ (sometimes with a coefficient ½ according to computational conventions). The explicit condition is

$$[\xi, \eta] + J[J\xi, \eta] + J[\xi, J\eta] - [J\xi, J\eta] = 0$$

where ξ, η are any real vector fields on X. This condition is due to Eckmann and Frölicher [23] (see also [22], [26]).

3 – ([31], [52]). Let $(T_{\mathbf{C}}^{(1,0)}, T_{\mathbf{C}}^{(0,1)})$ be the complex conjugate almost-product structure associated to the almost-complex structure on X and let P (or Q) be one of the corresponding projections. The integrability condition is again the vanishing of the torsion tensor $[P,P]$ (or $[Q,Q]$) associated to the almost-product structure. A more geometric condition is a special case of the complex Frobenius integrability condition: $[\xi,\eta]$ is a vector field of type $(1,0)$ (resp. $(0,1)$) whenever ξ and η are vector fields of type $(1,0)$ (resp. $(0,1)$) the bracket being the Lie bracket of complex vector fields. It can be shown that any of the above two conditions implies the other.

Let J be the almost-complex tensor and let d_J be the derivation of type d associated to the vector 1-form J. As in the case of real almost-product structures, the integrability condition can also be expressed by $d_J^2 = 0$, i.e., d_J defines a complex. It can be shown that integrability implies the exactness of the d_J-complex on the level of germs (local d_J-Poincaré lemma). The d_J-complex gives a de Rham type resolution of the constant sheaf $\mathbf{R}$ by the sheaves of (germs of) real exterior differential forms on X. The associated cohomology (on global sections) is isomorphic to the cohomology of X with real coefficients. The previous discussion has a complex analogue when J is replaced by $J_{\mathbf{C}}$ and real forms on TX by complex forms on $T_{\mathbf{C}}X$ (or simply real forms on TX by complex valued forms on TX keeping the same J). The resulting $d_{J_{\mathbf{C}}}$-complex is simply a complexification of the d_J-complex. It is more interesting to consider the operator $\bar{\partial} = d_Q$ (resp. $\partial = d_P$). The integrability condition is again expressed by $\bar{\partial}^2 = 0$ (resp. $\partial^2 = 0$). Integrability implies the exactness of the $\bar{\partial}$ (resp. ∂) complex on the level of germs (local $\bar{\partial}$ (resp. ∂)-Poincaré lemma). The $\bar{\partial}$-complex is the Dolbeault resolution of the sheaf Ω^p of germs of holomorphic complex valued p-forms by the fine sheaves of germs of complex valued differential forms of type (p,q). The associated cohomology is isomorphic to the cohomology of X with values in the sheaf Ω^p. The ∂-complex is a resolution of the sheaf $\overline{\Omega}^p$ of germs of anti-holomorphic p-forms (holomorphic for $-J$) by the sheaves of forms of type (q, p). It is the Dolbeault complex for the complex structure $-J$.

Ehresmann [24 (a)] observed that for real analytic data, the integrability condition is sufficient. The problem can be reduced to the Frobenius theorem with complex analytic data (see also [23], [26], [46 (a)]). In 1957, Newlander and Nirenberg [51] (see also [55]) solved the general problem. They proved the sufficiency of the integrability condition under the hypothesis that X is of class C^{2n+1} and J of class C^{2n} ($2n = \mathit{dim}\, X$), the complex analytic coordinates being of class C^{2n}. If J is assumed to be of class $C^{2n+\alpha}$ (α real, $\alpha > 0$, in the sense of Hölder), there exist complex analytic coordinates of class C^{2n+1}. The method of proof consists in dualizing the Cauchy-Riemann equations to obtain the operator $\bar{\partial} = Q \circ d$. The solutions of $\bar{\partial} f = 0$ are precisely the holomorphic functions on X, hence the problem consists in finding n independent solutions in the neighborhood of each point. The integrability condition is stated in terms of $\bar{\partial}$ and the analytic machinery involves potential theoretic estimates and elliptic linear partial differential equations. In 1963, Nijenhuis and Woolf [54] (see also [53]) improved the previous result by proving the sufficiency under the assumption that J is of class $C^{k+\alpha}$, $k \geq 1$, $0 < \alpha < 1$, hence X of class at least C^{k+1}. The complex analytic coordinates are then of class $C^{k+1+(\alpha/n)}$. The almost-complex structure is viewed as a complex conjugate almost-product structure with the corresponding integrability condition $[P, P] = 0$. In 1967, Malgrange [49 (b)] gave a very simple proof which has an extension to elliptic Lie structures [9(c), (d)]. His proof is based on a non-linear analogue of a method, due to H. Cartan, of solving a special type of elliptic linear equations. Other proofs were also given by Hörmander [35(a), (b)] and Kohn [40], the latter proof being a consequence of the solvability of the $\bar{\partial}$-Neumann problem.

Let us finally mention the complex 1-dimensional case, i.e., Riemann surfaces. In this case, the integrability condition is void, hence any almost-complex structure on a real 2-dimensional manifold is always integrable. The classical argument is the following. Let J be the almost-complex tensor on X and choose a J-invariant Riemann metric g. The next step is the construction of (local) oriented isothermal parameters, i.e., positively oriented local coordinates (u, v) which reduce g to $f(du^2 + dv^2)$, i.e., g is conformally equivalent to the Euclidean metric. The change of isothermal parameters is defined, on the overlap, by a local orientation preserving conformal map of $\mathbf{R}^2 = \mathbf{C}$, hence holomorphic. We infer that the functions $z = u + iv$ define a complex analytic structure on X which induces J since $\det J > 0$ and $g\left(\frac{\partial}{\partial u}, J\frac{\partial}{\partial u}\right) = 0$ imply that $\frac{\partial}{\partial v} = J\frac{\partial}{\partial u}$. The existence of isothermal parameters was proved by Korn [41] and Lichtenstein [48] and is obtained by integrating the classical Beltrami equation which is second order linear and elliptic. An elementary treatment was given by Chern [20(a)] with the assumption that J is of class C^{α}, $0 < \alpha < 1$ (see also [19], [21], [56]). If S is an oriented surface in $\mathbf{R}^3$, the canonical almost-complex (hence Riemann surface) structure on S is defined by the positively oriented rotation J of angle $\pi/2$ in each $T_x S$.

EXAMPLE (8): *The complex Frobenius theorem*

The classical real Frobenius theorem has been extended by Nirenberg [55(a)] to complex valued differential forms in such a way that the integrability condition for almost-complex structures is a special case of the general complex Frobenius theorem:

Let X be a real manifold of dimension N and let $\Omega = \Omega(y)$ be a K-dimensional subspace of the space of complex-valued forms defined at every point y (and varying in a C^∞ way) in a neighborhood of the origin. Set $\Lambda = \Omega \cap \overline{\Omega}$ and assume that $K' =$ dimension of Λ is constant and that Λ has a basis with C^∞ coefficients (we note that necessarily $K' \geq 2K - N$); set $L = K - K'$. Necessary and sufficient for the existence of new local coordinates x so that Ω is equivalent to the space spanned by

$$dx^a + i dx^{a+L}, \quad dx^\sigma, \quad a = 1, \dots, L; \quad \sigma = N - K' + 1, \dots, N,$$

are the conditions

(1) $$d\Omega \subset \text{ideal generated by } \Omega$$

(2) $$d\Lambda \subset \text{ideal generated by } \Lambda \,.$$

Other forms of the theorem, in terms of complex vector fields or first order partial differential equations, can also be found in [55(a)] [35(a)] and [49(a)]. The general theorem is derived from the above special case with the aid of the real Frobenius theorem. We remark finally that the C^∞-smoothness assumption can be weakened.

EXAMPLE (9): G-*structures*

The integrability of G-structures has been studied by several authors. The prevailing idea is the construction of invariants in the form of structure tensors with values in the second cohomology groups of the δ-complex (Section 6) defined in terms of the Lie algebra $\mathfrak{g}$ of G and its standard prolongations $\mathfrak{g}^{(k)}$. For every integer $k \geq 0$ a tensor c^k is defined and the integrability condition is again the vanishing of these tensors. The tensor c^0 was introduced by Ehresmann [24(b)] and was defined as the torsion associated to the elements of the first order prolongation. It has also been considered by Chern [20(b)] and Matsushima [50]. Bernard [18] and later Singer and Sternberg [58] gave alternate definitions of this tensor. Consider $(R^n)^* \otimes \mathfrak{g}$ imbedded canonically in $(R^n \otimes R^n)^* \otimes R^n$, $\mathfrak{g} \subset \mathfrak{gl}(n, R)$, and let ∂ be the skew-symmetrization operator. The first tensor is a function $c^0: H \to \Lambda^2(R^n)^* \otimes R^n/\partial(R^{n*} \otimes \mathfrak{g})$ which has tensor-like properties, i.e., transforms like a tensor under the right action of G on H and the standard action on the image space. It can be defined as the composite $c^0 = q \circ T$ where $T: H \to \Lambda^2(R^n)^* \otimes R^n$ is the torsion form associated to any H-connection and q is the quotient map. Let H be a G-structure defined by a field of endomorphisms h on X (i.e., a vector 1-form). Then c^0 is essentially the torsion tensor $[h, h]$ in the sense of Nijenhuis. For example, if H is an almost-complex structure defined by J or an almost-product structure defined by a projection π then the vanishing of c^0 is equivalent to the vanishing of $[J, J]$ or $[\pi, \pi]$ and therefore $c^0 = 0$ is a necessary and sufficient condition for integrability.

Let H be a G-structure with G the group of all transformations leaving invariant a p-dimensional subspace of R^n. H is associated to a field of contact elements of dimension p and an easy computation shows that $c^0 = 0$ is equivalent to the Frobenius integrability condition. Similarly, if $G = Sp(2n)$ is the symplectic group of R^{2n} then H is an almost-symplectic structure defined by a 2-form Ω and $c^0 = 0$ is equivalent to $d\Omega = 0$ (cf. [58]). Let us give an example where c^0 does not yield any information on the structure. If H is an O(n)-structure, i.e., Riemannian, then c^0 is identically zero (think of the torsion free Levi-Civita connection). The same thing holds for any group G defined by a non-degenerate symmetric 2-form. Structure tensors of higher order were introduced by Singer and Sternberg [58]. They first define the successive prolongations of (first-order) G-structures and then define c^k as being the tensor c^0 of the k-th prolongation $H^{(k)}$. Whereas the vanishing of c^0 (on H) is a necessary condition for the integrability of H, the same does not hold for the c^k, $k \geq 1$. In fact, H can be integrable without $H^{(k)}$ being so and c^k carries in itself, among other things, the Lie algebra structure of $\mathfrak{g}$ which need not be abelian. There is however a *meaningful* component in c^k which in the subsequent discussion we still denote by c^k. This latter c^k also has tensor-like properties and can be pushed down to the initial G-structure H, hence a tensor c^k on H. One proves that the vanishing of c^k for all k is not only a necessary condition for integrability but a suitable condition of integrability in the sense that it is equivalent to formal (power series) integrability. Guillemin [32] has redefined the structure tensors c^k by a direct approach which is based on the notion of k-th order structure preserving holonomic jets. A G-structure (or simply G) is of finite type, say k, if $\mathfrak{g}^{(k-1)} \neq 0$ and $\mathfrak{g}^{(k)} = 0$ hence $\mathfrak{g}^{(h)} = 0$ for $h > k$. This condition can also be stated in terms of G and its prolongations. Intuitively speaking, the finiteness condition means that the crucial invariants of the structure are determined by the partial derivatives of order $\leq k+1$ of the given data. As a consequence of this assumption one proves that the group of global automorphisms of H is a finite dimensional Lie group, i.e., the transformations only depend on a finite number of parameters (cf. [24(c)], [58]). Riemannian structures and more generally G-structure with G compact are of type 1. Conformal structures are of type 2 when *dim* $X > 2$. If G is of finite type k then the c^{k+h} vanish for $h > 0$. The integrability theorem of Guillemin [32] states that a G-structure of finite type is integrable if and only if all the tensors vanish. If the type is k then the vanishing of c^h, $h \leq k$, is required. If H is Riemannian then $c^0 \equiv 0$ and the vanishing of c^1 is equivalent to the vanishing of the curvature tensor (a second order invariant). If H is conformal, the non-vanishing tensors are also related to classical invariants (see [32]). The proof of the theorem involves an extensive algebraic machinery though the analytical tool is again the Frobenius theorem. It is shown that the integrability of H is reduced to the integrability of a parallelism which reflects the structure at order $k+1$. Another proof, using some of the techniques constructed in this monograph, follows from results of Spencer [59(d)] applied to pseudogroups of finite type.

A group G is of infinite type if $\mathfrak{g}^{(k)} \neq 0$ for all k. Not much is known in the latter case. For real analytic data the integrability theorem follows from the work of E. Cartan. Using the Cartan-Kähler theory of involutive systems one can prove that the vanishing of all c^k is equivalent to integrability. In particular, if G is involutive then the vanishing of c^0 is a necessary and sufficient condition for integrability (see [58]). Examples of involutive groups are $GL(n, \mathbb{R})$, $GL(n, \mathbb{C})$, $SL(n, \mathbb{R})$, $Sp(2n)$, the conformal group $Co(2)$, the group of a multifoliate structure and the contact groups. The involutive groups are of infinite type and so is the group $Csp(2n)$ of conformal symplectic transformations which is not involutive. The groups of finite type are not involutive. If $\mathfrak{g}$ contains an element of rank 1 then G is of infinite type. In the C^∞ case, integrability has been proved by Singer and Sternberg [58] under the assumption that G is completely reducible and by Malgrange, as a consequence of his theorem on integrability of elliptic Lie structures, when the group

G is elliptic. We observe that Malgrange's theorem also applies to G-structures of finite type.

Libermann [46 (d), (e)] extended the construction of structure tensors to G-structures of order k and studied the problem of formal integrability in the context of semi-holonomic jets (see also [62]). General methods for almost Γ-structures were developed by Spencer [13 (a)].

EXAMPLE (10): *Structures defined by fields of endomorphisms*

We now examine a class of G-structures whose integrability is characterized by the vanishing of the structure tensor c^0. It was observed in example 9 that c^0 is closely related to the Nijenhuis torsion tensor. The vanishing of c^0 is in general a stronger condition than the vanishing of the Nijenhuis tensor. We shall indicate some cases where the two conditions are equivalent, this being so in examples 6 and 7.

Let J_0 be an endomorphism of R^n (or a real $n \times n$ matrix) and let $G(J_0)$ be the subgroup of $GL(R^n)$ composed of all g which commute with J_0. A $G(J_0)$-structure (briefly a J_0-structure) H on a manifold X of dimension n is equivalent to the structure defined by a field of endomorphisms J on X (a tensor of type (1, 1)) such that for any $x \in X$ there exists a frame $h: R^n \to T_x X$ satisfying the condition $J_0 = h^{-1} \circ J_x \circ h$. In terms of matrices, J_x is the endomorphism of $T_x X$ whose matrix with respect to any frame in H_x is equal to J_0. Lehmann-Lejeune [45 (b)] proved that a necessary and sufficient condition for the integrability of a J_0-structure is the vanishing of the structure tensor c^0. The method of proof is an elaborate procedure of reduction to the Newlander-Nirenberg theorem. If c^0 vanishes then the Nijenhuis tensor $N_J = \frac{1}{2}[J, J]$ also vanishes. The converse is not true in general as was shown by Kobayashi [36] (see also [45 (b)]) who gave an example of a non-integrable almost-tangent structure $(J^2 = 0)$ with vanishing N_J. However, the converse holds in some special instances studied in [36] and [45 (b)]. In particular, this is the case when J_0 is completely reducible or when $\ker J_0 \cap \operatorname{im} J_0 = 0$. The latter condition holds for automorphisms, projections and endomorphisms with complete real spectrum; it does not hold for nilpotents other than 0. The converse statement also holds for the almost-tangent structures considered by Eliopoulos [25] ($\mathit{ker}\, J_0 = \mathit{im}\, J_0$).

Let d_J be the derivation of type d associated to the vector 1-form J and let $J^{\#}$ be the transpose of J considered as an endomorphism of the algebra of scalar exterior differential forms. We observe that $N_J = 0$ is equivalent to $d_J^2 = 0$. If $J_0 \in GL(R^n)$ then (cf. de Barros [17 (a)])

$$d_J\omega = J^{\#} \circ d \circ J^{\#-1}\omega + (J^{-1} \circ N_J) \barwedge \omega$$

or equivalently

$$(d_J \circ J^{\#} - J^{\#} \circ d)\omega = J^{\#}[(J^{-1} \circ N_J) \barwedge \omega]$$

where d is exterior differentiation. Moreover, $d_J^{\#} = J^{\#} \circ d \circ J^{\#-1}$ is a derivation of degree 1 which satisfies the local $d_J^{\#}$-Poincaré lemma. The previous formulas show that $d_J^{\#} = d_J$ if and only if $N_J = 0$. This being the case, we infer that d_J defines a de Rham type resolution of the constant sheaf R and that $J^{\#}$ defines a natural equivalence of the d_J-complex with the classical de Rham d-complex, hence induces an isomorphism on cohomology, i.e., the cohomology of the d_J-complex (on global sections) is isomorphic to the cohomology of X with real coefficients. More generally, assume that J_0 is not necessarily invertible, let Φ_J^q, $0 \leq q \leq n$, be the sheaf of germs of exterior q-forms ω such that $i(\xi)\omega = 0$ for any $\xi \in \mathit{ker}\, J$ and let Φ_J^{-1} be the sheaf of germs of functions f whose differential df vanishes on $\mathit{im}\, J$. If $N_J = 0$ we obtain the complex

$$0 \longrightarrow \Phi_J^{-1} \xrightarrow{\ i\ } \Phi_J^0 \xrightarrow{d_J} \Phi_J^1 \xrightarrow{d_J} \cdots \xrightarrow{d_J} \Phi_J^n \longrightarrow 0$$

which in the integrable case ($c^0 = 0$) is exact, i.e., the local d_J-Poincaré lemma holds and yields a resolution of the sheaf Φ_J^{-1} of germs of functions which are constant on the maximal integral leaves of the integrable field $\mathit{im}\, J$ (cf. [45 (a)]). Moreover, $J^{\#}$ yields a natural transformation of the de Rham complex into the d_J-complex which induces a natural transformation on the cohomologies. In general, the latter is neither injective nor surjective.

B. DEFORMATION THEORY OF LIE STRUCTURES

We begin by describing the main features of the general mechanism for the deformation of structures on manifolds defined by Lie equations. (For a treatment from a somewhat different point of view, namely perturbation of the Maurer-Cartan forms, see Griffiths [30].)

Let C be the category of topological spaces and continuous maps, and let M be a subcategory of C satisfying the conditions that any open subset of an object of M is an object of M and the restriction of any map of M to an open subset of an object is also a map of M. The maps of M will be called regular maps, and a map is biregular if it is a regular homeomorphism together with its inverse. Let X be a paracompact topological space. For each open set $U \subset X$, let

$$S(U) = \{\varphi \mid \varphi \text{ a homeomorphism from } U \text{ onto an object of } M\} .$$

If $U' \subset U$, we have $S(U) \to S(U')$ where $\varphi \in S(U)$ goes into $\varphi|_{U'}$. Next, let

$$G(U) = \{(\varphi, \psi) \mid \varphi, \psi \in S(U),\ \varphi \circ \psi^{-1} \quad \text{and} \quad \psi \circ \varphi^{-1} \in M\} .$$

We have the restriction map $G(U) \to G(U')$ where (φ, ψ) goes into $(\varphi|_{U'}, \psi|_{U'})$. An element of $G(U)$ is an identity if it is of the form (φ, φ), $\varphi \in S(U)$, and the product $(\varphi_1, \psi_1) \cdot (\varphi_2, \psi_2)$ is defined in $G(U)$ if and only if $\psi_1 = \varphi_2$ in which case

$$(\varphi_1, \psi_1) \cdot (\varphi_2, \psi_2) = (\varphi_1, \psi_2) .$$

The product clearly commutes with restriction so the restriction map is a homomorphism in the sense of groupoids. The assignments $U \mapsto G(U)$, $U \mapsto S(U)$ define presheaves which induce sheaves over X which we denote by $\mathcal{S}$, $\mathcal{G}$ respectively. We say that $\varphi, \psi \in S(U)$ are equivalent if $(\varphi, \psi) \in G(U)$ and we denote the equivalence classes of $S(U)$ modulo

G(U) by S(U)/G(U). Then $U \mapsto S(U)/G(U)$ induces a sheaf $\mathcal{S}/\mathcal{G}$ over X. It is easily verified that we have an isomorphism

$$H^0(X, \mathcal{S}/\mathcal{G}) \cong H^1(X, \mathcal{G})$$

(but this fact will not be used).

DEFINITION 1. *An M-structure on* X *is an element of* $H^0(X, \mathcal{S}/\mathcal{G})$, *i.e., it is a section of* $\mathcal{S}/\mathcal{G}$ *over* X.

It is convenient to enlarge the category *M* by adjoining to its objects all spaces with *M*-structures and to its maps all regular maps of *M*-spaces.

Let $\mathcal{R}_k$ be a linear Lie equation on R^n and denote by $\Gamma = \Gamma(\mathcal{R}_k)$ the corresponding non-linear equation. Consider the model category $M(\Gamma)$ whose objects are the open subsets of R^n with the structure induced by restriction and whose maps are the bidifferentiable maps between open subsets of R^n which are isomorphisms of the Lie equation Γ. An $M(\Gamma)$-structure on X will be called simply a Γ-structure. If Γ is the k-th order equation of a pseudogroup then the above notion reduces to the usual notion of pseudogroup structure.

Let B denote a *connected* differentiable manifold and let $(\mathcal{X}, \varpi, B)$ be a differentiable fibre bundle with base space B, total space $\mathcal{X}$, projection $\varpi: \mathcal{X} \to B$ and with fibre a (connected) differentiable manifold.

DEFINITION 2. *We say that* $\varpi: \mathcal{X} \to B$ *is a differentiable family of* Γ-*structures if each point of* $\mathcal{X}$ *has a neighborhood* U *together with a bidifferentiable map* $h: U \to \varpi(U) \times M$ *of* U *onto the product of an object* M *of* $M(\Gamma)$ *with* $\varpi(U)$ *whose restriction to* $U \cap \varpi^{-1}(t)$, $t \in \varpi(U)$, *is a biregular map, in the sense of* $M(\Gamma)$, *of* $U \cap \varpi^{-1}(t)$ *onto* $\{t\} \times M$.

Let $X_0 = \varpi^{-1}(0)$ be the fibre of $\mathcal{X}$ over a particular point $0 \in B$; then $X_t = \varpi^{-1}(t)$, $t \in B$, may be regarded as a deformation of the Γ-structure of X_0.

DEFINITION 3. *A differentiable family* $\varpi: \mathfrak{X} \to B$ *of* Γ*-structures is said to be locally trivial if and only if every point* $t_o \in B$ *has a neighborhood* U *together with a differentiable map* $h: \varpi^{-1}(U) \to X_{t_o}$ *whose restriction to each fibre* X_t *of* $\varpi^{-1}(U)$ *is a biregular map of* X_t *onto* X_{t_o} *in the sense of* $M(\Gamma)$.

Next, a local biregular transformation of a differentiable family $\varpi: \mathfrak{X} \to B$ of Γ-structures into another such family $\varpi': \mathfrak{X}' \to B'$ is a bidifferentiable map of an open set U of $\mathfrak{X}$ onto an open set U' of $\mathfrak{X}'$ which is a bundle mapping, i.e., it transforms fibres into fibres hence induces a map of $\varpi(U)$ onto $\varpi(U')$, and its restriction to each fibre is a biregular map in the sense of $M(\Gamma)$. In particular, among the local regular automorphisms of $\mathfrak{X}$ are those which induce the identity map on B. Let Π be the sheaf over $\mathfrak{X}$ of germs of infinitesimal regular automorphisms of $\mathfrak{X}$, Θ the subsheaf of Π of germs of infinitesimal automorphisms which leave the base space B fixed, and $\mathcal{T}$ the quotient sheaf, i.e., we have over $\mathfrak{X}$ the exact sequence of sheaves

$$0 \to \Theta \to \Pi \to \mathcal{T} \to 0 . \tag{1}$$

Let $\mathcal{T}(B)$ be the sheaf of germs of sections of the tangent bundle $T(B)$ of B; then $\mathcal{T}$ is isomorphic to the sheaf induced over $\mathfrak{X}$ from $\mathcal{T}(B)$ by the projection $\varpi: \mathfrak{X} \to B$. For a fixed t, let $\Theta_{(1),t}$ be the subsheaf of Θ composed of germs of vector fields which vanish on the fibre $X_t = \varpi^{-1}(t)$, $t \in B$; then we have the exact sequence

$$0 \to \Theta_{(1),t}|X_t \to \Theta|X_t \to \Theta_t \to 0$$

where $\Theta|X_t$ is a sheaf on $\mathfrak{X}$ vanishing outside the fibre X_t and Θ_t denotes the quotient sheaf. We denote by r_t the map (restriction) defined by the commutative triangle

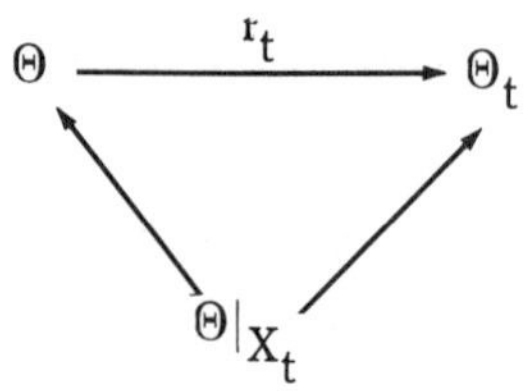

The map r_t induces a restriction map $r_t^*: H^*(\mathcal{X},\Theta) \to H^*(X_t, \Theta_t)$. In the same way we define restriction maps $r_r: \Pi \to \Pi_t$, $r_t: \mathcal{T} \to \mathcal{T}_t$, and we obtain the exact sequence of sheaves over X_t:

(2) $$0 \to \Theta_t \to \Pi_t \to \mathcal{T}_t \to 0\,.$$

Let Ξ be any sheaf of abelian groups over $\mathcal{X}$. Then we have the sheaf $\mathcal{H}^*(\Xi)$ over B of cohomology, namely the sheaf induced by the presheaf which assigns to each open set U of B the cohomology $H^*(\varpi^{-1}(U), \Xi)$. From (1) we obtain the following exact sequence of cohomology sheaves over B:

(3) $$0 \longrightarrow \mathcal{H}^0(\Theta) \longrightarrow \mathcal{H}^0(\Pi) \longrightarrow \mathcal{H}^0(\mathcal{T}) \xrightarrow{\partial^*} \mathcal{H}^1(\Theta) \longrightarrow \cdots$$

Since $\mathcal{H}^0(\mathcal{T}) \simeq \mathcal{T}(B)$, the coboundary map ∂^* induces a map

(4) $$\rho\colon \mathcal{T}(B) \to \mathcal{H}^1(\Theta)\,.$$

The following proposition may be proved (see [39 (a), (b)], [59 (b), (c)]):

PROPOSITION 1. *A differentiable family of compact Γ-manifolds without boundary is locally trivial if and only if the map ρ vanishes.*

From (2), and the fact that $H^0(X_t, \mathcal{T}_t) \simeq (T(B))_t$, we obtain the exact sequence

(5) $$0 \longrightarrow H^0(X_t,\Theta_t) \longrightarrow H^0(X_t,\Pi_t) \longrightarrow (T(B))_t \xrightarrow{\rho_t} H^1(X_t,\Theta_t) \longrightarrow \cdots$$

where

(6) $$\rho_t\colon (T(B))_t \to H^1(X_t, \Theta_t)$$

measures the infinitesimal deformation of the Γ-structure along the various tangent directions of B. However, the vanishing of ρ_t for all $t \in B$ does not imply that the family is locally trivial (see [39 (a), (b)]).

After these preliminaries we now describe how the techniques developed in this monograph apply to the deformation of Lie structure and the complexes of Chapter V constitute our starting point. For simplicity of notation, we write $\mathfrak{D}er^i_{\Sigma,k} = \mathcal{E}^i_k = \mathcal{E}^i$. Moreover, we drop the superscript "$\hat{\ }$" over D, $\mathfrak{D}$, $\mathfrak{D}_1$, since the operators defined earlier will not be used here. Let X be a manifold of dimension n with given Lie structure Γ. We restrict all the sheaves over X to the Lie equation $\mathcal{R} = \mathcal{R}(\Gamma)$ on X and we denote the restricted sheaves by $\mathcal{E}^i(\mathcal{R}) = \mathcal{E}^i_m(\mathcal{R})$, $\Gamma(\mathcal{R}) = \Gamma_{m+1}(\mathcal{R})$, etc., where m is an integer, $m \geq k$ (order of the Lie equation). Finally we extend the restricted sheaves to sheaves of germs of ℓ-parameter families depending on the *parameter* $t = (t^1, \ldots, t^\ell)$, where the t^j are real or, in the case of a complex Lie equation, complex and the germs are taken at $t = 0$ and any x. The sheaves of groupoids then become sheaves of groups since we assume that the transformations reduce to the identity at $t = 0$ and t lies in a germ of neighborhood of 0 in $\mathbf{R}^n$ (or $\mathbf{C}^n$ in the complex case). We attach the subscript "(1)" to $\mathcal{E}^i(\mathcal{R})$ (now assumed to be a sheaf of germs of ℓ-parameter families) to indicate that the germs vanish (to order 1) at $t = 0$. For example, if $u_t \in \mathcal{E}^i_{(1)}$, then $u_0 = 0$.

We have the following theorem (see [13 (c)]) where the notation conforms to that of Chapter V, with the modifications indicated above:

THEOREM 1.

(*i*) *The complex* (*over* X)

$$(7) \qquad 1 \longrightarrow \mathcal{A}ut(\mathcal{R}) \xrightarrow{\tilde{j}} \Gamma(\mathcal{R}) \xrightarrow{\mathfrak{D}} \mathfrak{D}(\Gamma(\mathcal{R})) \longrightarrow 0$$

is exact, where $\mathcal{A}ut(\mathcal{R})$, $\Gamma(\mathcal{R}) = \Gamma_{m+1}(\mathcal{R})$ *are sheaves of groups (corresponding respectively to* $\mathcal{A}ut$ (X), $\Gamma_{m+1}X$ *in Chapter* V – *e.g.,* $\mathcal{A}ut$ $(\mathcal{R})$ *is the sheaf of germs of* ℓ*-parameter families of regular automorphisms of* *the* Γ*-structure),* m *is a sufficiently large integer (to ensure stable range – i.e., the vanishing of the* δ*-cohomology), and* $\tilde{j} = \tilde{j}_{m+1}$. *The right operation of* $\Gamma(\mathcal{R})$ *on* $\mathfrak{D}(\Gamma(\mathcal{R}))$ *is given by the formula*

$$(8) \qquad (\mathfrak{D}F_1)^{F_2} = \mathfrak{D}(F_1 \bullet F_2) = \mathcal{A}dF_2 \bullet \mathfrak{D}F_1 + \mathfrak{D}F_2, \quad F_1, F_2 \in \Gamma(\mathcal{R}) .$$

In particular,

$$(9) \qquad \mathfrak{D}F^{-1} = -\mathcal{A}dF^{-1} \bullet \mathfrak{D}F, \qquad F \in \Gamma(\mathcal{R}) .$$

(ii) We have the non-linear complex

$$(10) \qquad 1 \longrightarrow \mathcal{A}ut(\mathcal{R}) \xrightarrow{\tilde{j}} \Gamma(\mathcal{R}) \xrightarrow{\mathfrak{D}} \mathcal{E}^1_{(1)}(\mathcal{R}) \xrightarrow{\mathfrak{D}_1} \mathcal{E}^2_{(1)}(\mathcal{R})$$

(where $\mathfrak{D}, \mathfrak{D}_1$ *correspond respectively to* $\hat{\mathfrak{D}}, \hat{\mathfrak{D}}_1$ *in Chapter* V) *and*

$$\mathcal{E}^q_{(1)}(\mathcal{R}) = \mathcal{E}^q_{(1),m}(\mathcal{R}), \quad q = 1, 2$$

and m *is the integer in part (i). The right operation of* $\Gamma(\mathcal{R})$ *on* $\mathcal{E}^1_{(1)}(\mathcal{R})$ *is defined, for* $F \in \Gamma(\mathcal{R})$ *and* $u \in \mathcal{E}^1_{(1)}(\mathcal{R})$, *by the formula*

$$(11) \qquad u^F = \mathcal{A}dF \bullet u + \mathfrak{D}F$$

where

$$(12) \qquad (u^{F_1})^{F_2} = u^{F_1 \bullet F_2}, \quad F_1, F_2 \in \Gamma(\mathcal{R}) ,$$

and

$$(13) \qquad \mathfrak{D}_1 u^F = \mathcal{A}dF \bullet \mathfrak{D}_1 u .$$

Hence $\ker \mathfrak{D}_1$ *is stable under the right operation of* F *on* $\mathcal{E}^1_{(1)}$, *and the cohomology of* (10) *at* $\mathcal{E}^1_{(1)}(\mathcal{R})$ *is the set of orbits of* $\ker \mathfrak{D}_1$ *under the operation of* $\Gamma(\mathcal{R})$*,*

(iii) Suppose that X *is compact without boundary. Then* $H^1(X, \mathcal{A}ut(\mathcal{R}))$ *is the set of germs of deformations of the* Γ*-structure of* X *(depending on* ℓ *parameters), with distinguished element* 1 *corresponding to the*

given Γ*-structure on* X. *Moreover, at least under the assumption that the Lie equation has constant rank – i.e., its equation is a bundle (in which case the* $\mathcal{E}^i(\mathcal{R})$ *are locally free), we have* $H^1(X, \Gamma(\mathcal{R})) = 1$ *and the cohomology sequence of* (7), *namely*

$$(14)\ 1 \longrightarrow H^0(X,\mathfrak{Aut}(\mathcal{R})) \longrightarrow H^0(X,\Gamma(\mathcal{R})) \xrightarrow{\mathfrak{D}} H^0(X,\mathfrak{D}(\Gamma(\mathcal{R}))) \xrightarrow{\partial^*} H^1(X,\mathfrak{Aut}(\mathcal{R})) \longrightarrow 1$$

is exact, i.e., two elements of $H^0(X, \mathfrak{D}(\Gamma(\mathcal{R}))$ *have the same image in* $H^1(X, \mathfrak{Aut}(\mathcal{R}))$ *if and only if they are congruent modulo the right operation of* $H^0(X, \Gamma(\mathcal{R}))$ *on* $H^0(X, \mathfrak{D}(\Gamma(\mathcal{R})))$.

The exactness of the complex (10) at $\mathcal{E}^1_{(1)}(\mathcal{R})$ has the following meaning. Let v_t be a section over X of $\mathcal{E}^1_{(1)}(\mathcal{R})$; then $v_0 = 0$ and v_t defines for each t on almost Γ-structure on X which, for $t = 0$, is the given Γ-structure. The family of almost Γ-structures on X is integrable if and only if

$$\mathfrak{D}_1 v_t = Dv_t - ½[v_t, v_t] = 0 .$$

Now suppose that (10) is exact at $\mathcal{E}^1_{(1)}(\mathcal{R})$ (hence exact everywhere by part (i)). Then there exists a locally finite covering $\{U_\alpha\}$ of X such that the equation $\mathfrak{D}F_t = v_t$ has a solution $F_{t,\alpha}$ on U_α, i.e., $\mathfrak{D}F_{t,\alpha} = v_t$ on U_α where $F_{t,\alpha}$ is a section of $\Gamma(\mathcal{R})$ over U_α. In the overlap $U_\alpha \cap U_\beta$ we also have $\mathfrak{D}F_{t,\beta} = v_t$ and hence, by the exactness of (7), we have

$$F_{t,\alpha} \circ F_{t,\beta}^{-1} = \tilde{j}f_{t,\alpha\beta} ,$$

i.e.,

$$F_{t,\alpha} = \tilde{j}f_{t,\alpha\beta} \circ F_{t,\beta} .$$

The system $\{f_{t,\alpha\beta}\}$ is a 1-cocycle representing a class in $H^1(X, \mathfrak{Aut}(\mathcal{R}))$, i.e., a *deformation* of the Lie structure of X.

The most important case is that in which Γ is the structure defined by a *transitive* (*continuous*) *pseudogroup* in the sense of Lie or, as we shall say shortly, Γ is a transitive Lie pseudogroup. If the transitive Lie pseudogroup is elliptic, the theorem of Malgrange [9(c), (d)] implies that (10) is exact at $\mathcal{E}^1_{(1)}(\mathcal{R})$. The exactness of (10) in the special case of complex transitive Lie pseudogroups (which are elliptic) was shown in [13(a)], Section 12, to be a consequence of the theorem of Newlander and Nirenberg [51]. For a general transitive Lie pseudogroup (not necessarily elliptic) the exactness of (10) remains an open question, although in special cases it can be established directly, as we shall see in the examples discussed below.

We shall now take up some simple examples, and we begin by listing the following classical examples of *primitive* transitive real Lie pseudogroups, of which the first four are simple, i.e., they contain no proper normal sub-pseudogroups different from the identity. (The deformation of the corresponding complex pseudogroup structures was first considered by Kodaira [37].)

(1) The pseudogroup of all differentiable transformations on $\mathbf{R}^n$ with non-vanishing jacobian determinants.

(2) The pseudogroup of all differentiable transformations on $\mathbf{R}^n$ with jacobian determinants equal to 1, i.e., the pseudogroup preserving the differential form $dx^1 \wedge dx^2 \wedge \dots \wedge dx^n$.

(3) (Symplectic structure). The pseudogroup of all differentiable transformations on $\mathbf{R}^{2m}$, $m \geq 2$, which leave invariant the exterior differential form $dx^1 \wedge dx^2 + dx^3 \wedge dx^4 + \dots + dx^{2m-1} \wedge dx^{2m}$.

(4) (Contact structure). The pseudogroup of all differentiable transformations on $\mathbf{R}^{2m+1}$, with variables $x^0, x^1, \dots, x^{2m}$, which leave invariant, up to a non-vanishing factor, the 1-form

$$dx^0 + x^1 dx^2 - x^2 dx^1 + \dots + x^{2m-1} dx^{2m} - x^{2m} dx^{2m-1} .$$

(5) The pseudogroup of all differentiable transformations on $\mathbf{R}^n$ with jacobian determinants which are constant (independent of $x = (x^1, x^2, \dots, x^n)$), i.e., the pseudogroup which leaves invariant, up to a constant factor, the differential form $dx^1 \wedge dx^2 \wedge \dots \wedge dx^n$.

(6) The pseudogroup of all differentiable transformations on $\mathbf{R}^{2m}$, $m \geq 2$, which leave invariant, up to a constant factor, the exterior differential form $dx^1 \wedge dx^2 + dx^3 \wedge dx^4 + \dots + dx^{2m-1} \wedge dx^{2m}$.

We shall discuss the deformation of the structures defined by the pseudogroups (2) and (3) (the pseudogroup (1) is locally rigid). Since these pseudogroups leave invariant given differential forms, we shall first recall briefly some facts about this type of pseudogroup.

We denote the coordinates on R^n by $x^1, x^2, \ldots, x^n$, and we write $x = (x^1, x^2, \ldots, x^n)$. Suppose that we are given on R^n a real-valued differential form ψ of degree p which, expressed in terms of the coordinates, has the form

$$\psi = \psi(x) = \sum_{i_1<\ldots<i_p} \psi_{i_1 \ldots i_p}(x)\, dx^{i_1} \wedge \ldots \wedge dx^{i_p} = \tag{15}$$

$$\frac{1}{p!} \sum_{i_1,\ldots,i_p} \psi_{i_1\ldots i_p}(x) dx^{i_1} \wedge \ldots \wedge dx^{i_p} .$$

Let $M(\psi)$ be the category whose objects are the open subsets of R^n and whose maps are all differentiable maps leaving ψ invariant. Let f_t be a 1-parameter family of local differentiable transformations of R^n. Then $\theta_t = \partial f_t/\partial t$ is a 1-parameter family of local vector fields and, as is easily verified, we have the formula

$$\frac{\partial}{\partial t} f_t^* \psi = f_t^*(\mathcal{L}(\theta_t)\psi) \tag{16}$$

where

$$\mathcal{L}(\theta_t)\psi = \theta_t \barwedge d\psi + d(\theta_t \barwedge \psi) \tag{17}$$

is, for each t, the Lie derivative of ψ along θ_t. Hence, if f_0 is the identity, we have $f_t^*\psi = \psi$ if and only if $\mathcal{L}(\theta_t)\psi = 0$ for all t. Since $\mathcal{L}(\theta)\psi$ involves only the vector field θ and its first-order partial derivatives, we can define $\mathcal{L}(\sigma)\psi$ for $\sigma \in \mathcal{J}_1 \otimes \underline{T}$ such that $\mathcal{L}(\sigma)\psi = \mathcal{L}(\theta)\psi$ for $\sigma = j_1\theta$. Then the Lie equation $\mathcal{R}_1 = \mathcal{R}_1(\psi)$ is the sheaf defined by

$$\mathcal{R}_1 = \{\sigma \in \mathcal{J}_1 \otimes \underline{T} \mid \mathcal{L}(\sigma)\psi = 0\} . \tag{18}$$

EXAMPLE (1): *Volume structure* (cf. [59(b)])

We consider the n-form

$$\psi = dx^1 \wedge dx^2 \wedge \ldots \wedge dx^n \tag{19}$$

on R^n. Let $f_t\colon y \to x = f_t(y)$ be a 1-parameter family of transformations (as above), and let

$$J_t(y) = det(\partial(f_t)/\partial(y))$$

be the jacobian determinant. Then we have

$$f_t^*\psi = J_t(y)\, dy^1 \wedge dy^2 \wedge \ldots \wedge dy^n$$

and hence

$$\frac{\partial}{\partial t}(f_t^*\psi) = \frac{\partial J_t(y)}{\partial t} dy^1 \wedge dy^2 \wedge \ldots \wedge dy^n .$$

On the other hand,

$$\mathcal{L}(\theta_t)\psi = (\nabla \bullet \theta_t)\, dx^1 \wedge dx^2 \wedge \ldots \wedge dx^n$$

where

$$\nabla \bullet \theta_t = \sum_{j=1}^{n} \partial\theta_t^j/\partial x^j$$

is the divergence of the vector field θ_t. Thus we obtain from (16) the formula

$$\partial J_t/\partial t = J_t \cdot (\nabla \bullet \theta_t) \tag{20}$$

or, if $J_t \neq 0$,

$$\partial\, log\, J_t/\partial t = \nabla \bullet \theta_t .$$

Thus J_t is independent of t (case of the pseudogroup (2)) if and only if

$$\nabla \bullet \theta_t = 0 . \tag{21}$$

More generally, the equation, $\nabla \bullet \theta_t$ = function of t, is equivalent to the condition that $J_t(y)$ is independent of y (case of the pseudogroup (5)) and $\nabla \bullet \theta$ is still invariant under this pseudogroup.

We now consider the deformation of Γ-structure defined by $M(\psi)$ where ψ is the n-form (19) (pseudogroup (2)), and we let $\mathcal{X} \to B$ be a differentiable family of compact $M(\psi)$-manifolds without boundary. The integral

$$(22) \qquad v(t) = \int_{X_t} dx^1 \wedge \ldots \wedge dx^n$$

over the fibre X_t then has an invariant meaning and will be called the volume of X_t. Denote by $\mathcal{O}_B$ the sheaf of germs of differentiable functions on B and, for any coordinate domain $U \subset B$, let $T_B(U) = H^0(U, \mathcal{T}(B))$ (module, over the functions, of vector fields on U). In view of (21), Θ is the sheaf of vector fields θ along the fibres satisfying $\nabla \bullet \theta = 0$.

PROPOSITION 2. *We have the isomorphism* $\mathcal{H}^1(\Theta) \simeq \mathcal{O}_B$ *and, for* $u \in T_B(U)$, $\rho(u) \in H^1(\mathcal{X}|_U, \Theta) \simeq H^0(U, \mathcal{O}_B)$ *may be identified with the function*

$$\mathcal{L}(u)v\colon t \mapsto \sum_{\nu=1}^{m} u^\nu(t)\,(\partial v(t)/\partial t^\nu)$$

where $u^1, \ldots, u^m$ *are the components of* u *referred to the coordinates* $t^1, \ldots, t^m$ *covering* U, *and* v *is defined by* (22).

Thus, in particular, we infer from Proposition 2 that the family $\mathcal{X} \to B$ is locally trivial if and only if all the fibres X_t of $\mathcal{X}$ have the same volume.

Finally, in the case of the pseudogroup (2), the complex (10) is exact at $\mathcal{E}^1_{(1)}(\mathcal{R})$. However, in this simple case, the general theory is unnecessary since deformations of the structure can be constructed directly. In fact, let X_0 be a compact manifold without boundary with structure defined by $M(\psi)$ where ψ is the form (19), and let φ be an arbitrary differential form of degree n. Then, if t is a sufficiently small real parameter, the form $\psi_t = \psi + t\varphi$ has maximal rank and hence each point of X_0 has a neighborhood with coordinate $x_t = (x_t^1, \ldots, x_t^n)$ depending differentiably on t such that, expressed in terms of this coordinate, $\psi_t = dx_t^1 \wedge dx_t^2 \wedge \ldots \wedge dx_t^n$. We thus obtain a differentiable family $\{X_t \mid |t| < \varepsilon\}$ of deformations of the $M(\psi)$-structure of X_0 which, by suitable choice of φ, will not be trivial in view of Proposition 2.

EXAMPLE (2): *Symplectic structure*

We consider the pseudogroup (3) of differentiable transformations, called canonical transformations, which leave invariant the form

$$\psi = dx^1 \wedge dx^2 + dx^3 \wedge dx^4 + \ldots + dx^{2m-1} \wedge dx^{2m} . \tag{23}$$

Let X_0 be a compact 2m-dimensional manifold without boundary and with $M(\psi)$-structure where ψ is the form (23). Since ψ is a 2-form of maximal rank, it establishes an isomorphism

$$\psi\colon T(X_0) \to T^*(X_0)$$

where the tangent vector θ is sent into $\psi(\theta) = \theta \barwedge \psi$. Let Θ_0 be the sheaf over X_0 of infinitesimal canonical transformations, i.e., the sheaf of germs of vector fields θ satisfying

$$\mathfrak{L}(\theta)\psi = d(\theta \barwedge \psi) \equiv 0 .$$

Since $\theta \barwedge \psi$ is a 1-form closed under d there exists, by the Poincaré lemma for d, a local function f such that $\theta \barwedge \psi = df$, i.e., $\theta = \psi^{-1}(df)$ and the vector fields which are locally of this form are precisely the infinitesimal canonical transformations. The above isomorphism established by ψ transforms Θ_0 into the sheaf Φ_0^1 over X_0 of closed 1-forms and hence we have the isomorphism

$$H^1(X_0, \Theta_0) \simeq H^1(X_0, \Phi_0^1) . \tag{24}$$

If $\mathcal{O}_0$ is the sheaf over X_0 of differentiable functions, we have the exact sequence

$$0 \longrightarrow R \longrightarrow \mathcal{O}_0 \xrightarrow{d} \Phi_0^1 \longrightarrow 0 \tag{25}$$

where, since $\mathcal{O}_0$ is a fine sheaf, we have $H^1(X_0, \mathcal{O}_0) = 0$. The cohomology sequence of (25) therefore establishes an isomorphism

$$H^1(X_0, \Phi_0^1) \simeq H^2(X_0, R)$$

and hence we obtain, by (24),

$$H^1(X_0, \Theta_0) \simeq H^2(X_0, R)\ . \tag{26}$$

Now let φ be a closed 2-form on X_0 representing a class $c \in H^2(X_0, R)$. Then, for sufficiently small t, the closed 2-form $\psi_t = \psi + t\varphi$ has maximal rank and hence, by the classical theorem of Darboux, each point of X_0 has a neighborhood with coordinate $x_t = (x_t^1, \ldots, x_t^{2m})$ depending differentiably on t such that, expressed in terms of this coordinate,

$$\psi_t = dx_t^1 \wedge dx_t^2 + dx_t^3 \wedge dx_t^4 + \ldots + dx_t^{2m-1} \wedge dx_t^{2m}\ .$$

Thus we obtain a differentiable family $\{X_t \mid |t| < \varepsilon\}$ of deformations of the $M(\psi)$-structure of X_0 and the infinitesimal deformation at time $t = 0$, namely $\rho_0(\partial/\partial t) \in H^1(X_0, \Theta_0)$ corresponds (up to sign) to the class $c \in H^2(X_0, R)$ by the isomorphism (26).

We observe finally that, since $H^0(X_0, \Theta_0) \simeq H^0(X_0, \Phi_0^1)$, the space $H^0(X_0, \Theta_0)$ is infinite dimensional (over R). We are therefore unable to construct a finite dimensional locally universal family of deformations of the symplectic structure on X_0, i.e., a finite dimensional family from which every sufficiently small deformation can be induced. (This difficulty disappears in the case of complex analytic symplectic structure on compact manifolds, since this structure is elliptic.) However, let $\varphi_1, \ldots, \varphi_\ell$ be closed 2-forms representing a basis of $H^2(X_0, R)$, and set

$$\sigma_t = \psi + t^1\varphi_1 + \ldots + t^\ell\varphi_\ell$$

where $t = (t^1, \ldots, t^\ell) \in R^\ell$. Then, for $|t| < \varepsilon$, σ_t is a closed 2-form of maximal rank which defines a family

$$\mathfrak{X} \to B = \{t \mid |t| < \varepsilon\}$$

of deformations of the symplectic structure of X_0. Let $\mathfrak{X}' \to B'$ be any family of deformations of X_0 where

$$B' = \{s \in R^{\ell'} \mid |s| < \eta\}.$$

Then, if η is sufficiently small, the family $\mathfrak{X}' \to B'$ is equivalent to a family of deformations $\mathfrak{X}'' \to B''$ which can be induced from $\mathfrak{X} \to B$.

EXAMPLE (3): Γ_G-*structure*

Let G be a linear group, $G \subset GL(n, R)$ or, in the complex case, $G \subset GL(n, C)$, and define Γ_G to be the transitive Lie pseudogroup on R^n (resp. $C^n \equiv R^{2n}$) of local differentiable transformations f whose jacobian matrices $\partial(f)/\partial(x)$ belong to G. It is easily seen that a vector field θ is an infinitesimal transformation of Γ_G if and only if the matrix

$$\partial(\theta)/\partial(x) = (\partial\theta^i/\partial x^j)_{i,j=1,2,\ldots,n}$$

belongs to the Lie algebra $\mathfrak{g}$ of G. Examples (1) and (2) above are Γ_G-structures with G respectively equal to $SL(n, R)$ (special linear group) and $Sp(2m, R)$ (symplectic group). A pseudogroup Γ_G is elliptic if and only if the Lie algebra of G contains no *real* sub-algebra generated by elements of rank 1; hence, in particular, every complex Γ_G (i.e., $G \subset GL(n, C)$) is elliptic and therefore the complex (10) is exact for complex Γ_G-structure.

EXAMPLE (4): *Multifoliate structure*

Multifoliate structure is an interesting type of Γ_G-structure since, in particular, it is characterized by sheaves of regular functions and is the only Lie structure which is so characterized (see [39 (c)], Section 10). *Foliate structure* (real or complex), *differentiable structure* (pseudogroup (1) above) and *complex analytic structure* provide the commonest examples of multifoliate structure.

Multifoliate structure is Γ_G-structure corresponding to a multifoliate group G defined as follows, where $\mathfrak{gl}(n, R)$, $\mathfrak{gl}(n, C)$ denote respectively the Lie algebras of $GL(n, R)$, $GL(n, C)$.

DEFINITION 4. *The group* G, *with Lie algebra* $\mathfrak{g} \subset \mathfrak{gl}(n, R)$ *or, in the complex case,* $\mathfrak{g} \subset \mathfrak{gl}(n, C)$, *is multifoliate if the following conditions are satisfied*:

(i) $\mathfrak{g}$ *is an associative algebra with Lie algebra bracket defined by commutators, and* G *is the subset of invertible elements of* $\mathfrak{g}$.

(ii) $\mathfrak{g}$ *is generated by elements of rank* 1.

A different definition is given in [39 (c)]; the above definition is due to C. Buttin.

In view of (ii), a real multifoliate pseudogroup is never elliptic; nevertheless, the complex (10) (for multifoliate structure) is exact (see [39 (c)]). In contrast with examples (1) and (2), the deformation of multifoliate structure requires an essential part of the mechanism of the general theory. Lack of space prevents us from giving here a description of it, and we refer the reader to the paper [39 (c)] for details.

EXAMPLE (5): *Cohen-Macaulay structure*

Suppose that X is a manifold of dimension n with Lie structure Γ, and assume that the equations $\mathcal{R}_{k+\ell}$, $\ell \geq 0$, are all locally free, where k is the order of the Lie equation, i.e., for $\ell \geq 0$, $\mathcal{R}_{k+\ell}$ is the sheaf $\underline{R}_{k+\ell}$ of sections of a vector bundle $R_{k+\ell}$. Then the linear complex of Chapter V, restricted to $\underline{R}_{k+\ell}$ where ℓ is sufficiently large (to ensure stable range), has the form

$$(27) \qquad 0 \longrightarrow \Theta \longrightarrow \underline{E}^0(R) \xrightarrow{D} \underline{E}^1(R) \xrightarrow{D} \cdots \xrightarrow{D} \underline{E}^n(R) \longrightarrow 0$$

where the $E^j(R)$ are vector bundles over X and Θ is the sheaf over X of infinitesimal automorphisms of the Γ-structure. We remark that the assumption of constant rank which we have just made is satisfied by all the examples considered and, in any case, few results have so far been proved without it.

For simplicity, we denote the complex (27) by $\{D, E(R)\}$ where $E(R) = \oplus E^j(R)$. The characteristic ideal $\mathcal{J}$ of $\{D, E(R)\}$ is the set of all polynomial functions on the complexified cotangent bundle $T^*_{\mathbf{C}} = T^*(X) \otimes \mathbf{C}$ of X which are defined as follows. If $(x, \zeta) \in T^*_{\mathbf{C}}$, $\zeta \neq 0$, we have the complexified (principal) symbol complex

$$(28)\quad 0 \longrightarrow E^0_{\mathbf{C}}(R)_x \xrightarrow{\sigma(D)(x,\zeta)} E^1_{\mathbf{C}}(R)_x \longrightarrow \cdots \xrightarrow{\sigma(D)(x,\zeta)} E^n_{\mathbf{C}}(R)_x \longrightarrow 0$$

where $E^j(R)_x$ denotes the fibre of $E^j(R)$ over the point $x \in X$. Consider all homotopy operators

$$a(x,\zeta)\colon E_{\mathbf{C}}(R)_x \to E_{\mathbf{C}}(R)_x$$

of degree -1, i.e.,

$$E^n_{\mathbf{C}}(R)_x \xrightarrow{a(x,\zeta)} E^{n-1}_{\mathbf{C}}(R)_x \longrightarrow \cdots \xrightarrow{a(x,\zeta)} E^1_{\mathbf{C}}(R)_x \xrightarrow{a(x,\zeta)} E^0_{\mathbf{C}}(R)_x$$

where the $a(x,\zeta)$'s are polynomial mappings (in ζ) satisfying

$$\sigma(D)(x,\zeta) \circ a(x,\zeta) + a(x,\zeta) \circ \sigma(D)(x,\zeta) = p(\zeta) \cdot I$$

where I is the identity map and $p(\zeta)$ is a polynomial in ζ. The set of all polynomials $p = p(\zeta)$ obtained in this way constitutes the characteristic ideal $\mathcal{J}_x$ at the point $x \in X$. Let V be the (complex) characteristic variety of $\{D, E(R)\}$ defined by $\mathcal{J}$, i.e., V_x, the characteristic variety at x, is the set on which $\mathcal{J}_x$ vanishes.

Since we have assumed constant rank, *dim* V_x is independent of $x \in X$ (see [33]) and we let q be the codimension of V_x. Following Guillemin and Sternberg [33], we say that a point $(x,\zeta) \in V_x$ is a *Cohen-Macaulay* point if (28) is exact at $E^{q+1}(R)_x, \ldots, E^n(R)_x$. We remark that the Cohen-Macaulay points of V_x form an open set in the Zariski topology.

DEFINITION 5. *We say that a Lie equation is a Cohen-Macaulay equation, or that the Lie structure is Cohen-Macaulay, if it has constant rank and every point of the characteristic variety V is a Cohen-Macaulay point.*

For a Cohen-Macaulay Lie equation, it follows from a theorem of Sweeney [61] that the complex (27) is exact at $\underline{E}^q(R), \ldots, \underline{E}^n(R)$ (and the *coercive estimate* for the Neumann problem holds at these positions – see Sweeney, *loc. cit.*) or, as we shall say, the complex (27) has homological length q (the codimension of the characteristic variety).

In the case of a Cohen-Macaulay Lie equation it is therefore natural to ask the question: can be complex (27) be replaced by one whose length is equal to its homological length? In the local sense the answer to this question is affirmative, namely (see [33]) if $(x, \zeta) \in V$ is a Cohen-Macaulay point, then there exists a conic neighborhood of (x, ζ) in $T^*_{\mathbf{C}}$ and a *Poincaré complex* of length q, constructed from q commuting *formal* pseudodifferential operators on this conic neighborhood, whose characteristic ideal coincides with the ideal $\mathcal{J}$ of $\{D, E(R)\}$. By a formal pseudo-differential operator we mean the total symbol of such an operator defined on a conic neighborhood of the complexified cotangent bundle or, more precisely, the (classical) pseudodifferential operators on X modulo those whose wave-front sets (in the sense of Hörmander [35(c)]) do not meet the conic neighborhood. Given q pair-wise commuting pseudo-differential operators $A_1, \ldots, A_q$ (which may be formal) mapping the sections of a vector bundle F into sections of F, a Poincaré complex is constructed as follows in terms of a given q-dimensional vector space W with basis $w_1, \ldots, w_q$. Let

$$\Lambda^*(w) = \bigoplus_{j=0}^{q} \Lambda^j(w)$$

denote the exterior algebra of the w_i's with constant coefficients, set $F^i = F \otimes \Lambda^i(w)$, and let $D_A: \underline{F}^i \to \underline{F}^{i+1}$ be defined by the formula

$$D_A(f \otimes \omega) = \sum_{i=1}^{q} A_i f \otimes (w_i \wedge \omega)$$

where f is a section of F and $\omega \in \Lambda^*(w)$. It follows that we have a complex $\{D_A, F\}$, i.e., $D_A^2 = 0$ since $A_i \circ A_j = A_j \circ A_i$ and $w_i \wedge w_j = -w_j \wedge w_i$.

EXAMPLE (6): *Complex analytic structure*

Complex analytic structure is elliptic Γ_G-structure, $G = GL(n, \mathbb{C})$, and it is multifoliate. Moreover, as we now show, it is Cohen-Macaulay and the complex (27) (for complex analytic structure) can be replaced by the well-known Dolbeault complex which is a Poincaré complex with $w_i = d\bar{z}^i$ and $A_i = \partial/\partial\bar{z}^i$, where $z^1, \ldots, z^n$ are complex analytic coordinates.

Let X be a complex analytic manifold, $\dim_{\mathbb{C}} X = n$. Then the complexified tangent bundle $T_{\mathbb{C}}$ of X splits: $T_{\mathbb{C}} = T \oplus \bar{T}$ where T is the holomorphic tangent bundle of X and $\bar{T}$ is the conjugate bundle ($T_{\mathbb{C}}^{(1,0)}$ and $T_{\mathbb{C}}^{(0,1)}$ in the notation of (b), example 7). Let $H_1(T)$ denote the vector bundle of holomorphic 1-jets of the holomorphic tangent bundle T, $\bar{H}_1(\bar{T})$ the conjugate bundle. Then

$$J_1(T_{\mathbb{C}}) \supset H_1(T) \oplus \bar{H}_1(\bar{T})$$

and the Lie equation $\mathcal{R}_1$ is equal to $\underline{R}_1$ where R_1 is the real subbundle of $H_1(T) \oplus \bar{H}_1(\bar{T})$ (composed of the real jets). However, it is obviously more convenient to use the complex representation, in which case R_1 is identified with $H_1(T)$. If $T_{\mathbb{C}}^* = T^* \oplus \bar{T}^*$ is the corresponding splitting of the complex cotangent bundle $T_{\mathbb{C}}^*$ of X, then the complex characteristic variety V is T^*, the holomorphic cotangent bundle; therefore the codimension of V is n. It may be verified that all points of V are Cohen-Macaulay points; hence complex analytic structure is Cohen-Macaulay. Moreover, we have the following exact, commutative diagram:

$$(29)\qquad \begin{array}{ccccccccccccccc}
0 \to \Theta \to & \underline{E}^0(R) & \xrightarrow{D} & \underline{E}^1(R) & \xrightarrow{D} \cdots \xrightarrow{D} & \underline{E}^n(R) & \xrightarrow{D} \cdots \xrightarrow{D} \underline{E}^{2n}(R) \to 0 \\
\| \qquad & \downarrow{\scriptstyle \pi} & & \downarrow{\scriptstyle \pi} & & \downarrow{\scriptstyle \pi} & \\
0 \to \Theta \to & \underline{T} & \xrightarrow{\bar\partial} & \underline{\bar{T}}^* \otimes \underline{T} & \xrightarrow{\bar\partial} \cdots \xrightarrow{\bar\partial} & \Lambda^n \underline{\bar{T}}^* \otimes \underline{T} \to 0 &
\end{array}$$

where Θ is the sheaf of holomorphic vector fields, $\pi\colon \underline{E}^p(R) \longrightarrow \Lambda^p \underline{\bar{T}}^* \otimes \underline{T}$ is a projection, $0 \leq p \leq n$, and, expressed in terms of local coordinates,

$$\bar{\partial} = \sum_{j=1}^{n} d\bar{z}^j \wedge \frac{\partial}{\partial \bar{z}^j} .$$

The second line of (29) is the Dolbeault complex.

Next, we consider briefly the deformation of complex analytic structure. Let $\mathfrak{L}^{0,p}$ be the sheaf of germs of ℓ-parameter families of sections of $\Lambda^p \bar{T}^* \otimes T$ and $\mathfrak{E}^p(\mathfrak{R})$ the sheaf of ℓ-parameter families of sections of $E^p(R)$. Then (compare [13(b)], pp. 400-401) we have the following exact, commutative diagram:

$$\text{(30)}\quad \begin{array}{ccccccccccccccc} 0 \to \Xi & \to & \mathfrak{E}^0(\mathfrak{R}) & \xrightarrow{D} & \mathfrak{E}^1(\mathfrak{R}) & \xrightarrow{D} & \cdots & \xrightarrow{D} & \mathfrak{E}^n(\mathfrak{R}) & \xrightarrow{D} & \cdots & \xrightarrow{D} & \mathfrak{E}^{2n}(\mathfrak{R}) & \to & 0 \\ \| & & \downarrow \pi & & \downarrow \pi & & & & \downarrow \pi & & & & & & \\ 0 \to \Xi & \to & \mathfrak{L}^{0,0} & \xrightarrow{\bar{\partial}} & \mathfrak{L}^{0,1} & \xrightarrow{\bar{\partial}} & \cdots & \xrightarrow{\bar{\partial}} & \mathfrak{L}^{0,n} & \to & 0 & & & & \end{array}$$

where Ξ is the sheaf of ℓ-parameter families of holomorphic vector fields and $\pi: \mathfrak{E}^p(\mathfrak{R}) \to \mathfrak{L}^{0,p}$ is a projection, $0 \le p \le n$.

An analogous projection can also be defined for the non-linear complex (10). In fact, $\mathfrak{Aut}(\mathfrak{R})$ in (10) (for complex analytic structure) is the sheaf of groups of germs of ℓ-parameter families of biholomorphic transformations which reduce to the identity for $t = 0$ (where $t = (t^1, \ldots, t^\ell) \in \mathbf{C}^\ell$). Let Φ denote the sheaf of groups of germs of ℓ-parameter families of bidifferentiable transformations of X which are the identity at $t = 0$. Then we have the exact, commutative diagram:

$$\text{(31)}\quad \begin{array}{ccccccccc} 1 & \to & \mathfrak{Aut}(\mathfrak{R}) & \to & \Gamma(\mathfrak{R}) & \xrightarrow{\mathfrak{D}} & \mathfrak{E}^1_{(1)}(\mathfrak{R}) & \xrightarrow{\mathfrak{D}_1} & \mathfrak{E}^2_{(1)}(\mathfrak{R}) \\ & & \| & & \downarrow \pi & & \downarrow \pi & & \downarrow \pi \\ 1 & \to & \mathfrak{Aut}(\mathfrak{R}) & \to & \Phi & \xrightarrow{\mathfrak{D}} & \mathfrak{L}^{0,1}_{(1)} & \xrightarrow{\mathfrak{D}_1} & \mathfrak{L}^{0,2}_{(1)} \end{array}$$

Here

$$\mathfrak{D}: \Phi \to \mathfrak{L}^{0,1}_{(1)}$$

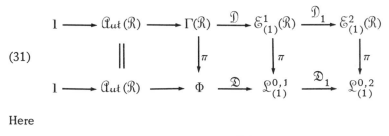

sends f_t into $\mathfrak{D} f_t = f_t^{-1} \bullet \bar{\partial} f_t$ where, in terms of a complex analytic coordinate system $(z^1, \ldots, z^n)$, $f_t^{-1} \bullet \bar{\partial} f_t$ has the components

$$(f_t^{-1} \bullet \bar{\partial} f_t)^i = \sum_{j=1}^{n} [\partial (f_t^{-1})^i / \partial z^j] \, \bar{\partial} f_t^j, \quad i = 1, 2, \ldots, n \ .$$

Finally,

$$\mathfrak{D}_1 : \mathfrak{L}^{0,1}_{(1)} \to \mathfrak{L}^{0,2}_{(1)}$$

maps φ_t into

$$\mathfrak{D}_1 \varphi_t = \bar{\partial} \varphi_t - \tfrac{1}{2} [\varphi_t, \varphi_t]$$

where $[\varphi_t, \varphi_t]$ is the Nijenhuis bracket (see [38]).

The deformation of complex analytic structure has become a rather large subject, and we do not give further details here but refer the reader to the literature (which is too large to list – but see, for example, M. Kuranishi [44], and numerous papers of K. Kodaira which have appeared in the *American Journal of Mathematics* during the last ten years).

CHAPTER I

JET SHEAVES AND DIFFERENTIAL EQUATIONS

1. *Notation*

Throughout these notes X will be a Hausdorff C^∞-manifold of dimension n and *differentiable* will mean of class C^∞. Since the results are local, one can replace X by an open set in R^n. Let $E \to X$ be a vector bundle (always locally trivial, C^∞ and of finite rank) and $\underline{E}$ the sheaf of germs of sections of E. We recall that local sections of $\underline{E}$ identify with local sections of E. For the trivial line bundle $E = X \times R$ we obtain the structure sheaf $\underline{E} = \mathcal{O}_X$ (or simply $\mathcal{O}$) of the manifold X. $\mathcal{O}$ is the sheaf, of local rings, of germs of differentiable functions on X and sections of $\mathcal{O}$ are differentiable functions on X. For any vector bundle E, the sheaf $\underline{E}$ is a sheaf of $\mathcal{O}$-modules, the structure being defined by extending to germs the linear operations of E. This structure will be called natural to distinguish it from other $\mathcal{O}$-module structures to be introduced later. Let $\mathcal{E}$ be a sheaf of $\mathcal{O}$-modules on X (briefly, an $\mathcal{O}$-module). For $a \in X$, denote by $\mathcal{E}(a)$ the stalk of $\mathcal{E}$ over the point a. $\mathcal{E}$ is said to be locally free of finite rank if every point $a \in X$ has a neighborhood U_a and a finite number of sections σ_a^i over U_a which induce at every point $b \in U_a$ an $\mathcal{O}(b)$-basis of $\mathcal{E}(b)$. The sheaf $\underline{E}$ associated to a vector bundle E is locally free of finite rank (equal to the rank of E). Moreover any vector bundle map $\phi : E \to F$ extends to an $\mathcal{O}$-linear sheaf map $\underline{\phi} : \underline{E} \to \underline{F}$. The functor $\phi \mapsto \underline{\phi}$ is exact and defines an equivalence of the category of vector bundles over X with the category of sheaves of $\mathcal{O}$-modules over X which are locally free of finite rank. In other words, any locally free sheaf of $\mathcal{O}$-modules (of finite rank) is isomorphic to the sheaf of sections of a vector bundle and any $\mathcal{O}$-linear map of such sheaves is induced by a vector bundle map. One should observe however, that non-exact bundle sequences can transform into exact sheaf sequences (e.g., $0 \to \mathbf{R} \times \mathbf{R} \xrightarrow{\phi} \mathbf{R} \times \mathbf{R}$, $\phi(t, v) = (t, tv)$). A subsheaf $\mathcal{F}$ of $\mathcal{E}$ is said to be locally a direct factor if for any $x \in X$ there is a neighborhood U of x such that $\mathcal{F}|_U$ is a direct factor of $\mathcal{E}|_U$. From this we also derive the notion of local splitting. The exactness of a vector bundle sequence can be verified on the sheaf level as is shown by the

LEMMA. *The following conditions are equivalent:*

1) $E \xrightarrow{\phi} F \xrightarrow{\psi} G$ *is exact;*
2) $\underline{E} \xrightarrow{\phi} \underline{F} \xrightarrow{\psi} \underline{G}$ *is locally split exact and the image of* $\underline{\psi}$ *is locally split;*
3) $\underline{E} \xrightarrow{\phi} \underline{F} \xrightarrow{\psi} \underline{G}$ *is split exact in each stalk and* $im\underline{\psi}$ *splits in each stalk.*

Proof: A projective module over a local ring being free, we apply this property to a stalk over x and observe that $im\underline{\phi}$ as well as $im\underline{\psi}$ are locally finitely generated.

In particular we derive the

COROLLARY. *Let* $\mathcal{F}$ *be a subsheaf of* $\underline{E}$. *The following conditions are equivalent:*

1) $\mathcal{F} = \underline{F}$ *with* F *a sub-bundle of* E;
2) $\mathcal{F}$ *is locally finitely generated and locally a direct factor of* $\underline{E}$;
3) $\mathcal{F}$ *is locally finitely generated and a direct factor in each stalk;*
4) $\mathcal{F}$ *is locally a direct factor of* $\underline{E}$.

Any of the above conditions implies that $\mathcal{F}$ is locally free of finite rank. This property alone does not imply that $\mathcal{F} = \underline{F}$ (cf. the previous example). If X is paracompact then local can be replaced by global in the lemma and its corollary.

2. *Jet bundles*

Let E be a vector bundle and denote by I_a the maximal ideal of $\mathcal{O}(a)$ (germs vanishing at the point a). I_a^{k+1} is the ideal of germs flat to order

$k+1$ at a (all partial derivatives of order $\leq k$ with respect to some, hence any coordinate system, vanish at a) and $I_a^{k+1}\,\underline{E}(a)$ is the $\mathcal{O}(a)$-submodule of $\underline{E}(a)$ composed of germs flat to order $k+1$ at a. Moreover,

$$(J_kE)_a = \underline{E}(a)/I_a^{k+1}\,\underline{E}(a)$$

is a finite dimensional vector space over $\mathbf{R}$. If $\{e_\lambda\}_{1 \leq \lambda \leq N}$ is a local basis of E and $x = (x_1, \dots, x_n)$ a system of local coordinates around a, then (Taylor's expansion) given any $A \in (J_kE)_a$ there is a unique family of polynomials $P_\lambda \in \mathbf{R}[X_1, \dots, X_n]$ such that A is the equivalence class of $\sum_\lambda P_\lambda(x-x(a))\,e_\lambda$. The equivalence classes of $\frac{1}{\alpha!}(x-x(a))^\alpha e_\lambda$, $|\alpha| \leq k$, and $1 \leq \lambda \leq N$ (for the notation cf. Section 5), form a basis of $(J_kE)_a$. Set

$$J_kE = \bigcup_{a \in X} (J_kE)_a$$

and define, for any local section σ of E, the section $j_k\sigma$ of J_kE by $j_k\sigma(a) =$ class of σ in $(J_kE)_a$. There is a unique vector bundle structure on J_kE which induces the linear structure $(J_kE)_a$ on each fibre and makes differentiable all the sections $j_k\sigma$. We call J_kE the bundle of k-jets of E. $J_k(X \times \mathbf{R})$ will be denoted by J_k. Each fibre of J_k inherits a ring structure: $j_kf(a) \cdot j_kg(a) = j_k(fg)(a)$ and two permuting $\mathcal{O}(a)$-module structures: the *left* structure given by $f \cdot j_kg(a) = j_k(f(a)g)(a)$ and the *right* structure given by $(j_kg(a)) \cdot f = j_k(fg)(a)$, $f \in \mathcal{O}(a)$, in such a way that $(J_k)_a$ is a left and right $\mathcal{O}(a)$-algebra. Since $\mathcal{O}(a)$ is commutative, the terminology as well as the notation are only intended to distinguish the two structures. Similarly each fibre $(J_kE)_a$ inherits a $(J_k)_a$-module structure and left, right $\mathcal{O}(a)$-module structures in such a way that $(J_kE)_a$ becomes a module over the left, right $\mathcal{O}(a)$-algebra $(J_k)_a$. Extending all these structures to germs, $\underline{J_k}$ becomes a sheaf of left, right $\mathcal{O}$-algebras and $\underline{J_kE}$ becomes a sheaf of modules over the left, right $\mathcal{O}$-algebra $\underline{J_k}$. Since $f \cdot A = f(a)A$, $A \in (J_kE)_a$, it is clear that the left $\mathcal{O}$-module structure on $\underline{J_kE}$ coincides with the natural one. The map $j_k\colon \underline{E} \to \underline{J_kE}$, $\sigma \mapsto j_k\sigma$, is right $\mathcal{O}$-linear and injective. In particular, $j_k\colon \mathcal{O} \to \underline{J_k}$ is an injective $\mathcal{O}$-algebra morphism for the right structure of $\underline{J_k}$. The map $i_k : \mathcal{O} \to \underline{J_k}$, $i_k f : x \mapsto j_k(f(x))(x)$, is an injective $\mathcal{O}$-algebra morphism for the left structure of $\underline{J_k}$. Moreover the image of i_k is a direct factor for the left $\mathcal{O}$-module structure of $\underline{J_k}$ since i_k is the extension, to germs, of the injective bundle map $X \times R \to J_k$, $(x, \lambda) \mapsto j_k\lambda(x)$. We shall see later that the image of j_k splits for the right $\mathcal{O}$-module structure. The sheaf $\underline{J_k E}$ is locally free of finite rank for the left (= natural) $\mathcal{O}$-module structure. Also we shall see later that the same thing is true for the right structure. On $(J_0)_a$ the left and right $\mathcal{O}(a)$-module structures coincide and $i_0 = j_0 : \mathcal{O} \to \underline{J_0}$ is an algebra isomorphism. We identify J_0 with $X \times R$ and $\underline{J_0}$ with $\mathcal{O}$. Similarly, $J_0 E$ identifies with E and $\underline{J_0 E}$ with $\underline{E}$ (all the structures coincide with the natural one).

We mention, for later use, the following bundle maps: the projection $\rho_h : J_k E \to J_h E$, $j_k\sigma(a) \mapsto j_h\sigma(a)$ with $h \leq k$, the injection $\nu_{k+\ell} : J_{k+\ell}E \to J_\ell(J_k E)$, $j_{k+\ell}\sigma(a) \mapsto j_\ell(j_k\sigma)(a)$, and the bundle map $J_k\phi : J_k E \to J_k F$ associated to a bundle map $\phi : E \to F$, $j_k\sigma(a) \mapsto j_k(\phi \circ \sigma)(a)$. The functor $J_k : \phi \mapsto J_k\phi$ is exact.

In the next section we redefine jet sheaves in a way more suitable to our purpose. The different structures as well as the maps i_k and j_k will appear more naturally. In fact, left and right structures turn out to be essentially the same thing, all depending on how we choose to look at jets. The present section was intended for the reader who is familiar with the classical definitions of Ehresmann [2]. Later we shall transcribe some of the results in terms of these.

3. *The prolongation space* $X_{(k)}$

Let Δ be the diagonal of $X^2 = X \times X$, π_1 (resp. π_2) the projection on the first (resp. second) factor and denote by x (resp. y) the variable in the first (resp. second) factor of X^2. Denote by $\Delta : X \to X^2$ the canonical injection $x \mapsto (x, x)$. Vector bundles and sheaves on X can be transported to corresponding objects on the diagonal Δ via the map Δ and vice-versa. Nevertheless we shall not identify X with Δ since X will play different roles according to whether it is considered as the horizontal (first) component or the vertical (second) component of X^2. These roles will in fact be unsymmetrical since we shall be forced later to give preference to one of the components. For some obscure reason preference will be given to the horizontal component.

Loosely speaking, we wish to extend (or prolong) the structure sheaf $\mathcal{O}_\Delta \simeq \mathcal{O}$ of the diagonal to an *infinitesimal neighborhood of order* k of Δ in X^2. Let $\mathcal{O}_{X^2}$ be the structure sheaf of X^2 and $\mathcal{I}$ the ideal of $\mathcal{O}_{X^2}$ of (germs of) functions vanishing on the diagonal Δ (defining ideal of the diagonal). $\mathcal{I}^{k+1}$ is the ideal of (germs of) functions flat to order $k+1$ on Δ (i.e., $j_k f(p) = 0$, $\forall p \in \Delta$). Following Malgrange [9], define the sheaf of rings

$$\mathcal{J}_k = \mathcal{O}_{X^2} / \mathcal{I}^{k+1} .$$

It vanishes outside the diagonal; hence, we shall consider $\mathcal{J}_k$ as a sheaf on Δ and $\mathcal{O}_{X^2}$ will usually mean the restriction of this sheaf to the diagonal. $\mathcal{O}$ can be imbedded as a subsheaf of $\mathcal{J}_k$ in essentially two different ways. A function f on X can be lifted to X^2 either by π_1 or π_2. Lifting by π_1 gives rise to a (sheaf of) ring morphism $i: \mathcal{O} \to \mathcal{O}_{X^2}$, $f \mapsto f \circ \pi_1$ and lifting by π_2 gives rise to $j: \mathcal{O} \to \mathcal{O}_{X^2}$, $f \mapsto f \circ \pi_2$. These maps impart to $\mathcal{O}_{X^2}$ two permuting $\mathcal{O}$-module structures: the *left* structure defined by i and the *right* structure defined by j in such a way that $\mathcal{O}_{X^2}$ becomes a left, right $\mathcal{O}$-algebra. The left, right terminology and the corresponding notation serve only to distinguish the two structures. If $f \in \mathcal{O}$ and $g \in \mathcal{O}_{X^2}$ then $(f \cdot g)(x, y) = f(x)\, g(x, y)$ and $(g \cdot f)(x, y) = g(x, y)\, f(y)$. The sheaf $\mathcal{I}^{k+1}$ is a left and right ideal hence the algebra structures pass to the quotient and define left, right $\mathcal{O}$-algebra structures on $\mathcal{J}_k$. It is clear that both structures coincide when restricted to constants. The maps i and j composed with the quotient map yield injective ring morphisms $i_k, j_k: \mathcal{O} \to \mathcal{J}_k$ which are the two imbeddings of $\mathcal{O}$ in $\mathcal{J}_k$. If $f \in \mathcal{O}$ and $A \in \mathcal{J}_k$ then $(i_k f) \cdot A = f \cdot A$ and $(j_k f) \cdot A = A \cdot f$ where left hand terms are ring products and right hand terms are left, right scalar multiplications. It shows that i_k is left $\mathcal{O}$-linear and j_k is right $\mathcal{O}$-linear.

Moreover the map $\beta_k : \mathcal{J}_k \to \mathcal{O}$ induced by $f \in \mathcal{O}_{X^2} \to f \circ \Delta \in \mathcal{O}$ is a left, right algebra morphism and $\beta_k \circ i_k = \beta_k \circ j_k = \mathrm{Id}$. Hence i_k is left split injective, j_k is right split injective and, for both imbeddings, $\mathcal{J}_k$ is an inessential extension of the subalgebra $\mathcal{O}$ by the ideal $\ker \beta_k = \mathcal{J}/\mathcal{J}^{k+1}$. For $k = 0$, the left and right structures coincide and for $k > 0$ they are isomorphic. To see this, let

$$r : \mathcal{O}_{X^2} \to \mathcal{O}_{X^2}$$

be the map induced by the reflection on the diagonal, i.e., $r(f)(x, y) = f(y, x)$. This map is an involutive $(r^2 = \mathrm{Id})$ $\mathcal{O}$-algebra isomorphism when terms are given opposite structures. Since $\mathcal{J}^{k+1}$ is invariant, r factors to the quotient and defines an involutive $\mathcal{O}$-algebra isomorphism $r_k : \mathcal{J}_k \to \mathcal{J}_k$ which permutes left and right structures. The following relations hold: $r_k \circ j_k = i_k$, $r_k \circ i_k = j_k$ and $r_0 = \mathrm{Id}$. $\mathcal{J}_k$ contains nilpotent elements for $k > 0$ and $\mathcal{J}_0$ identifies with $\mathcal{O}$. In fact, $i_0 = j_0 : \mathcal{O} \to \mathcal{J}_0$ is an $\mathcal{O}$-algebra isomorphism whose inverse is β_0.

$\mathcal{J}_k$ is called the sheaf of k-jets (of functions) and the manifold Δ with the structure sheaf $\mathcal{J}_k$ is called a *prolongation space of order* k of X or a *generalized differentiable manifold with nilpotent elements.* We denote the prolongation space by $X_{(k)}$ so that $\mathcal{O}_{X_{(k)}} = \mathcal{J}_k$.

We shall now relate $\mathcal{J}_k$ with $\underline{J_k}$. Let g be a function on X^2 defined in a neighborhood of a point $(a, a) \in \Delta$ and for each x near a, let g_x be the function defined in a neighborhood of $x \in X$ by $g_x(y) = g(x, y)$. If $[g]_{(a,a)}$ represents the germ of g at (a, a), there is a well defined sheaf map $\mathcal{O}_{X^2} \to \underline{J_k}$, $[g]_{(a,a)} \mapsto [x \mapsto j_k g_x(x)]_a$, which factors to

$$\Pi_2 : \mathcal{J}_k \to \underline{J_k}\ .$$

It is easy to show that Π_2 is a left, right $\mathcal{O}$-algebra isomorphism. Similarly, if we define g_y by $g_y(x) = g(x, y)$ then $[g] \mapsto [y \mapsto j_k g_y(y)]$ factors to an $\mathcal{O}$-algebra isomorphism

$$\Pi_1 : \mathcal{J}_k \to \underline{J_k}$$

when terms are given opposite structures. In other words, Π_2 identifies $\mathcal{J}_k$ with $\underline{J_k}$ while Π_1 permutes left and right structures. It follows that $\mathcal{J}_k$ is locally free of finite rank for both left and right $\mathcal{O}$-module structures since both are isomorphic to the natural structure of $\underline{J_k}$. Write

$$\bar{r}_k = \Pi_2 \circ \Pi_1^{-1} : \underline{J_k} \to \underline{J_k} \, .$$

The commutativity of the diagram

$$\begin{array}{ccc} \mathcal{J}_k & \xrightarrow{\ r\ } & \mathcal{J}_k \\ {\scriptstyle \Pi_1}\downarrow & & \downarrow{\scriptstyle \Pi_2} \\ \underline{J_k} & \xrightarrow{\ \mathrm{Id}\ } & \underline{J_k} \end{array}$$

implies the commutativity of

$$\begin{array}{ccc} \mathcal{J}_k & \xrightarrow{\ r_k\ } & \mathcal{J}_k \\ {\scriptstyle \Pi_1}\downarrow & & \downarrow{\scriptstyle \Pi_1} \\ \underline{J_k} & \xrightarrow{\ \bar{r}_k\ } & \underline{J_k} \end{array} \qquad \begin{array}{ccc} \mathcal{J}_k & \xrightarrow{\ r_k\ } & \mathcal{J}_k \\ {\scriptstyle \Pi_2}\downarrow & & \downarrow{\scriptstyle \Pi_2} \\ \underline{J_k} & \xrightarrow{\ \bar{r}_k\ } & \underline{J_k} \end{array}$$

The map $\bar{r}_k$ is an involutive $\mathcal{O}$-algebra isomorphism which permutes left and right structures. The following relations hold: $\bar{r}_k \circ j_k = i_k$, $\bar{r}_k \circ i_k = j_k$, $\Pi_2 \circ i_k = i_k$, $\Pi_2 \circ j_k = j_k$, $\Pi_1 \circ i_k = j_k$ and $\Pi_1 \circ j_k = i_k$.

4. *Prolongation of sheaves*

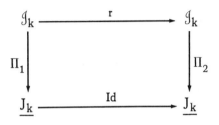

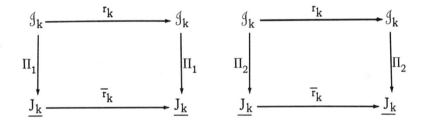

Let E be a vector bundle on X. $\underline{E}$ is an $\mathcal{O}$-module and we want to extend it to a $\mathcal{J}_k$-module. There are several (though equivalent) ways of achieving this and we shall start by generalizing the extension procedure of $\mathcal{O}$, namely extending $\underline{E}$ to an *infinitesimal neighborhood of order* k of the diagonal.

First observe that $\mathcal{O}_{X^2}$ is the sheaf of germs of sections of the trivial line bundle $X^2 \times R$. If E is trivial, i.e., $E = X \times V$ with V a finite dimensional real vector space, then Section 3 carries over on replacing R by V, hence $\mathcal{O}_{X^2}$ by $\underline{X^2 \times V} = \underline{F}$. The sheaf $\underline{F}$ is a module over the left, right $\mathcal{O}$-algebra $\mathcal{O}_{X^2}$ and $\mathcal{J}^{k+1}\underline{F}$ is the submodule of (germs of) sections which are flat to order $k+1$ on the diagonal Δ. The quotient sheaf

$$\mathcal{J}_k E = \underline{F}/\mathcal{J}^{k+1}\underline{F}$$

vanishes outside the diagonal, hence will be considered as a sheaf on Δ. It carries a module structure over the left, right $\mathcal{O}$-algebra $\mathcal{O}_{X^2}$. Since $\mathcal{J}^{k+1}$ is the annihilator of $\mathcal{J}_k E$, this structure factors to $\mathcal{J}_k$ hence $\mathcal{J}_k E$ becomes a module over the left, right $\mathcal{O}$-algebra $\mathcal{J}_k$. We define similarly split injective morphisms $i_k: \underline{E} \to \mathcal{J}_k E$ (left $\mathcal{O}$-linear), $j_k: \underline{E} \to \mathcal{J}_k E$ (right $\mathcal{O}$-linear) and an r_k-semilinear isomorphism $r_k: \mathcal{J}_k E \to \mathcal{J}_k E$ (the first r_k refers to $\mathcal{J}_k \to \mathcal{J}_k$) which permutes left and right structures and i_k with j_k. We also define the Π_2-semilinear isomorphism

$$\Pi_2: \mathcal{J}_k E \to \underline{J_k E}$$

which preserves left and right structures and the Π_1-semilinear isomorphism

$$\Pi_1: \mathcal{J}_k E \to \underline{J_k E}$$

which permutes left and right structures. The $\mathcal{O}$-module $\underline{E}$ is thus extended to the $\mathcal{J}_k$-module $\mathcal{J}_k E$ in two different ways, one corresponding to i_k (hence to the left structure of $\mathcal{J}_k E$) and the other to j_k (right structure). These two imbeddings of $\underline{E}$ are permuted by r_k.

Consider now a vector bundle F on X^2. Trivializing F along the diagonal we obtain a local situation analogous to the one discussed above. The intrinsic description is as follows. The sheaf

$$\mathcal{G}_k F = \underline{F}/\mathcal{J}^{k+1}\underline{F}$$

is again a module over the left, right $\mathcal{O}$-algebra $\mathcal{J}_k$ whose support is the diagonal Δ. Taking representatives $\sigma \in \underline{F}$ and $f \in \mathcal{O}$, the left and right $\mathcal{O}$-module structures are given respectively by $f \cdot \sigma = \tau$ with $\tau(x, y) = f(x)\sigma(x, y)$ and $\sigma \cdot f = \nu$ with $\nu(x, y) = f(y)\sigma(x, y)$. We will show that the left and right $\mathcal{O}$-module structures are isomorphic to the natural structure of a vector bundle (sheaf of sections). For any $x \in X$ (first factor of X^2) let $F_x = F|\pi_1^{-1}(x)$ and for any $y \in X$ (second factor) let $F_y = F|\pi_2^{-1}(y)$. Define

$$F_k = \bigcup_{x \in X} (J_k F_x)_x \quad \text{and} \quad F'_k = \bigcup_{y \in X} (J_k F_y)_y \,.$$

Since F is locally trivial it follows easily that F_k and F'_k are locally trivial vector bundles with base space X and, using similar definitions as in Section 2, the sheaves $\underline{F_k}$ and $\underline{F'_k}$ carry module structures over the left, right $\mathcal{O}$-algebra $\underline{J_k}$ (the left $\mathcal{O}$-module structure coincides with the natural one). Furthermore, imitating Section 3 (for any section σ of F consider σ_x and σ_y), we define the Π_2-semilinear isomorphism

$$\Pi_2 : \mathcal{G}_k F \to \underline{F_k}$$

which preserves left and right $\mathcal{O}$-module structures and the Π_1-semilinear isomorphism

$$\Pi_1 : \mathcal{G}_k F \to \underline{F'_k}$$

which permutes left and right $\mathcal{O}$-module structures.

Let now E be a vector bundle on X. E can be lifted to X^2 in two different ways: $\pi_1^* E$ and $\pi_2^* E$. We set

$$\mathcal{J}'_k E = \mathcal{G}_k \pi_1^* E = \underline{\pi_1^* E} / \mathcal{J}^{k+1} \underline{\pi_1^* E}$$

and

$$\mathcal{J}_k E = \mathcal{G}_k \pi_2^* E \,.$$

$\mathcal{J}_k E$ is a module over the left, right $\mathcal{O}$-algebra $\mathcal{J}_k$, $(\pi_2^* E)_k = J_k E$ and $(\pi_2^* E)'_k$ is the vector bundle on X whose fibre over $a \in X$ is the vector space $\{j_k \sigma(a) | \sigma : X \to E_a\}$. It follows that $\mathcal{J}_k E$ identifies with $\underline{J_k E}$ since

$$\Pi_2 : \mathcal{J}_k E \to \underline{J_k E}$$

is an isomorphism for all the structures, whereas

$$\Pi_1 : \mathcal{J}_k E \to \underline{(\pi_2^* E)'_k}$$

permutes the $\mathcal{O}$-module structures. Since a section σ of E can be lifted to a section $j\sigma = \sigma \circ \pi_2$ of $\pi_2^* E$, this yields an injective right $\mathcal{O}$-linear morphism

$$j_k : \underline{E} \to \mathcal{J}_k E$$

which is transformed by Π_2 into $j_k : \underline{E} \to \underline{J_k E}$ and by Π_1 into the left $\mathcal{O}$-linear injective map $i_k : \underline{E} \to \underline{(\pi_2^* E)'_k}$, $\sigma \mapsto \{x \mapsto j_k(\sigma(x))(x)\}$ which is the extension, to germs, of the injective bundle map $E \to (\pi_2^* E)'_k$, $v \mapsto j_k \sigma(x)$, where $v \in E_x$ and $\sigma : X \to E_x$ is the constant map with value v. The map

$$\beta_k : \mathcal{J}_k E \to \underline{E}$$

induced by $\sigma \in \underline{\pi_2^* E} \mapsto \sigma|_\Delta \in \underline{E}$ is β_k-semilinear $(\beta_k : \mathcal{J}_k \to \mathcal{O})$ and $\beta_k \circ j_k = \mathrm{Id}$, hence j_k is right split injective.

Similarly, $\mathcal{J}'_k E$ is a module over the left, right $\mathcal{O}$-algebra $\mathcal{J}_k$ and

$$\Pi_1 : \mathcal{J}'_k E \to \underline{J_k E}$$

is an isomorphism of $\mathcal{J}_k \simeq \underline{J_k}$-modules which permutes left and right $\mathcal{O}$-module structures $(\mathcal{J}_k \simeq \underline{J_k}$ via $\Pi_1)$. The map

$$\Pi_2 : \mathcal{J}'_k E \to \underline{(\pi_1^* E)_k} = \underline{(\pi_2^* E)'_k}$$

is an isomorphism for all the structures. Lifting sections by π_1, we define the injective left $\mathcal{O}$-linear morphism

$$i_k : \underline{E} \to \mathcal{J}'_k E$$

which is transformed by Π_1 into j_k and by Π_2 into $i_k : \underline{E} \to \underline{(\pi_1^* E)_k} = \underline{(\pi_2^* E)'_k}$. We also define the β_k-semilinear map

$$\beta_k : \mathcal{J}'_k E \to \underline{E} \, ,$$

$\beta_k \circ i_k = \mathrm{Id}$ and i_k is left split injective. The $\mathcal{O}$-module $\underline{E}$ is thus extended via i_k (hence the left structure) to the $\mathcal{J}_k$-module $\mathcal{J}'_k E$. The reflection on the diagonal induces an r_k-semilinear isomorphism

$$r_k : \mathcal{J}'_k E \to \mathcal{J}_k E$$

which permutes left, right structures and the maps i_k, j_k. The map $\Pi_2 \circ r_k \circ \Pi_1^{-1}$ is the identity of $\underline{J_k E}$ and $\Pi_1 \circ r_k \circ \Pi_2^{-1}$ is the identity of $\underline{(\pi_1^* E)_k}$. For $k = 0$, the left and right $\mathcal{O}$-module structures and the $\mathcal{J}_0 = \mathcal{O}$-module structures on $\mathcal{J}'_0 E$ and $\mathcal{J}_0 E$ all coincide. Since $r_0 = \mathrm{Id} : \mathcal{J}_0 \to \mathcal{J}_0$, the map $r_0 : \mathcal{J}'_0 E \to \mathcal{J}_0 E$ is an $\mathcal{O}$-linear isomorphism so that we can identify $\mathcal{J}'_0 E$ with $\mathcal{J}_0 E$. Moreover the maps $\beta_0 : \mathcal{J}'_0 E \to \underline{E}$ and $\beta_0 : \mathcal{J}_0 E \to \underline{E}$ are $\mathcal{O}$-linear isomorphisms inverse to i_0 and j_0 respectively. The composition

$$\underline{E} \xrightarrow{i_0} \mathcal{J}'_0 E \xrightarrow{r_0} \mathcal{J}_0 E \longrightarrow \underline{E}$$

is the identity.

Let $\mathcal{E}$ be a sheaf of $\mathcal{O}$-modules on X. We want to extend it to a $\mathcal{J}_k$-module. The above discussed procedure which applies to locally free sheaves of finite rank does not extend to $\mathcal{E}$ since, for example, $\underline{\pi_2^* E}$ is different from $\pi_2^* \underline{E}$. The latter is the subsheaf of the former whose elements are the germs of sections $\sigma(x, y)$ of $\pi_2^* E$ which are independent of the first variable. Let us agree that in tensor products of $\mathcal{O}$-modules containing a bi-module as a factor (e.g., $\mathcal{J}_k$, $\underline{J_k}$) the position of this factor will indicate the structure to be used. Define

$$\mathcal{J}_k \mathcal{E} = \mathcal{J}_k \otimes_{\mathcal{O}} \mathcal{E}$$

where the tensor product is taken with respect to the right structure of $\mathcal{J}_k$. The left, right $\mathcal{O}$-algebra structure of $\mathcal{J}_k$ induces on $\mathcal{J}_k\mathcal{E}$ a module structure over the left, right algebra $\mathcal{J}_k$ (extension of scalars via j_k). Let $1 \in \mathcal{J}_k$ be the equivalence class of $1 \in \mathcal{O}_{X^2}$ $(1 = j_k 1 = i_k 1)$ and define

$$j_k : \mathcal{E} \to \mathcal{J}_k\mathcal{E}$$

by $\mu \mapsto 1 \otimes \mu$. This map is right $\mathcal{O}$-linear hence $j_k(f\mu) = (j_k f) \otimes \mu$. It is split injective (right structure) since $j_k = j_k \otimes \mathrm{Id} : \mathcal{O} \otimes_{\mathcal{O}} \mathcal{E} \to \mathcal{J}_k \otimes_{\mathcal{O}} \mathcal{E}$ and $j_k : \mathcal{O} \to \mathcal{J}_k$ is split injective for the right structure. If $\mathcal{E}$ is locally free, $\mathcal{E} = \underline{E}$, then the map

$$\mathcal{J}_k \otimes_{\mathcal{O}} \underline{E} \xrightarrow{\psi_k} \mathcal{J}_k E$$

defined by $A \otimes \eta \mapsto A \cdot j_k \eta$ (standard extension of $j_k : \underline{E} \to \mathcal{J}_k E$) is an isomorphism of $\mathcal{J}_k$-modules which commutes with the maps j_k. The map ψ_k is induced by the isomorphism $f \otimes \sigma \in \mathcal{O}_{X^2} \otimes_{\mathcal{O}} \underline{E} \mapsto f(j\sigma) \in \underline{\pi_2^* E}$ and bijectivity results from the exactness of

$$0 \to \mathcal{J}^{k+1} \otimes_{\mathcal{O}} \underline{E} \to \mathcal{O}_{X^2} \otimes_{\mathcal{O}} \underline{E} \to \mathcal{J}_k \otimes_{\mathcal{O}} \underline{E} \to 0 .$$

We can as well make the extension of scalars via i_k. In this case we define the $\mathcal{J}_k$-module

$$\mathcal{J}'_k\mathcal{E} = \mathcal{E} \otimes_{\mathcal{O}} \mathcal{J}_k$$

and the split injective left $\mathcal{O}$-linear map

$$i_k : \mathcal{E} \to \mathcal{J}'_k\mathcal{E}$$

by $\mu \to \mu \otimes 1$. If $\mathcal{E} = \underline{E}$ then $i_k : \underline{E} \to \mathcal{J}'_k E$ extends to a $\mathcal{J}_k$-module isomorphism

$$\underline{E} \otimes_{\mathcal{O}} \mathcal{J}_k \xrightarrow{\psi'_k} \mathcal{J}'_k E$$

which commutes with the maps i_k.

The two extensions differ by the r_k-semilinear isomorphism

$$r_k : \mathcal{E} \otimes_{\mathcal{O}} \mathcal{J}_k \to \mathcal{J}_k \otimes_{\mathcal{O}} \mathcal{E}$$

defined by $\eta \otimes A \mapsto (r_k A) \otimes \eta$ which permutes left, right $\mathcal{O}$-module structures and the maps i_k, j_k. The diagram

$$\begin{array}{ccc} \underline{E} \otimes_{\mathcal{O}} \mathcal{J}_k & \xrightarrow{\;r_k\;} & \mathcal{J}_k \otimes_{\mathcal{O}} \underline{E} \\ \downarrow{\scriptstyle \psi'_k} & & \downarrow{\scriptstyle \psi_k} \\ \mathcal{J}'_k E & \xrightarrow{\;r_k\;} & \mathcal{J}_k E \end{array}$$

is commutative.

We can finally combine the two extension procedures and consider the sheaf

$$\mathcal{E} \otimes_{\mathcal{O}} \mathcal{J}_k \otimes_{\mathcal{O}} \mathcal{F}$$

which carries a module structure over the left, right $\mathcal{O}$-algebra $\mathcal{J}_k$. If $\mathcal{E} = \underline{E}$ and $\mathcal{F} = \underline{F}$ then $\underline{E} \otimes_{\mathcal{O}} \mathcal{J}_k \otimes_{\mathcal{O}} \underline{F}$ is isomorphic to

$$\mathcal{G}_k H = \underline{H} / \mathcal{J}^{k+1} \underline{H}$$

with $H = \pi_1^* E \otimes \pi_2^* F$. The $\mathcal{O}_{X^2}$-isomorphism

$$\underline{E} \otimes_{\mathcal{O}} \mathcal{O}_{X^2} \otimes_{\mathcal{O}} \underline{F} \to \underline{H}$$

defined by $\sigma \otimes f \otimes \tau \mapsto f\, i(\sigma) \otimes j(\tau)$, factors to the desired $\mathcal{J}_k$-isomorphism.

5. *Coordinates*

Let $x = (x_1, \ldots, x_n)$ be a system of local coordinates on X considered as first factor of X^2 and denote by $y = (y_1, \ldots, y_n)$ the same system of coordinates on X considered as second factor. Then

$$(x,y) = (x_1, \ldots, x_n, y_1, \ldots, y_n)$$

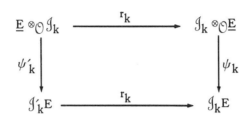

is a system of coordinates on X^2 and the diagonal, in this chart, is given by the equations $y_i - x_i = 0$. Let $\alpha = (\alpha_1, \dots, \alpha_n)$ be a multi-index, α_i integers ≥ 0, and recall the usual notation:

$$|\alpha| = \Sigma\alpha_i,\ \alpha! = \alpha_1! \dots \alpha_n!,\ 1_i = (0, \dots, 0, \alpha_i = 1, 0, \dots, 0),\ (y-x)^\alpha =$$

$$(y_1 - x_1)^{\alpha_1} \dots (y_n - x_n)^{\alpha_n},\ D_x^\alpha f(x, y) = \frac{\partial^{|\alpha|} f}{\partial x_1^{\alpha_1} \dots \partial x_n^{\alpha_n}} (x, y)$$

(partial derivative with respect to the x coordinates) and similarly for $D_y^\alpha f(x, y)$. Let $f(x, y)$ be a differentiable function. By Taylor's expansion with respect to the y coordinates along the diagonal, we infer that

$$f(x, y) \equiv \sum_{|\alpha| \leq k} D_y^\alpha f(x, x) \frac{(y-x)^\alpha}{\alpha!} \ \mathrm{mod}\ \mathcal{J}^{k+1}\ .$$

Moreover, two polynomial functions

$$\sum_{|\alpha| \leq k} a_\alpha(x)(y-x)^\alpha \qquad \text{and} \qquad \sum_{|\alpha| \leq k} b_\alpha(x)(y-x)^\alpha$$

are equivalent $\mathrm{mod}\ \mathcal{J}^{k+1}$ if and only if $a_\alpha = b_\alpha$ for all α. It follows that any element $A \in \mathcal{J}_k(a, a)$ is determined by a unique germ of polynomial

$$\sum_{|\alpha| \leq k} a_\alpha(x) \frac{(y-x)^\alpha}{\alpha!}$$

at the point (a, a) or, equivalently, by the germs of the coefficients $a_\alpha(x)$ at the point a. If $g \in \mathcal{O}$ then

$$g \cdot A = \sum_{|\alpha| \leq k} g(x)\, a_\alpha(x) \frac{(y-x)^\alpha}{\alpha!},\ A \cdot g = \sum_{|\alpha| \leq k} \left(\sum_{\beta+\gamma=\alpha} \frac{\alpha!}{\beta!\,\gamma!} a_\beta(x)\, D^\gamma g(x) \right) \frac{(y-x)^\alpha}{\alpha!}$$

and

$$A \cdot B = \sum_{|\alpha| \leq k} \left(\sum_{\beta+\gamma=\alpha} \frac{a_\beta\, b_\gamma}{\beta!\,\gamma!} \right) (y-x)^\alpha.$$

Left multiplication shows that $\mathcal{J}_k$ is locally (left) generated by the monomials $\frac{(y-x)^\alpha}{\alpha!}$ and these are linearly independent, hence form a local left basis. Π_2 transforms the section

$$\sum_{|\alpha|\leq k} a_\alpha(x) \frac{(y-x)^\alpha}{\alpha!}$$

into the section of J_k whose jet coordinates at x are the $a_\alpha(x)$. Applying the isomorphism r_k (or expanding with respect to x) we have similar conclusions when x is interchanged with y and Π_2 with Π_1. Since $(x-y)^\alpha = (-1)^{|\alpha|}(y-x)^\alpha$ it follows that $\frac{(y-x)^\alpha}{\alpha!}$ is a left, right basis of $\mathcal{J}_k$. Call $i_k\mathcal{O} \subset \mathcal{J}_k$ the left holonomic elements of $\mathcal{J}_k$ and $j_k\mathcal{O}$ the right holonomic elements. Expanding $\frac{(y-x)^\alpha}{\alpha!}$ we also obtain a left basis $\frac{y^\alpha}{\alpha!}$ and a right basis $\frac{x^\alpha}{\alpha!}$ which shows that $\mathcal{J}_k$ is left $\mathcal{O}$-generated by its right holonomic elements and right $\mathcal{O}$-generated by its left holonomic elements.

Let $\mathcal{E}$ be locally free of finite rank with local basis $\{\phi_\mu\}$. The sheaf $\mathcal{J}_k \otimes_{\mathcal{O}} \mathcal{E}$ is locally free of finite rank for both left and right structures and $\frac{(y-x)^\alpha}{\alpha!} \otimes \phi_\mu$ is a left, right basis. Assume further that $\mathcal{E} = \underline{E}$ and trivialize E with respect to ϕ_μ. Then

$$\sum_{|\alpha|\leq k} a_\alpha^\mu(x) \frac{(y-x)^\alpha}{\alpha!} \otimes \phi_\mu$$

is mapped by ψ_k into

$$\sum_{|\alpha|\leq k} a_\alpha^\mu(x) \frac{(y-x)^\alpha}{\alpha!} \phi_\mu(y) \ \mathit{mod}\ \mathcal{J}^{k+1} \underline{\pi_2^* E}$$

and this is mapped by Π_2 into the section of $J_k E$ whose jet coordinates (with respect to the given trivialization) are the a_α^μ. If we replace (locally) E by its trivialization $U \times R^N$ and correspondingly $\pi_2^* E$ by $U^2 \times R^N$ then, for any section $\tau(x,y)$ of $U^2 \times R^N$, Taylor's expansion with respect to y yields

$$\tau(x,y) \equiv \sum_{|\alpha|\leq k} b_\alpha(x) \frac{(y-x)^\alpha}{\alpha!} \ \mathit{mod}\ \mathcal{J}^{k+1} \underline{U^2 \times R^N}$$

where the b_α's are functions with values in R^N and $b_\alpha = \Sigma\, a_\alpha^\mu\, e_\mu$ where e_μ is the standard basis of R^N. On the other hand, since $\pi_2^* E$ is trivial in the horizontal direction (y fixed), Taylor's expansion with respect to x is well defined; hence any $A \in \mathcal{J}_k E$ is uniquely represented by a (germ of) polynomial section

$$\sum_{|\alpha| \leq k} b_\alpha(y) \frac{(x-y)^\alpha}{\alpha!}$$

of $\pi_2^* E$ where the b_α's are sections of E. This corresponds to the unique representation

$$\sum_{|\alpha| \leq k} \frac{(x-y)^\alpha}{\alpha!} \otimes b_\alpha$$

of elements in $\mathcal{J}_k \otimes_{\mathcal{O}} \mathcal{E}$.

Similar results hold for $\mathcal{E} \otimes_{\mathcal{O}} \mathcal{J}_k$ and $\mathcal{J}'_k E$. For later use we write the local expression of $r_k : \mathcal{J}_k \to \mathcal{J}_k$, namely

$$\sum_{|\alpha| \leq k} \frac{a_\alpha(x)}{\alpha!} (y-x)^\alpha \mapsto \sum_{|\alpha+\beta| \leq k} (-1)^{|\alpha|} \frac{D^\beta a_\alpha(x)}{\alpha!\, \beta!} (y-x)^{\alpha+\beta} .$$

6. *The first linear complex*

We shall now be mainly concerned with sheaves of jets of tensor bundles on X. For computational reasons we introduce some identifications. Let T be the tangent bundle of X and T^* the cotangent bundle. TX^2 is the direct sum of two sub-bundles $TX^2 = T_H \oplus T_V$ where T_H is the bundle of *horizontal* vectors (annihilated by π_{2*}) and T_V is the bundle of *vertical* vectors (annihilated by π_{1*}). We denote by $H: TX^2 \to T_H \subset TX^2$ and $V: TX^2 \to T_V \subset TX^2$ the corresponding projections. By duality, $T^*X^2 = T_H^* \oplus T_V^*$, where T_H^* is the bundle of *horizontal* forms (annihilator of T_V in T^*X^2) and T_V^* the bundle of *vertical* forms (annihilator of T_H in T^*X^2). The corresponding horizontal and vertical projections are the transposed maps H^* and V^*. We can identify π_1^*T with T_H, π_2^*T with T_V, $\pi_1^*T^*$ with T_H^* and $\pi_2^*T^*$ with T_V^* so that the jet sheaves $\mathcal{J}'_kT$, $\mathcal{J}_kT$, etc., are now defined as quotient sheaves of $\underline{T_H}$, $\underline{T_V}$, etc. The morphisms i and j are obviously defined: vector fields on X are lifted to horizontal or vertical fields on X^2 and differential forms are pulled back by the projections π_i. The derivative of the reflection $(x, y) \to (y, x)$ and its transpose yield the isomorphism r_k. The map r_0 is an $\mathcal{O}$-linear isomorphism and the inverses of i_0 and j_0 are the quotients of the maps $\theta \mapsto \pi_{i*}(\theta|_\Delta)$ for vector fields and $\omega \mapsto \Delta^*\omega$ for differential forms where $\Delta: X \to X^2$ is the diagonal inclusion. In this context, $\pi_1^* \Lambda^p T^*$ identifies with the sub-bundle $\Lambda^p T_H^*$ of $\Lambda^p T^* X^2$ whose elements are the forms ω such that $\omega(v_1, \ldots, v_p) = \omega(Hv_1, \ldots, Hv_p)$, hence sections of $\pi_1^* \Lambda^p T^*$ identify with exterior differential p-forms ω on X^2 satisfying $\omega(\theta_1, \ldots, \theta_p) = \omega(H\theta_1, \ldots, H\theta_p)$ for any vector fields θ_i. Similarly, $\pi_2^* \Lambda^p T^*$ identifies with the sub-bundle $\Lambda^p T_V^*$ whose elements are the forms ω such that $\omega(v_1, \ldots, v_p) = \omega(Vv_1, \ldots, Vv_p)$.

Let

$$d_H: \underline{\Lambda^p T_H^*} \to \underline{\Lambda^{p+1} T_H^*}$$

be the exterior differential operator with respect to the x variable (alternatively, the covariant differential operator with respect to the trivial horizontal connection on X^2), namely

$$d_H\omega\,(\theta_1, \ldots, \theta_{p+1}) = d\omega(H\theta_1, \ldots, H\theta_{p+1}) .$$

The operator d_H can also be expressed in the following way. Let $(x, y) \in X^2$ and let ω_y be the restriction of ω to the submanifold $X \times \{y\} \subset X^2$. Then

$$(d_H\omega)_{(x,y)} = (d\omega_y) \circ H_{(x,y)} .$$

It is clear that d_H is right $\mathcal{O}$-linear and $d_H(\mathcal{J}^{k+1}\, \underline{\Lambda^p T_H^*}) \subset \mathcal{J}^k\, \underline{\Lambda^{p+1} T_H^*}$, hence it induces a right $\mathcal{O}$-linear quotient map

$$D : \mathcal{J}'_k \wedge^p T^* \to \mathcal{J}'_{k-1} \wedge^{p+1} T^* .$$

Recalling the isomorphism

$$\underline{\wedge^p T^*} \otimes_{\mathcal{O}} \mathcal{J}_k \xrightarrow{\psi'_k} \mathcal{J}'_k \wedge^p T^*$$

induced by $\omega \otimes f \mapsto f\pi_1^*\omega$, we obtain for any $p \geq 0$ a right $\mathcal{O}$-linear map

$$D : \underline{\wedge^p T^*} \otimes_{\mathcal{O}} \mathcal{J}_k \to \underline{\wedge^{p+1} T^*} \otimes_{\mathcal{O}} \mathcal{J}_{k-1} ,$$

hence a complex ($d_H \circ d_H = 0$ since the connection is integrable and $d_H(f(y)) = 0$)

$$(6.1)_k \quad 0 \longrightarrow \mathcal{O} \xrightarrow{j_k} \mathcal{J}_k \xrightarrow{D} \underline{T}^* \otimes \mathcal{J}_{k-1} \xrightarrow{D} \cdots \longrightarrow \underline{\wedge^n T^*} \otimes \mathcal{J}_{k-n} \longrightarrow 0$$

where it is understood that $\mathcal{J}_\ell = 0$ whenever $\ell < 0$ and the subscript $\mathcal{O}$ is omitted in $\otimes$. Tensoring on the right the above (right linear) complex by an $\mathcal{O}$-module $\mathcal{E}$, we obtain the first linear complex for $\mathcal{E}$

$$(6.2)_k \quad 0 \longrightarrow \mathcal{E} \xrightarrow{j_k} \mathcal{J}_k \otimes \mathcal{E} \xrightarrow{D} \underline{T}^* \otimes \mathcal{J}_{k-1} \otimes \mathcal{E} \xrightarrow{D} \cdots \xrightarrow{D} \underline{\wedge^n T^*} \otimes \mathcal{J}_{k-n} \otimes \mathcal{E} \longrightarrow 0$$

where each operator $D = D \otimes \mathrm{Id}$ is right $\mathcal{O}$-linear. We claim that $(6.2)_k$ is split exact (splitting with respect to the right structure). Before proving this we shall make some very naive computations and show exactness in two special cases.

Let $\omega \otimes f \in \underline{\wedge^p T^*} \otimes \mathcal{O}_{X^2}$ represent a decomposable element of $\underline{\wedge^p T^*} \otimes \mathcal{J}_k$. Then $\omega \otimes f$ identifies with $f\pi_1^*\omega \in \underline{\wedge^p T_H^*}$ and

$$d_H(f\pi_1^*\omega) = f\pi_1^* d\omega + \sum_{1 \leq i \leq n} \frac{\partial f}{\partial x_i} \pi_1^*(dx_i \wedge \omega)$$

hence

$$d_H(\omega \otimes f) = d\omega \otimes f + \sum_i (dx_i \wedge \omega) \otimes \frac{\partial f}{\partial x_i}.$$

Setting

$$f \equiv \sum_{|\alpha| \leq k} a_\alpha(x) \frac{(y-x)^\alpha}{\alpha!} \mod \mathcal{J}^{k+1}$$

then

$$s = \omega \otimes \sum_{|\alpha| \leq k} a_\alpha(x) \frac{(y-x)^\alpha}{\alpha!} \in \underline{\Lambda^p T^*} \otimes \mathcal{J}_k$$

is mapped into

$$Ds = d\omega \otimes \sum_{|\alpha| < k-1} a_\alpha(x) \frac{(y-x)^\alpha}{\alpha!} +$$

$$\sum_{1 \leq i \leq n} (dx_i \wedge \omega) \otimes \sum_{|\alpha| \leq k-1} \left(\frac{\partial a_\alpha}{\partial x_i} - a_{\alpha+1_i} \right)(x) \frac{(y-x)^\alpha}{\alpha!}$$

In particular

$$D\left(\sum_{|\alpha| \leq k} a_\alpha(x) \frac{(y-x)^\alpha}{\alpha!} \right) = \sum_{1 \leq i \leq n} dx_i \otimes \sum_{|\alpha| \leq k-1} \left(\frac{\partial a_\alpha}{\partial x_i} - a_{\alpha+1_i} \right)(x) \frac{(y-x)^\alpha}{\alpha!}$$

hence $\ker(\mathcal{J}_k \xrightarrow{D} \underline{T}^* \otimes \mathcal{J}_{k-1})$ is the subsheaf of $\mathcal{J}_k$ composed of the elements

$$\sum_{|\alpha| \leq k} a_\alpha(x) \frac{(y-x)^\alpha}{\alpha!}$$

with $a_\alpha = D^\alpha a_0$, i.e., the subsheaf $j_k(\mathcal{O})$. Alternatively, the relation $d_H f \equiv 0 \mod \mathcal{J}^k \underline{T}^*_H$ means that $\partial f / \partial x_i \in \mathcal{J}^k$, hence $f(x,y) - g(y) \in \mathcal{J}^{k+1}$ with $g(x) = f(x,x)$, so that the class of f is equal to $j_k g$. The complex $(6.1)_k$ is exact at $\mathcal{J}_k$. It is also trivial to check that

$$(6.1)_1 \qquad 0 \longrightarrow \mathcal{O} \xrightarrow{j_1} \mathcal{J}_1 \longrightarrow \underline{T}^* \otimes \mathcal{J}_0 \simeq \underline{T}^* \longrightarrow 0$$

is exact. In fact, if $\omega = \sum_i a_i(x)\, dx_i$ is a section of $\underline{T}^*$ then

$$\omega = D(\sum_i a_i(y)\, x_i) = D(\sum_i a_i(x)(x_i - y_i)) .$$

To prove split exactness of $(6.2)_k$ it is enough to prove split exactness of $(6.1)_k$. Since d_H is horizontal differentiation it is more suitable in this case to represent the elements of $\mathcal{J}_k$ by their Taylor expansion with respect to x. Let β denote a multi-index such that $|\beta| = p$ and $\beta_i \leq 1$. Then $dx^{\wedge\beta} = dx_1^{\beta_1} \wedge \ldots \wedge dx_n^{\beta_n}$ (delete all terms $dx_i^{\beta_i}$ with $\beta_i = 0$), $|\beta| = p$, is a local basis for $\underline{\Lambda^p T^*}$ and $dx^{\wedge\beta} \otimes \frac{(x-y)^\alpha}{\alpha!}$ is a right local basis for $\underline{\Lambda^p T^*} \otimes \mathcal{J}_k$. The element

$$s = \sum_{|\beta|=p} dx^{\wedge\beta} \otimes \sum_{|\alpha|\leq k} a_{\alpha\beta}(y) \frac{(x-y)^\alpha}{\alpha!} \in \underline{\Lambda^p T^*} \otimes \mathcal{J}_k$$

is mapped into

$$Ds = \sum_{|\beta|=p} \sum_{1\leq i\leq n} dx_i \wedge dx^{\wedge\beta} \otimes \sum_{|\alpha|\leq k-1} a_{\alpha+1_i,\beta}(y) \frac{(x-y)^\alpha}{\alpha!} \in \underline{\Lambda^{p+1} T^*} \otimes \mathcal{J}_{k-1} .$$

If we fix $y = x_0$ and write $s(x_0)$ as an exterior differential form with polynomial coefficients (of degree $\leq k$) in the variable x, i.e.,

$$s(x_0) = \sum_{|\alpha|\leq k,\, \beta} a_{\alpha\beta}(x_0) \frac{(x-x_0)^\alpha}{\alpha!} dx^{\wedge\beta}$$

then the above expression for D shows that D_{x_0} is simply exterior differentiation:

$$(Ds)(x_0) = \sum_{1\leq i\leq n} \sum_{|\alpha|\leq k-1,\beta} a_{\alpha+1_i,\beta}(x_0) \frac{(x-x_0)^\alpha}{\alpha!} dx_i \wedge dx^{\wedge\beta} = d(s(x_0)) .$$

The intrinsic description is as follows: the complex $(6.1)_k$ is transformed by

$$\Pi_1 : \underline{\Lambda^p T^*} \otimes \mathcal{J}_{k-p} \to \underline{J_{k-p}\Lambda^p T^*}$$

into a complex of sheaves and left $\mathcal{O}$-linear operators which is simply the extension to germs of the following complex of vector bundles

$$(6.3)_k \qquad 0 \longrightarrow X \times R \xrightarrow{i_k} J_k \xrightarrow{d} J_{k-1}T^* \xrightarrow{d} \ldots \xrightarrow{d} J_{k-n}\Lambda^n T^* \longrightarrow 0$$

where

$$d : J_h \Lambda^p T^* \to J_{h-1} \Lambda^{p+1} T^*$$

is exterior differentiation in each fibre, namely $j_h\omega(x) \mapsto j_{h-1}d\omega(x)$. This complex is exact (Poincaré lemma for differential forms with polynomial coefficients) hence its extension to germs is left split exact. Applying Π_1^{-1} we obtain the right split exactness of $(6.1)_k$.

The above interpretation of D as exterior differentiation of polynomial differential forms is not intrinsic. In fact we have to take coordinates in order to represent k-jets by polynomials. However, we shall see later that restricting D to the kernel of

$$\underline{\Lambda^p T^*} \otimes \mathcal{J}_h \xrightarrow{\mathrm{Id}\otimes\rho_{h-1}} \underline{\Lambda^p T^*} \otimes \mathcal{J}_{h-1}$$

i.e., restricting D to differential p-forms with homogeneous polynomial coefficients, the description becomes intrinsic.

Let $\rho_h : \mathcal{J}_k \to \mathcal{J}_h$, $h \le k$, be the map induced by the identity on $\mathcal{O}_{X^2}$. It is a surjective left, right $\mathcal{O}$-algebra morphism, the sequence

$$0 \longrightarrow \mathcal{I}^k / \mathcal{I}^{k+1} \longrightarrow \mathcal{J}_k \xrightarrow{\rho_{k-1}} \mathcal{J}_{k-1} \longrightarrow 0$$

is left, right split exact (in each stalk), the left, right $\mathcal{O}$-module structures on $\mathcal{k}er\, \rho_{k-1}$ coincide and the $\mathcal{J}_k$-module structure reduces to the $\mathcal{O}$-module structure via the map $\beta_k : \mathcal{J}_k \to \mathcal{O}$. We want to identify $\mathcal{k}er\, \rho_{k-1}$ with $\underline{S^k T^*}$. Remark first that the map $f \in \mathcal{I} \mapsto d_V f \in \underline{T_V^*}$, where d_V is exterior differentiation with respect to the variable y, is left $\mathcal{O}$-linear and factors to an $\mathcal{O}$-linear isomorphism $\varepsilon_1^{-1} : \mathcal{I}/\mathcal{I}^2 \to \mathcal{J}_0 T^* \simeq \underline{T}^*$ (identification via j_0). In fact, $d_V f \equiv 0 \; \mathit{mod}\; \mathcal{I}\underline{T_V^*}$ means that $\partial f/\partial y_i \in \mathcal{I}$ hence $f \in \mathcal{I}^2$. Surjectivity is given by the relation $d_V(\sum_i a_i(x)(y_i - x_i)) \equiv \sum_i a_i(y)\, dy_i \; \mathit{mod}\; \mathcal{I}\underline{T_V^*}$ where the second term identifies with $\sum a_i(x)\, dx_i \in \underline{T}^*$. In the same manner, $f \in \mathcal{I} \mapsto d_H f \in \underline{T_H^*}$ factors to an $\mathcal{O}$-linear isomorphism $\zeta_1^{-1} : \mathcal{I}/\mathcal{I}^2 \to \mathcal{J}'_0 T^* \simeq \underline{T}^*$ (identification via i_0) and $\varepsilon_1^{-1} \circ \zeta_1 = -\,\mathrm{Id}$.

Since $\underline{S^kT^*} = S^k\underline{T}^*$, we define the map $\varepsilon_k : \underline{S^kT^*} \to \mathcal{J}^k/\mathcal{J}^{k+1}$ by $d_Vf_1 \vee \ldots \vee d_Vf_k \mapsto f_1 \ldots f_k \mod \mathcal{J}^{k+1}$ where $f_i \in \mathcal{J}$ and $\vee$ denotes the symmetric product. ε_k is well defined since $d_Vf \equiv d_Vg \mod \mathcal{J}\underline{T^*_V}$ means that $f-g \in \mathcal{J}^2$ and it is clearly $\mathcal{O}$-linear since right and left structures coincide on these spaces. The sequence

$$0 \longrightarrow \underline{S^kT^*} \xrightarrow{\varepsilon_k} \mathcal{J}_k \xrightarrow{\rho_{k-1}} \mathcal{J}_{k-1} \longrightarrow 0$$

is exact since $\ker \rho_{k-1}$ is the subsheaf of elements $\sum_{|\alpha|=k} a_\alpha(x)\frac{(y-x)^\alpha}{\alpha!}$, $[d_V(y_1-x_1)]^{\alpha_1} \vee \ldots \vee [d_V(y_n-x_n)]^{\alpha_n}$ maps into $(y-x)^\alpha$ and the class of $d_V(y_i-x_i) \mod \mathcal{J}\underline{T^*_V}$ identifies with $dx_i \in \underline{T}^*$.

Similarly we define the $\mathcal{O}$-linear map $\zeta_k : \underline{S^kT^*} \to \mathcal{J}^k/\mathcal{J}^{k+1}$ by $d_Hf_1 \vee \ldots \vee d_Hf_k \mapsto f_1 \ldots f_k \mod \mathcal{J}^{k+1}$ with $f_i \in \mathcal{J}$. The sequence

$$0 \longrightarrow \underline{S^kT^*} \xrightarrow{\zeta_k} \mathcal{J}_k \xrightarrow{\rho_{k-1}} \mathcal{J}_{k-1} \longrightarrow 0$$

is exact and the relation $\varepsilon_1^{-1} \circ \zeta_1 = -\,\mathrm{Id}$ implies that $\varepsilon_k^{-1} \circ \zeta_k = (-1)^k \mathrm{Id}$. Applying Π_1 and Π_2 to $\mathcal{J}_k \xrightarrow{\rho_{k-1}} \mathcal{J}_{k-1}$ one infers that the above sequences are left, right split exact (a fact which can also be verified directly by taking polynomial representatives).

It is worthwhile to give a second description of ε_k and ζ_k. If $f \in \mathcal{J}^k$ then the kth order Fréchet derivative (with respect to some coordinate system) $D^{(k)}f(x,x)$ at a point of the diagonal defines a symmetric form of degree k on $T_{(x,x)}X^2$ which is independent of the coordinate system and only depends on $f \mod \mathcal{J}^{k+1}$. Moreover, $D^{(k)}f(x,x)(w_1, \ldots, w_k) = 0$ whenever some vector w_i is diagonal. Denote diagonal vectors by δ_i, horizontal vectors by h_i and vertical vectors by v_i. There exists a unique horizontal symmetric k-form σ_x on $T_{(x,x)}X^2$ (horizontal means $\sigma_x(w_1, \ldots, w_k) = \sigma_x(Hw_1, \ldots, Hw_k)$) and a unique vertical symmetric k-form σ'_x on $T_{(x,x)}X^2$ satisfying $D^{(k)}f(x,x)(\delta_1 + h_1, \ldots, \delta_k + h_k) = \sigma_x(h_1, \ldots, h_k)$ and $D^{(k)}f(x,x)(\delta_1 + v_1, \ldots, \delta_k + v_k) = \sigma'_x(v_1, \ldots, v_k)$. It is clear that σ_x is the kth order (partial) Fréchet derivative of f with respect to x and similarly for σ'_x (with respect to y). Hence $D^{(k)}f(x,x)$ is determined either by the horizontal form σ_x or the vertical form σ'_x and further, $\sigma'_x(v_1,\dots,v_k) = (-1)^k\,\sigma_x(h_1,\dots,h_k)$ whenever v_i is the horizontal vector h_i thought of as vertical, i.e., $h_i + v_i$ is diagonal. If we identify σ_x and σ'_x with elements of $(S^kT^*)_x$ via π_1^* and π_2^* respectively, then ε_k^{-1} is equal to $f \in \mathcal{J}^k/\mathcal{J}^{k+1} \mapsto \mathit{germ}\,\{x \mapsto \sigma'_x\} \in \underline{S}^kT^*$ and ζ_k^{-1} is equal to $f \in \mathcal{J}^k/\mathcal{J}^{k+1} \mapsto \mathit{germ}\,\{x \mapsto \sigma_x\} \in \underline{S}^kT^*$. It follows, in particular, that the restriction of r_k to $\mathit{ker}\,\rho_{k-1}$ is equal to $(-1)^k\mathrm{Id}$ (which is rather obvious).

Define $\underline{\delta}$ to be the restriction of D to the kernel of

$$\underline{\Lambda^pT^*} \otimes \mathcal{J}_h \to \underline{\Lambda^{p+1}T^*} \otimes \mathcal{J}_{h-1}\,.$$

The sheaf map $\underline{\delta}$ is $\mathcal{O}$-linear (left and right structures coincide on the kernel) and the following diagram is commutative (tensor product with respect to $\mathcal{O}$ and $S^\ell T^* = 0$ for $\ell < 0$).

$(6.4)_k$

$$\begin{array}{ccccccccccccc}
 & & & & 0 & & 0 & & & & 0 & & \\
 & & & & \downarrow & & \downarrow & & & & \downarrow & & \\
 & & 0 & \longrightarrow & \underline{S^kT^*} & \xrightarrow{\underline{\delta}} & \underline{T}^* \otimes \underline{S^{k-1}T^*} & \xrightarrow{\underline{\delta}} & \cdots & \xrightarrow{\underline{\delta}} & \underline{\Lambda^nT^*} \otimes \underline{S^{k-n}T^*} & \longrightarrow & 0 \\
 & & \downarrow & & \downarrow{\scriptstyle \zeta_k} & & \downarrow{\scriptstyle \mathrm{Id}\otimes\zeta_{k-1}} & & & & \downarrow{\scriptstyle \mathrm{Id}\otimes\zeta_{k-n}} & & \\
0 & \longrightarrow & \mathcal{O} & \xrightarrow{j_k} & \mathcal{J}_k & \xrightarrow{D} & \underline{T}^* \otimes \mathcal{J}_{k-1} & \xrightarrow{D} & \cdots & \xrightarrow{D} & \underline{\Lambda^nT^*} \otimes \mathcal{J}_{k-n} & \longrightarrow & 0 \\
 & & \downarrow & & \downarrow & & \downarrow & & & & \downarrow & & \\
0 & \longrightarrow & \mathcal{O} & \xrightarrow{j_{k-1}} & \mathcal{J}_{k-1} & \xrightarrow{D} & \underline{T}^* \otimes \mathcal{J}_{k-2} & \xrightarrow{D} & \cdots & \xrightarrow{D} & \underline{\Lambda^nT^*} \otimes \mathcal{J}_{k-n-1} & \longrightarrow & 0 \\
 & & \downarrow & & \downarrow & & \downarrow & & & & \downarrow & & \\
 & & 0 & & 0 & & 0 & & & & 0 & &
\end{array}$$

Since the two bottom lines are right split exact and the columns are left, right split exact it follows that the first line is split exact (for $k \geq 1$).

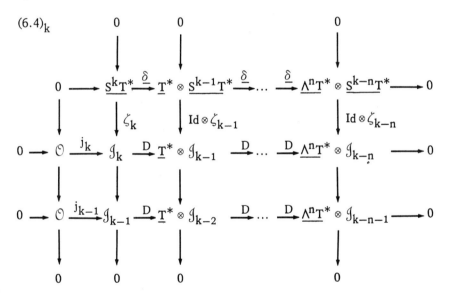

Let $\omega \otimes (s_1 \vee \ldots \vee s_h) \in \underline{\Lambda^p T^*} \otimes \underline{S^h T^*}$ be a decomposable element and represent each s_i by $d_H f_i$ with $f_i \in \mathcal{I}$ ($\pi_1^* s_i \equiv d_H f_i$ *mod* $\mathcal{I}\underline{T_H^*}$). Computing modulo $\mathcal{I}^h \underline{\Lambda^{p+1} T_H^*}$ we have

$$\underline{\delta}(\omega \otimes s_1 \vee \ldots \vee s_h) = d_H(f_1 \ldots f_h \pi_1^* \omega) = d_H(f_1 \ldots f_h) \wedge \pi_1^* \omega =$$

$$\sum_i (f_1 \ldots \hat{f}_i \ldots f_h) d_H f_i \wedge \pi_1^* \omega = \sum_i (f_1 \ldots \hat{f}_i \ldots f_h) \pi_1^*(s_i \wedge \omega) =$$

$$\sum_i (s_i \wedge \omega) \otimes s_1 \vee \ldots \vee \hat{s}_i \vee \ldots \vee s_h;$$

hence

$$\underline{\delta}(\omega \otimes s_1 \vee \ldots \vee s_h) = \sum_i (s_i \wedge \omega) \otimes s_1 \vee \ldots \vee \hat{s}_i \vee \ldots \vee s_h \,. \tag{6.5}$$

It is now clear that $\underline{\delta}$ is the $\mathcal{O}$-linear sheaf map induced by the bundle morphism $\Lambda^p T^* \otimes S^h T^* \xrightarrow{\delta} \Lambda^{p+1} T^* \otimes S^{h-1} T^*$ whose definition is analogous to (6.5) (replace germs by vectors). Taking coordinates (x, y) in X^2 and writing $dx^{\vee\alpha} = dx_1^{\alpha_1} \vee \ldots \vee dx_n^{\alpha_n}$, then

$$\omega \otimes \sum_{|\alpha|=h} a_\alpha(x) \frac{dx^{\vee\alpha}}{\alpha!} \in \underline{\Lambda^p T^*} \otimes \underline{S^h T^*}$$

identifies (via $\mathrm{Id} \otimes \zeta_h$) with

$$s = \omega \otimes \sum_{|\alpha|=h} a_\alpha(y) \frac{(x-y)^\alpha}{\alpha!}$$

and

$$\underline{\delta} s = Ds = \sum_{1 \leq i \leq n} (dx_i \wedge \omega) \otimes \sum_{|\alpha|=h-1} a_{\alpha+1_i}(y) \frac{(x-y)^\alpha}{\alpha!} \,.$$

Fixing $y = x_0$ and writing $s(x_0)$ as a differential form with polynomial coefficients we find again that $\underline{\delta}_{x_0}$ is exterior differentiation on differential forms with homogeneous polynomial coefficients. An intrinsic description is as follows. Let $\mathcal{V}$ be a real vector space of finite dimension and identify each tangent space $T_x\mathcal{V}$ with $\mathcal{V}$. An exterior differential form ω of degree p on $\mathcal{V}$ identifies with a map $\omega : \mathcal{V} \to \Lambda^p\mathcal{V}^*$. We say that ω is polynomial of degree h if this map is a homogeneous polynomial function of degree h (induced by an h-multilinear map). Since polynomial functions of degree h, $\mathcal{V} \to \Lambda^p\mathcal{V}^*$, identify with h-multilinear symmetric maps $\mathcal{V}^h \to \Lambda^p\mathcal{V}^*$, the vector space of exterior differential p-forms on $\mathcal{V}$ which are polynomial of degree h is equal to $\Lambda^p\mathcal{V}^* \otimes S^h\mathcal{V}^*$ and exterior differentiation induces an R-linear map $\Lambda^p\mathcal{V}^* \otimes S^h\mathcal{V}^* \xrightarrow{d} \Lambda^{p+1}\mathcal{V}^* \otimes S^{h-1}\mathcal{V}^*$ which is precisely

$$\omega \otimes s_1 \vee \ldots \vee s_h \mapsto \sum_i s_i \wedge \omega \otimes s_1 \vee \ldots \vee \hat{s}_i \vee \ldots \vee s_h .$$

The Poincaré lemma implies that the complex

$$0 \longrightarrow S^k\mathcal{V}^* \xrightarrow{d} \mathcal{V}^* \otimes S^{k-1}\mathcal{V}^* \xrightarrow{d} \ldots \xrightarrow{d} \Lambda^n\mathcal{V}^* \otimes S^{k-n}\mathcal{V}^* \longrightarrow 0$$

is exact. A homotopy operator is also given by

$$\omega_1 \wedge \cdots \wedge \omega_p \otimes s \mapsto \sum_{1 \le i \le p} \frac{(-1)^{i+1}}{p+h} \omega_1 \wedge \ldots \wedge \hat{\omega}_i \wedge \ldots \wedge \omega_p \otimes \omega_i \vee s .$$

The complex

$$(6.5)_k \qquad 0 \longrightarrow S^kT^* \xrightarrow{\delta} T^* \otimes S^{k-1}T^* \xrightarrow{\delta} \ldots \xrightarrow{\delta} \Lambda^nT^* \otimes S^{k-n}T^* \longrightarrow 0$$

is exact and δ is exterior differentiation on each fibre. It follows again that the first line of diagram $(6.4)_k$ is split exact.

Tensoring diagram $(6.4)_k$ on the right with the $\mathcal{O}$-module $\mathcal{E}$ we obtain the commutative diagram $(\underline{\delta} = \underline{\delta} \otimes \mathrm{Id})$

$(6.6)_k$

$$
\begin{array}{ccccccccccc}
 & & & & 0 & & 0 & & & & 0 \\
 & & & & \downarrow & & \downarrow & & & & \downarrow \\
 & & 0 & \longrightarrow & \underline{S^kT^*}\otimes\mathcal{E} & \xrightarrow{\underline{\delta}} & \underline{T^*}\otimes\underline{S^{k-1}T^*}\otimes\mathcal{E} & \xrightarrow{\underline{\delta}} \cdots \xrightarrow{\underline{\delta}} & \underline{\Lambda^nT^*}\otimes\underline{S^{k-n}T^*}\otimes\mathcal{E} & \longrightarrow & 0 \\
 & & \downarrow & & \downarrow{\scriptstyle \zeta_k\otimes \mathrm{Id}} & & \downarrow{\scriptstyle \mathrm{Id}\otimes\zeta_{k-1}\otimes\mathrm{Id}} & & \downarrow{\scriptstyle \mathrm{Id}\otimes\zeta_{k-n}\otimes\mathrm{Id}} & & \\
0 & \longrightarrow & \mathcal{E} & \xrightarrow{j_k} & \mathcal{J}_k\otimes\mathcal{E} & \xrightarrow{D} & \underline{T^*}\otimes\mathcal{J}_{k-1}\otimes\mathcal{E} & \xrightarrow{D} \cdots \xrightarrow{D} & \underline{\Lambda^nT^*}\otimes\mathcal{J}_{k-n}\otimes\mathcal{E} & \longrightarrow & 0 \\
 & & \downarrow & & \downarrow & & \downarrow & & \downarrow & & \\
0 & \longrightarrow & \mathcal{E} & \xrightarrow{j_{k-1}} & \mathcal{J}_{k-1}\otimes\mathcal{E} & \xrightarrow{D} & \underline{T^*}\otimes\mathcal{J}_{k-2}\otimes\mathcal{E} & \xrightarrow{D} \cdots \xrightarrow{D} & \underline{\Lambda^nT^*}\otimes\mathcal{J}_{k-n-1}\otimes\mathcal{E} & \longrightarrow & 0 \\
 & & \downarrow & & \downarrow & & \downarrow & & \downarrow & & \\
 & & 0 & & 0 & & 0 & & 0 & &
\end{array}
$$

whose lines and columns are right split exact (the first line has only one $\mathcal{O}$-module structure). In the literature one always finds diagram $(6.6)_k$ with ζ_h replaced by ε_h and $\underline{\delta}$ by $-\underline{\delta}$ (since $\zeta_h = (-1)^h \varepsilon_h$).

Assume now that $\mathcal{E} = \underline{E}$ is locally free and consider $F = \Lambda^p T^*_H \otimes_R \pi_2^* E$ which is a vector bundle on X^2. $\underline{F}$ is the sheaf of germs of horizontal exterior differential p-forms on X^2 with values in $\pi_2^* E$. One checks easily that $\underline{\Lambda^p T^*} \otimes \mathcal{J}_k \otimes \underline{E}$ identifies with the sheaf $\underline{F}/\mathcal{J}^{k+1}\underline{F}$ (cf. end of Section 4) and that D is the exterior differentiation with respect to the variable x (which makes sense since $\pi_2^* E$ is trivial in the horizontal direction). The first line of diagram $(6.6)_k$ is the extension to germs of the following exact vector bundle complex

$$(6.7)_k \quad 0 \longrightarrow S^kT^* \otimes E \xrightarrow{\delta} T^* \otimes S^{k-1}T^* \otimes E \xrightarrow{\delta} \cdots \xrightarrow{\delta} \Lambda^nT^* \otimes S^{k-n}T^* \otimes E \longrightarrow 0$$

where $\delta = \delta \otimes \mathrm{Id}$ and δ_x is exterior differentiation of homogeneous polynomial exterior differential forms on T_x with values in E_x. The second line of $(6.6)_k$ is transformed by Π_1 into a complex of sheaves and left $\mathcal{O}$-linear operators which is the extension to germs of the following exact complex of vector bundles:

$$(6.8)_k \quad 0 \longrightarrow E \xrightarrow{i_k} G_k(E) \xrightarrow{d} G_{k-1}(T^*, E) \xrightarrow{d} \cdots \xrightarrow{d} G_{k-n}(\Lambda^nT^*, E) \longrightarrow 0$$

where $G_h(\Lambda^p T^*, E)$ is the bundle on X whose fibre over the point a is the vector space of h-jets of exterior differential p-forms with values in E_a and d is exterior differentiation in each fibre $j_h\omega(a) \mapsto j_{h-1}d\omega(a)$.

Remark finally that $\underline{\Lambda T^*} \otimes \mathcal{J}_k \otimes \mathcal{E}$ has a natural left $\underline{\Lambda T^*}$-module structure (extension of scalars) and the direct sum of the operators D defines an operator $D: \underline{\Lambda T^*} \otimes \mathcal{J}_k \otimes \mathcal{E} \to \underline{\Lambda T^*} \otimes \mathcal{J}_{k-1} \otimes \mathcal{E}$. The operator D is the unique differential operator ($\mathbf{R}$-linear sheaf map) satisfying the two properties:

1) $D \circ j_k = 0$
2) $D(\omega \wedge \tau) = d\omega \wedge \rho_{k-1}\tau + (-1)^{\deg\omega}\omega \wedge D\tau$ for any $\tau \in \underline{\Lambda T^*} \otimes \mathcal{J}_k \otimes \mathcal{E}$ and any homogeneous ω.

In particular $\underline{\delta}(\omega \wedge \tau) = (-1)^{\deg\omega}\omega \wedge \underline{\delta}(\tau)$ for any $\tau \in \underline{\Lambda T^*} \otimes \underline{S^k T^*} \otimes \mathcal{E}$, hence $\underline{\delta}$ is, up to a sign, $\underline{\Lambda T^*}$-linear. A similar formula holds for δ.

7. *The second linear complex*

Consider the three consecutive columns of diagram (6.6)

$$
\begin{array}{ccccccccc}
 & & 0 & & 0 & & 0 & & \\
 & & \downarrow & & \downarrow & & \downarrow & & \\
\cdots \to & & \underline{\Lambda^{p-1}T^*} \otimes \underline{S^{k+2}T^*} \otimes \mathcal{E} & \xrightarrow{\underline{\delta}_{p-1,\,k+2}} & \underline{\Lambda^p T^*} \otimes \underline{S^{k+1}T^*} \otimes \mathcal{E} & \xrightarrow{\underline{\delta}_{p,\,k+1}} & \underline{\Lambda^{p+1}T^*} \otimes \underline{S^k T^*} \otimes \mathcal{E} & \to \cdots \\
 & & \downarrow \zeta_{p-1,\,k+2} & & \downarrow \zeta_{p,\,k+1} & & \downarrow \zeta_{p+1,\,k} & & \\
\cdots \to & & \underline{\Lambda^{p-1}T^*} \otimes \mathcal{J}_{k+2} \otimes \mathcal{E} & \xrightarrow{D_{p-1,\,k+2}} & \underline{\Lambda^p T^*} \otimes \mathcal{J}_{k+1} \otimes \mathcal{E} & \xrightarrow{D_{p,\,k+1}} & \underline{\Lambda^{p+1}T^*} \otimes \mathcal{J}_k \otimes \mathcal{E} & \to \cdots \\
 & & \downarrow \rho_{k+1} & & \downarrow \rho_k & & \downarrow \rho_{k-1} & & \\
\cdots \to & & \underline{\Lambda^{p-1}T^*} \otimes \mathcal{J}_{k+1} \otimes \mathcal{E} & \xrightarrow{D_{p-1,\,k+1}} & \underline{\Lambda^p T^*} \otimes \mathcal{J}_k \otimes \mathcal{E} & \xrightarrow{D_{p,\,k}} & \underline{\Lambda^{p+1}T^*} \otimes \mathcal{J}_{k-1} \otimes \mathcal{E} & \to \cdots \\
 & & \downarrow & & \downarrow & & \downarrow & & \\
 & & 0 & & 0 & & 0 & &
\end{array}
$$

We define the quotient sheaves

$$
\mathcal{C}_k^{p+1}\mathcal{E} = \underline{\Lambda^{p+1}T^*} \otimes \mathcal{J}_k \otimes \mathcal{E} / \zeta_{p+1,\,k} \circ \underline{\delta}_{p,\,k+1}(\underline{\Lambda^p T^*} \otimes \underline{S^{k+1}T^*} \otimes \mathcal{E})
$$

for $p \geq 0$ and $\mathcal{C}^0_k\mathcal{E} = \mathcal{J}_k \otimes \mathcal{E}$. $\mathcal{C}^q_k\mathcal{E}$ inherits a module structure over the left and right $\mathcal{O}$-algebra $\mathcal{J}_k$ and $\mathcal{C}^q_k\mathcal{E} = 0$ for $q > n$. The relation

$$\zeta_{p+1,k} \circ \underline{\delta}_{p,\,k+1} = D_{p,\,k+1} \circ \zeta_{p,\,k+1}$$

implies that $D_{p,k+1}$ factors to a right $\mathcal{O}$-linear operator

$$D^1_{p,\,k+1}: \underline{\Lambda^p T^*} \otimes \mathcal{J}_k \otimes \mathcal{E} \to \mathcal{C}^{p+1}_k\mathcal{E}$$

and $D_{p,k+1} \circ D_{p-1,k+2} = 0$ implies that $D^1_{p,k+1} \circ D_{p-1,k+1} = 0$, hence $D^1_{p,k+1}$ factors to a right $\mathcal{O}$-linear operator

$$D'_{p,\,k}: \mathcal{C}^p_k\mathcal{E} \to \mathcal{C}^{p+1}_k\mathcal{E}\,.$$

It also follows from $D^1_{p,k+1} \circ D_{p-1,k+1} = 0$ that $D'_{p,k} \circ D'_{p-1,k} = 0$; hence we obtain the second linear complex

$$(7.1)_k \qquad 0 \longrightarrow \mathcal{E} \xrightarrow{j_k} \mathcal{C}^0_k\mathcal{E} \xrightarrow{D'} \mathcal{C}^1_k\mathcal{E} \xrightarrow{D'} \cdots \xrightarrow{D'} \mathcal{C}^n_k\mathcal{E} \longrightarrow 0$$

which by a simple diagram chase is shown to be exact. Since the lines and columns of diagram (6.6) are right split exact one shows actually that $(7.1)_k$ is right split exact. Moreover, since

$$0 \longrightarrow \underline{\Lambda^p T^*} \otimes \underline{S^{k+1}T^*} \xrightarrow{\zeta_{p+1,k} \circ \underline{\delta}_{p,k+1}} \underline{\Lambda^{p+1}T^*} \otimes \mathcal{J}_k \longrightarrow \mathcal{C}^{p+1}_k \longrightarrow 0$$

is right split exact ($\mathcal{C}^{p+1}_k = \mathcal{C}^{p+1}_k\mathcal{O}$), it follows that $(7.1)_k$ is the complex obtained by tensoring on the right with $\mathcal{E}$ the right split exact complex

$$(7.2)_k \qquad 0 \longrightarrow \mathcal{O} \xrightarrow{j_k} \mathcal{C}^0_k \xrightarrow{D'} \mathcal{C}^1_k \xrightarrow{D'} \cdots \xrightarrow{D'} \mathcal{C}^n_k \longrightarrow 0\,.$$

If $\mathcal{E}$ is locally free of finite rank then $\mathcal{C}^q_k\mathcal{E}$ is also locally free of finite rank over the left and right $\mathcal{O}$-module structures. If $\mathcal{E} = \underline{E}$ then $\mathcal{C}^q_k\underline{E}$ is the sheaf (of germs) of sections of the quotient vector bundle

$$\Lambda^q T^* \otimes J_k E / \zeta_{q,\,k} \circ \delta_{q-1,\,k+1}(\Lambda^{q-1}T^* \otimes S^{k+1}T^* \otimes E).$$

Furthermore ($\mathcal{E}$ arbitrary), the last formula of Section 6 restricted to $\underline{\Lambda T}^* \otimes \underline{S}^{k+1}T^* \otimes \mathcal{E}$ shows that the left $\underline{\Lambda T}^*$-module structure of $\underline{\Lambda T}^* \otimes \mathcal{J}_k \otimes \mathcal{E}$ factors to $\mathcal{C}_k^{\bullet}\mathcal{E} = \oplus_q \mathcal{C}_k^q\mathcal{E}$. It follows that D' has a characterization analogous to the one given for D at the end of Section 6.

8. *Homogeneous linear partial differential equations*

We begin this section by introducing some natural maps of jet sheaves. Let $\mathcal{E}$ and $\mathcal{F}$ be two sheaves of $\mathcal{O}$-modules and $\mathcal{E} \xrightarrow{\phi} \mathcal{F}$ an $\mathcal{O}$-linear morphism. The map $\mathrm{Id} \otimes \phi : \mathcal{J}_k \otimes \mathcal{E} \to \mathcal{J}_k \otimes \mathcal{F}$ is $\mathcal{J}_k$-linear (hence left, right $\mathcal{O}$-linear). It is the unique left $\mathcal{O}$-linear map which makes commutative the diagram

(8.1)
$$\begin{array}{ccc} \mathcal{J}_k \otimes \mathcal{E} & \xrightarrow{\ \mathrm{Id} \otimes \phi\ } & \mathcal{J}_k \otimes \mathcal{F} \\ {\scriptstyle j_k}\uparrow & & \uparrow{\scriptstyle j_k} \\ \mathcal{E} & \xrightarrow{\ \phi\ } & \mathcal{F} \end{array}$$

The functor $\mathcal{J}_k \bullet = \mathcal{J}_k \otimes \bullet$ is exact since the sheaf $\mathcal{J}_k$ is locally free to the right, hence flat. If $\mathcal{E} = \underline{E}$ and $\mathcal{F} = \underline{F}$ then $\phi = \underline{\psi}$ with $\psi : E \to F$ a bundle morphism and $\mathrm{Id} \otimes \phi = \underline{J_k \psi}$ (cf. Section 2).

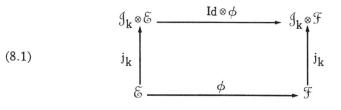

The map $\mathcal{O} \otimes_{\mathbf{R}} \mathcal{O} \to \mathcal{O}_{X^2}$, $f \otimes g \mapsto i(f)\, j(g)$, is an injective left, right $\mathcal{O}$-algebra morphism (with the obvious structure on $\mathcal{O} \otimes_{\mathbf{R}} \mathcal{O}$). We identify $\mathcal{O} \otimes_{\mathbf{R}} \mathcal{O}$ with its image. Since $\mathcal{J}_k$ is left generated by its right holonomic elements $j_k\mathcal{O}$, it follows that the restriction of the quotient map $\mathcal{O}_{X^2} \to \mathcal{J}_k$ to $\mathcal{O} \otimes_{\mathbf{R}} \mathcal{O}$ is surjective hence $\mathcal{J}_k = \mathcal{O} \otimes_{\mathbf{R}} \mathcal{O} / (\mathcal{O} \otimes_{\mathbf{R}} \mathcal{O} \cap \mathcal{J}^{k+1})$. We define $\iota : \mathcal{O} \otimes_{\mathbf{R}} \mathcal{O} \to \mathcal{J}_\ell \otimes \mathcal{J}_k$, $f \otimes g \mapsto i_\ell f \otimes j_k g$, which factors to an injective map $\iota_{k+\ell} : \mathcal{J}_{k+\ell} \to \mathcal{J}_\ell \otimes \mathcal{J}_k$. Observe now that $\mathcal{J}_\ell \otimes_{\mathcal{O}} \mathcal{J}_k$ has three permuting $\mathcal{O}$-module structures namely the left, *middle* and right structures induced respectively by the left structure of $\mathcal{J}_\ell$, the right structure of $\mathcal{J}_\ell$ (or the left one of $\mathcal{J}_k$) and the right structure of $\mathcal{J}_k$ (the middle structure corresponds to the previously defined right structure on $\mathcal{J}_\ell \otimes_{\mathcal{O}} \mathcal{E}$) as well as a ring structure which makes it into a three fold $\mathcal{O}$-algebra. The map $\iota_{k+\ell}$ is a left, right $\mathcal{O}$-algebra morphism and a split injection (in each stalk) for both left and right $\mathcal{O}$-module structures. Tensoring $\iota_{k+\ell}$ on the right with $\mathrm{Id}_{\mathcal{E}}$ we obtain an injective left, right $\mathcal{O}$-linear map still noted by $\iota_{k+\ell}$: $\mathcal{J}_{k+\ell} \otimes_{\mathcal{O}} \mathcal{E} \to \mathcal{J}_\ell \otimes_{\mathcal{O}} \mathcal{J}_k \otimes_{\mathcal{O}} \mathcal{E}$ which splits to the right (the right structure of $\mathcal{J}_\ell \otimes_{\mathcal{O}} \mathcal{J}_k \otimes_{\mathcal{O}} \mathcal{E}$ induced by the right structure of $\mathcal{J}_k$ should not be confused with the right structure of $\mathcal{J}_\ell \otimes_{\mathcal{O}} (\mathcal{J}_k \otimes_{\mathcal{O}} \mathcal{E})$). It is the unique left $\mathcal{O}$-linear map which makes commutative the diagram

(8.2)
$$\begin{array}{ccc} \mathcal{J}_{k+\ell} \otimes_{\mathcal{O}} \mathcal{E} & \xrightarrow{\iota_{k+\ell}} & \mathcal{J}_\ell \otimes_{\mathcal{O}} \mathcal{J}_k \otimes_{\mathcal{O}} \mathcal{E} \\ \uparrow{\scriptstyle j_{k+\ell}} & & \uparrow{\scriptstyle j_\ell} \\ \mathcal{E} & \xrightarrow{j_k} & \mathcal{J}_k \otimes_{\mathcal{O}} \mathcal{E} \end{array}$$

$\iota_{k+\ell}$ is a natural transformation of the functor $\mathcal{J}_{k+\ell}{}^\bullet$ to the functor $\mathcal{J}_\ell \circ \mathcal{J}_k{}^\bullet$. If $\mathcal{E} = \underline{E}$ then $\iota_{k+\ell} = \underline{\nu_{k+\ell}} : \underline{J_{k+\ell} E} \to \underline{J_\ell J_k E}$ (cf. Section 2). Iteration of $\iota_{k+\ell}$ yields a commutative diagram

(8.3)
$$\begin{array}{ccccc} & & \mathcal{J}_{\ell+h} \otimes \mathcal{J}_k \otimes \mathcal{E} & & \\ & \nearrow^{\iota_{k+(\ell+h)}} & & \searrow^{\iota_{\ell+h} \otimes \mathrm{Id}} & \\ \mathcal{J}_{k+\ell+h} \otimes \mathcal{E} & & \xrightarrow{\quad \iota_{k+\ell+h} \quad} & & \mathcal{J}_h \otimes \mathcal{J}_\ell \otimes \mathcal{J}_k \otimes \mathcal{E} \\ & \searrow_{\iota_{(k+\ell)+h}} & & \nearrow_{\mathrm{Id} \otimes \iota_{k+\ell}} & \\ & & \mathcal{J}_h \otimes \mathcal{J}_{k+\ell} \otimes \mathcal{E} & & \end{array}$$

where all the arrows are right split injections and $\iota_{k+\ell+h}$ is defined by tensoring on the right, with $\mathrm{Id}_{\mathcal{E}}$, the left, right split injection $\iota_{k+\ell+h}$: $\mathcal{J}_{k+\ell+h} \to \mathcal{J}_h \otimes \mathcal{J}_\ell \otimes \mathcal{J}_k$ quotient of $f \otimes g \mapsto i_h f \otimes 1 \otimes j_k g$. Imbedding all the sheaves into $\mathcal{F} = \mathcal{J}_h \otimes \mathcal{J}_\ell \otimes \mathcal{J}_k \otimes \mathcal{E}$, one can show that

$$(\mathcal{J}_{\ell+h} \otimes \mathcal{J}_k \otimes \mathcal{E}) \cap (\mathcal{J}_h \otimes \mathcal{J}_{k+\ell} \otimes \mathcal{E}) = \mathcal{J}_{k+\ell+h} \otimes \mathcal{E}$$

whenever $\ell > 0$ (first prove that $(J_{\ell+h} J_k) \cap (J_h J_{k+\ell}) = J_{k+\ell+h}$, then use splitting properties). We infer that

$$\mathcal{J}_{k+\ell+h} \otimes \mathcal{E} = (\mathcal{J}_{\ell+h} \otimes \mathcal{J}_k \otimes \mathcal{E}) \times_{\mathcal{F}} (\mathcal{J}_h \otimes \mathcal{J}_{k+\ell} \otimes \mathcal{E})$$

where the fibre product is taken with respect to the maps $\iota_{\ell+h} \otimes \mathrm{Id}$ and $\mathrm{Id} \otimes \iota_{k+\ell}$.

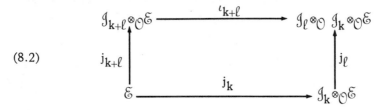

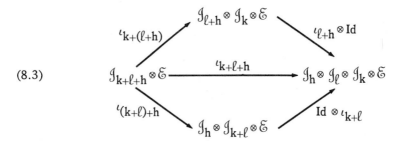

We finally consider a left $\mathcal{O}$-linear morphism $\mathcal{J}_k \otimes_{\mathcal{O}} \mathcal{E} \xrightarrow{\phi} \mathcal{F}$ and define

$$p_\ell \phi = (\mathrm{Id} \otimes \phi) \circ \iota_{k+\ell} : \mathcal{J}_{k+\ell} \otimes \mathcal{E} \to \mathcal{J}_\ell \otimes \mathcal{J}_k \otimes \mathcal{E} \to \mathcal{J}_\ell \otimes \mathcal{F}$$

which is only left $\mathcal{O}$-linear. $p_\ell \phi$ is called the ℓth prolongation of ϕ and is the unique left $\mathcal{O}$-linear map which makes commutative the diagram

$$\text{(8.4)} \qquad \begin{array}{ccccc} & \nearrow^{j_{k+\ell}} & \mathcal{J}_{k+\ell} \otimes \mathcal{E} & \xrightarrow{p_\ell \phi} & \mathcal{J}_\ell \otimes \mathcal{F} \\ \mathcal{E} & & & & \uparrow{\scriptstyle j_\ell} \\ & \searrow_{j_k} & \mathcal{J}_k \otimes \mathcal{E} & \xrightarrow{\phi} & \mathcal{F} \end{array}$$

If ϕ is injective then $p_\ell \phi$ is injective. We observe that $p_\ell \phi = \mathrm{Id} \otimes \phi$ for $k = 0$, $p_\ell \mathrm{Id}_k = \iota_{k+\ell}$ where $\mathrm{Id}_k : \mathcal{J}_k \otimes \mathcal{E} \to \mathcal{J}_k \otimes \mathcal{E}$ and iteration of p_ℓ yields the commutative diagram

$$\text{(8.5)} \qquad \begin{array}{ccc} & & 0 \\ & & \downarrow \\ \mathcal{J}_{k+\ell+h} \otimes \mathcal{E} & \xrightarrow{p_{\ell+h} \phi} & \mathcal{J}_{\ell+h} \otimes \mathcal{F} \\ & \searrow^{p_h(p_\ell \phi)} & \downarrow{\scriptstyle p_h \mathrm{Id}_\ell} \\ & & \mathcal{J}_h \otimes \mathcal{J}_\ell \otimes \mathcal{F} \end{array}$$

where the column is exact. If $\mathcal{E} = \underline{E}$ and $\mathcal{F} = \underline{F}$ then $p_\ell \phi = \underline{p_\ell \psi}$ where $\phi = \underline{\psi}$ and the maps $\psi : J_k E \to F$ and $p_\ell \psi = (J_\ell \psi) \circ \nu_{k+\ell} : J_{k+\ell} E \to J_\ell F$ are bundle morphisms. Moreover, if ϕ is split injective then $p_\ell \phi$ is left split injective.

Let $\mathcal{E}$ be a sheaf of $\mathcal{O}$-modules on X. A homogeneous linear partial differential equation of order k on $\mathcal{E}$ is by definition a left $\mathcal{O}$-submodule $\mathcal{R}_k$ of $\mathcal{J}_k \otimes_{\mathcal{O}} \mathcal{E}$. If $\mathcal{E} = \underline{E}$ and $\mathcal{R}_k$ is a locally free direct factor (in each stalk) then $\mathcal{R}_k$ identifies with the sheaf of germs of sections of a subbundle $R_k \subset J_k E$. The sheaf of (germs of) solutions of $\mathcal{R}_k$ is the subsheaf

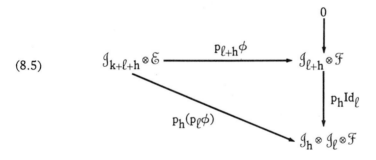

$\mathcal{S} = j_k^{-1}(\mathcal{R}_k) \subset \mathcal{E}$, hence a germ of solution of $\mathcal{R}_k$ is an element $\sigma \in \mathcal{E}$ such that $j_k\sigma \in \mathcal{R}_k$. A local solution of $\mathcal{R}_k$ is a local section τ of $\mathcal{S}$ or, equivalently, a local section τ of $\mathcal{E}$ such that $j_k \circ \tau$ is a local section of $\mathcal{R}_k$. The sheaf $\mathcal{S}$ is only an R-submodule of $\mathcal{E}$ and the restricted map $j_k : \mathcal{S} \to \mathcal{J}_k \otimes \mathcal{E}$ is an R-linear isomorphism of $\mathcal{S}$ onto the subsheaf $j_k\mathcal{S}$ of $\mathcal{J}_k \otimes \mathcal{E}$ composed of the right holonomic element contained in $\mathcal{R}_k$. The equation is said to be integrable if it is left $\mathcal{O}$-generated by $j_k\mathcal{S}$. If $\mathcal{E} = \underline{E}$ and $\mathcal{R}_k$ is a direct factor, this is equivalent to saying that each element in R_k is the k-jet of a solution. In this section we shall give a brief outline of formal integrability which, loosely speaking, is the condition of compatibility for the equation (or the existence of formal solutions) hence a first step towards integrability. Further developments can be found reviewed in Spencer [13(d)] and in the references given in that paper, especially the references to H. Goldschmidt (see, in particular, Goldschmidt [5(a), (b)]). An interesting aspect can also be found in Lehmann [8(a)].

The definitions adopted are those that fit best into the present framework. However, it should be pointed out that, up to now, the proof of deeper results in the theory requires some restrictions, namely one has to assume some *regularity* conditions.

The ℓth prolongation of $\mathcal{R}_k$ is the differential equation of order $k+\ell$ defined by $\mathcal{R}_{k+\ell} = \iota_{k+\ell}^{-1}(\mathcal{J}_\ell \otimes \mathcal{R}_k)$ where $\mathcal{J}_\ell \otimes \mathcal{R}_k$ is identified with a subsheaf of $\mathcal{J}_\ell \otimes \mathcal{J}_k \otimes \mathcal{E}$. Imbedding $\mathcal{J}_{k+\ell} \otimes \mathcal{E}$ into $\mathcal{J}_\ell \otimes \mathcal{J}_k \otimes \mathcal{E}$, we can write $\mathcal{R}_{k+\ell} = (\mathcal{J}_{k+\ell} \otimes \mathcal{E}) \cap (\mathcal{J}_\ell \otimes \mathcal{R}_k)$. We observe that $\mathcal{R}_{(k+\ell)+h} = \mathcal{R}_{k+(\ell+h)} = \iota_{k+\ell+h}^{-1}(\mathcal{J}_h \otimes \mathcal{J}_\ell \otimes \mathcal{R}_k)$, hence $\mathcal{R}_{k+\ell}$ can be defined inductively as $\mathcal{R}_{k+\ell} = \mathcal{R}_{k+1+1+\ldots+1}$ by iterating 1-prolongations ℓ times. To prove this, consider again the commutative diagram

$$
\begin{array}{ccccc}
 & & \mathcal{J}_{\ell+h} \otimes \mathcal{R}_k & & \\
 & & \cap & & \\
\mathcal{R}_{k+(\ell+h)} \subset \mathcal{J}_{k+\ell+h} \otimes \mathcal{E} & \xrightarrow{\iota_{k+(\ell+h)}} & \mathcal{J}_{\ell+h} \otimes \mathcal{J}_k \otimes \mathcal{E} & & \\
 & & & \searrow^{\iota_{\ell+h} \otimes \mathrm{Id}} & \\
\Vert & & & & \mathcal{J}_h \otimes \mathcal{J}_\ell \otimes \mathcal{J}_k \otimes \mathcal{E} \supset \mathcal{J}_h \otimes \mathcal{J}_\ell \otimes \mathcal{R}_k \\
 & & & \nearrow_{\mathrm{Id} \otimes \iota_{k+\ell}} & \\
\mathcal{R}_{(k+\ell)+h} \subset \mathcal{J}_{k+\ell+h} \otimes \mathcal{E} & \xrightarrow{\iota_{(k+\ell)+h}} & \mathcal{J}_h \otimes \mathcal{J}_{k+\ell} \otimes \mathcal{E} & & \\
 & & \cup & & \\
 & & \mathcal{J}_h \otimes \mathcal{R}_{k+\ell} & &
\end{array}
$$

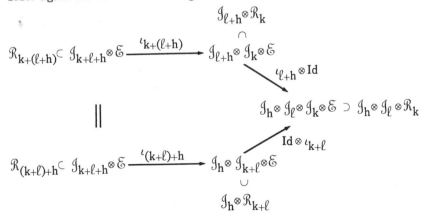

Since $\mathcal{R}_{k+(\ell+h)} = \iota^{-1}_{k+(\ell+h)}(\mathcal{J}_{\ell+h} \otimes \mathcal{R}_k)$ and $\mathcal{R}_{(k+\ell)+h} = \iota^{-1}_{(k+\ell)+h}(\mathcal{J}_h \otimes \mathcal{R}_{k+\ell})$ it is enough to prove that $\mathcal{J}_{\ell+h} \otimes \mathcal{R}_k = (\iota_{\ell+h} \otimes \mathrm{Id})^{-1}(\mathcal{J}_h \otimes \mathcal{J}_\ell \otimes \mathcal{R}_k)$ and $\mathcal{J}_h \otimes \mathcal{R}_{k+\ell} = (\mathrm{Id} \otimes \iota_{k+\ell})^{-1}(\mathcal{J}_h \otimes \mathcal{J}_\ell \otimes \mathcal{R}_k)$. The first contention follows from the right splitting of $\mathit{im}\, \iota_{\ell+h}$ in $\mathcal{J}_h \otimes \mathcal{J}_\ell$. The second contention is equivalent to the exactness of

$$\mathcal{J}_h \otimes \mathcal{R}_{k+\ell} \to \mathcal{J}_h \otimes \mathcal{J}_{k+\ell} \otimes \mathcal{E} \to \mathcal{J}_h \otimes \mathcal{J}_\ell \otimes \mathcal{J}_k \otimes \mathcal{E} / \mathcal{J}_h \otimes \mathcal{J}_\ell \otimes \mathcal{R}_k .$$

But

$$\mathcal{R}_{k+\ell} \to \mathcal{J}_{k+\ell} \otimes \mathcal{E} \to \mathcal{J}_\ell \otimes \mathcal{J}_k \otimes \mathcal{E} / \mathcal{J}_\ell \otimes \mathcal{R}_k$$

is exact, hence

$$\mathcal{J}_h \otimes \mathcal{R}_{k+\ell} \to \mathcal{J}_h \otimes \mathcal{J}_{k+\ell} \otimes \mathcal{E} \to \mathcal{J}_h \otimes (\mathcal{J}_\ell \otimes \mathcal{J}_k \otimes \mathcal{E} / \mathcal{J}_\ell \otimes \mathcal{R}_k)$$

is exact. On the other hand

$$0 \to \mathcal{J}_\ell \otimes \mathcal{R}_k \to \mathcal{J}_\ell \otimes \mathcal{J}_k \otimes \mathcal{E} \xrightarrow{q} \mathcal{J}_\ell \otimes \mathcal{J}_k \otimes \mathcal{E} / \mathcal{J}_\ell \otimes \mathcal{R}_k \to 0$$

yields the exact sequence

$$0 \to \mathcal{J}_h \otimes \mathcal{J}_\ell \otimes \mathcal{R}_k \to \mathcal{J}_h \otimes \mathcal{J}_\ell \otimes \mathcal{J}_k \otimes \mathcal{E} \xrightarrow{\mathrm{Id} \otimes q} \mathcal{J}_h \otimes (\mathcal{J}_\ell \otimes \mathcal{J}_k \otimes \mathcal{E} / \mathcal{J}_\ell \otimes \mathcal{R}_k) \to 0$$

and the diagram

$$\begin{array}{ccccc}
\mathcal{J}_h \otimes \mathcal{R}_{k+\ell} & \to & \mathcal{J}_h \otimes \mathcal{J}_{k+\ell} \otimes \mathcal{E} & \to & \mathcal{J}_h \otimes \mathcal{J}_\ell \otimes \mathcal{J}_k \otimes \mathcal{E} / \mathcal{J}_h \otimes \mathcal{J}_\ell \otimes \mathcal{R}_k \\
\| & & \| & & \simeq \big\downarrow \mathrm{Id} \otimes q \\
\mathcal{J}_h \otimes \mathcal{R}_{k+\ell} & \to & \mathcal{J}_h \otimes \mathcal{J}_{k+\ell} \otimes \mathcal{E} & \to & \mathcal{J}_h \otimes (\mathcal{J}_\ell \otimes \mathcal{J}_k \otimes \mathcal{E} / \mathcal{J}_\ell \otimes \mathcal{R}_k)
\end{array}$$

is commutative. This completes the proof.

Imbedding all the sheaves of the first diagram into $\mathcal{J}_h \otimes \mathcal{J}_\ell \otimes \mathcal{J}_k \otimes \mathcal{E}$ we infer that

$$\mathcal{R}_{k+\ell+h} = (\mathcal{J}_{k+\ell+h} \otimes \mathcal{E}) \cap (\mathcal{J}_h \otimes \mathcal{J}_\ell \otimes \mathcal{R}_k) = (\mathcal{J}_{\ell+h} \otimes \mathcal{R}_k) \cap (\mathcal{J}_h \otimes \mathcal{R}_{k+\ell})$$

$(\ell > 0)$ hence

$$\mathcal{R}_{k+\ell+h} = (\mathcal{J}_{\ell+h} \otimes \mathcal{R}_k) \times_{\mathcal{F}} (\mathcal{J}_h \otimes \mathcal{R}_{k+\ell}) \text{ with } \mathcal{F} = \mathcal{J}_h \otimes \mathcal{J}_\ell \otimes \mathcal{R}_k .$$

We now relate the first prolongation to the operator D. Consider the diagram

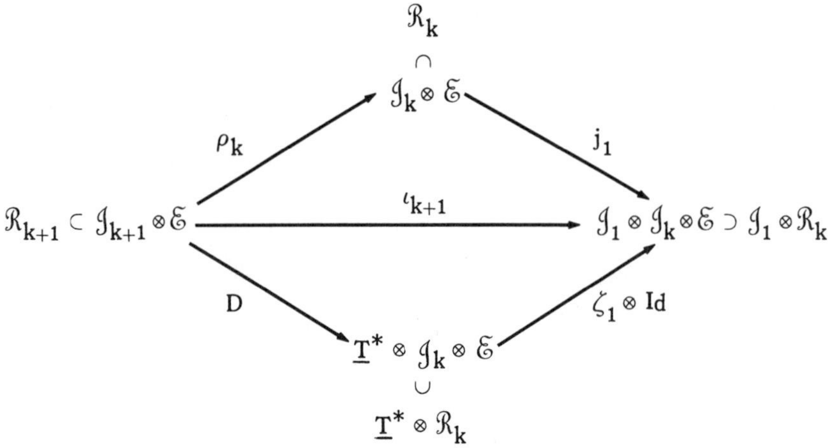

A straightforward computation shows that $(\zeta_1 \otimes \mathrm{Id}) \circ D = \iota_{k+1} - j_1 \circ \rho_k$. Since $\zeta_1 \otimes \mathrm{Id}$ is injective and $\zeta_1 : \underline{T}^* \to \mathcal{J}_1$ splits to the right, it follows that

$$\mathcal{R}_{k+1} = \{\tau \in \mathcal{J}_{k+1} \otimes \mathcal{E} \mid \rho_k \tau \in \mathcal{R}_k,\ D\tau \in \underline{T}^* \otimes \mathcal{R}_k\}$$

and this together with $D(\omega \wedge \tau) = d\omega \wedge \rho_{k-1}\tau + (-1)^{\deg \omega} \omega \wedge D\tau$ (cf. end of Section 6) implies that $D : \underline{\Lambda T}^* \otimes \mathcal{J}_{k+1} \otimes \mathcal{E} \to \underline{\Lambda T}^* \otimes \mathcal{J}_k \otimes \mathcal{E}$ restricts to

$$D : \underline{\Lambda T}^* \otimes \mathcal{R}_{k+1} \to \underline{\Lambda T}^* \otimes \mathcal{R}_k.$$

We finally observe that $\mathcal{R}_k$ and $\mathcal{R}_{k+\ell}$ have the same sheaf of solutions (use diagram (8.2)). We can now restrict the complex $(6.2)_\ell$ to

$$(8.6)_\ell \qquad 0 \longrightarrow \mathcal{S} \xrightarrow{j_\ell} \mathcal{R}_\ell \xrightarrow{D} \underline{T}^* \otimes \mathcal{R}_{\ell-1} \xrightarrow{D} \cdots \xrightarrow{D} \underline{\Lambda^p T}^* \otimes \mathcal{R}_{\ell-p} \xrightarrow{D} \cdots$$

where $\mathcal{S}$ is the sheaf of solutions of the given equation $\mathcal{R}_k$ and $\mathcal{R}_\ell = \mathcal{J}_\ell \otimes \mathcal{E}$ for $\ell < k$. Let $\mathcal{G}_\ell$ be the kernel of $\mathcal{R}_\ell \xrightarrow{\rho_{\ell-1}} \mathcal{R}_{\ell-1}$. The sheaf $\underline{\Lambda^p T}^* \otimes \mathcal{G}_\ell$ identifies via ζ_ℓ with an $\mathcal{O}$-submodule of $\underline{\Lambda^p T}^* \otimes \underline{S^\ell T}^* \otimes \mathcal{E}$ and diagram $(6.6)_\ell$ restricts to

$$(8.7)_\ell \qquad \begin{array}{ccccccccccc}
 & & & & 0 & & 0 & & & 0 & \\
 & & & & \downarrow & & \downarrow & & & \downarrow & \\
 & & 0 & \longrightarrow & \mathcal{G}_\ell & \xrightarrow{\underline{\delta}} & \underline{T}^*\otimes\mathcal{G}_{\ell-1} & \xrightarrow{\underline{\delta}} \cdots & \xrightarrow{\underline{\delta}} & \underline{\Lambda^p T^*}\otimes\mathcal{G}_{\ell-p} & \longrightarrow \cdots \\
 & & \downarrow & & \downarrow & & \downarrow & & & \downarrow & \\
0 & \longrightarrow & \mathcal{S} & \xrightarrow{j_\ell} & \mathcal{R}_\ell & \xrightarrow{D} & \underline{T}^*\otimes\mathcal{R}_{\ell-1} & \xrightarrow{D} \cdots & \xrightarrow{D} & \underline{\Lambda^p T^*}\otimes\mathcal{R}_{\ell-p} & \longrightarrow \cdots \\
 & & \| & & \downarrow & & \downarrow & & & \downarrow & \\
0 & \longrightarrow & \mathcal{S} & \xrightarrow{j_{\ell-1}} & \mathcal{R}_{\ell-1} & \xrightarrow{D} & \underline{T}^*\otimes\mathcal{R}_{\ell-2} & \xrightarrow{D} \cdots & \xrightarrow{D} & \underline{\Lambda^p T^*}\otimes\mathcal{R}_{\ell-p-1} & \longrightarrow \cdots
\end{array}$$

Let $\mathcal{G} = \bigoplus_{\ell\geq 0} \mathcal{G}_\ell$. The complex

$$(8.8) \qquad 0 \longrightarrow \mathcal{G} \xrightarrow{\underline{\delta}} \underline{T}^*\otimes\mathcal{G} \xrightarrow{\underline{\delta}} \underline{\Lambda^2 T^*}\otimes\mathcal{G} \xrightarrow{\underline{\delta}} \cdots \xrightarrow{\underline{\delta}} \underline{\Lambda^n T^*}\otimes\mathcal{G} \longrightarrow 0$$

is called the $\underline{\delta}$-complex for the equation $\mathcal{R}_k$. Let $H^q_m = H^q_m\mathcal{G}$ denote the cohomology of

$$\underline{\Lambda^{q-1}T^*}\otimes\mathcal{G}_{m+1} \to \underline{\Lambda^q T^*}\otimes\mathcal{G}_m \to \underline{\Lambda^{q+1}T^*}\otimes\mathcal{G}_{m-1}.$$

We say that $\mathcal{G}_\ell$ (or $\mathcal{R}_\ell$) is p-acyclic if $H^q_m = 0$ for $m \geq \ell$ and $0 \leq q \leq p$. $\mathcal{G}_\ell$ is acyclic or involutive if it is p-acyclic for all p. It is clear that $\mathcal{G}_\ell$ is 0-acyclic for any $\ell \geq 1$ and we claim that it is 1-acyclic for $\ell \geq k$. In fact,

$$\mathcal{G}_{\ell+1} = \{\tau \in \underline{S^{\ell+1}T^*} \otimes \mathcal{E} \mid \underline{\delta}\tau \in \underline{T}^*\otimes\mathcal{G}_\ell\}$$

and any cocycle $\sigma \in \underline{T}^*\otimes\mathcal{G}_\ell$ is cobounded by an element $\tau \in \underline{S^{\ell+1}T^*}\otimes\mathcal{E}$.

DEFINITION. *The equation* $\mathcal{R}_k$ *is formally integrable if* $\mathcal{R}_{\ell+1} \xrightarrow{\rho_\ell} \mathcal{R}_\ell \longrightarrow 0$ *is exact for* $\ell \geq k$.

The following criterion is an improved version, due to Goldschmidt [5(a)] and Malgrange [9(a), (b)], of a theorem of Quillen [12]. An elegant proof, due to R. Bkouche, can be found in [8(b)].

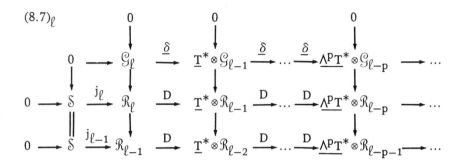

THEOREM. *Let* $\mathcal{R}_k \subset \mathcal{J}_k \otimes \mathcal{E}$ *be a linear differential equation of order* k *and assume that*

1) $\mathcal{R}_{k+1} \xrightarrow{\rho_k} \mathcal{R}_k \longrightarrow 0$ *is exact,*
2) $\mathcal{G}_k$ *is 2-acyclic.*

Then $\mathcal{R}_k$ *is formally integrable.*

The rest of this section is devoted to discussing condition (2) of the theorem. Not much can be said in the general setting, hence we shall introduce some regularity conditions. To start with, assume that $\mathcal{E} = \underline{E}$ is locally free and let $\mathcal{R}_k \subset \mathcal{J}_k \otimes \underline{E}$ be a linear differential equation. Denote by $g_\ell \subset S^\ell T^* \otimes E$ the subset induced by $\mathcal{G}_\ell \subset \underline{S^\ell T^*} \otimes \underline{E}$: each germ $\sigma \in \mathcal{G}_\ell(x)$ induces a vector $\sigma(x) \in (g_\ell)_x$. The fibre $(g_\ell)_x$ is a vector subspace of $S^\ell T^*_x \otimes E_x$ though g_ℓ is not in general a sub-bundle (locally trivial). Since g_ℓ is induced by a subsheaf it follows that $\dim(g_\ell)_x$ is lower semi-continuous. We call g_ℓ a family of vector spaces on X (more precisely a family of subspaces of $S^\ell T^* \otimes E$). The complex (8.8) induces a complex of families of vector spaces on X

(8.9) $$0 \longrightarrow g \xrightarrow{\delta} T^* \otimes g \xrightarrow{\delta} \Lambda^2 T^* \otimes g \longrightarrow \dots \longrightarrow \Lambda^n T^* \otimes g \longrightarrow 0$$

with $g = \bigoplus_{\ell \geq 0} g_\ell$ and $g_\ell = S^\ell T^* \otimes E$ for $\ell < k$. Denote by $H^q_m g$ the cohomology of

$$\Lambda^{q-1} T^* \otimes g_{m+1} \to \Lambda^q T^* \otimes g_m \to \Lambda^{q+1} T^* \otimes g_{m-1}.$$

$H^q_m g$ is a family of vector spaces on X,

$$H^q_m g = \bigcup_{x \in X} H^q_m g_x$$

where $H^q_m g_x$ is the cohomology of

$$\Lambda^{q-1} T^*_x \otimes (g_{m-1})_x \to \Lambda^q T^*_x \otimes (g_m)_x \to \Lambda^{q+1} T^*_x \otimes (g_{m-1})_x.$$

If g_m is a sub-bundle then $H^q_m g = 0$ implies that $H^q_m \mathcal{G} = 0$.

PROPOSITION. *The following assertions are equivalent:*

a) *The subsheaf* $\mathcal{G}_m$ *is a locally free direct factor (in each stalk) of* $\underline{S^mT^* \otimes E}$.

b) *There exists a sub-bundle* $F \subset S^mT^* \otimes E$ *such that* $\mathcal{G}_m = \underline{F}$

c) g_m *is a sub-bundle of* $S^mT^* \otimes E$.

We say that $\mathcal{R}_k$ satisfies the hypothesis h_ℓ if $\mathcal{G}_m$ satisfies the above equivalent conditions for $m \geq \ell$. A criterion for h_ℓ, due to Goldschmidt [5(a)], is given by the

LEMMA. *If* g_ℓ *and* $g_{\ell+1}$ *are sub-bundles and* g_ℓ *is 2-acyclic then* $g_{\ell+m}$ *is a sub-bundle for* $m \geq 0$.

Consider now a vector bundle E on X and let g_ℓ, $\ell \geq 0$, be a family of vector subspaces of $S^\ell T^* \otimes E$ such that $\delta(g_{\ell+1}) \subset T^* \otimes g_\ell$. This implies the more general relation $\delta(\Lambda^p T^* \otimes g_{\ell+1}) \subset \Lambda^{p+1} T^* \otimes g_\ell$, hence we can define a complex analogous to (8.9) by setting $g = \bigoplus_{\ell \geq 0} g_\ell$.

δ-POINCARE LEMMA.

1) *There exists an integer valued function* $k_o : X \to \mathbb{N}$ *such that* $(g_{k_o(x)})_x$ *is involutive* $(H^q_m g_x = 0,\ m \geq k_o(x),\ q \geq 0)$.

2) *If* g_{m_o} *is 1-acyclic for some* m_o, *then* k_o *can be chosen bounded from above and there is an upper bound* K_o *which depends only on* $n = \dim X$, m_o *and* $\operatorname{rank} E = \dim E_x$.

3) *If* g_m, $m \geq \ell_o$, *are sub-bundles then* k_o *can be chosen upper semi-continuous.*

For a proof we refer the reader to [14]. We remark that, if the g_ℓ's arise from a regular equation $\mathcal{R}_k$ (i.e., $\mathcal{R}_{k+m} = \underline{R}_{k+m}$ with R_{k+m} a sub-bundle of $J_{k+m}(E)$ and $g_\ell = g_{k+m}$ a sub-bundle of $S^{k+m}T^* \otimes E$), then 1-acyclicity automatically holds.

Returning to a differential equation $\mathcal{R}_k \subset \mathcal{J}_k \otimes \underline{E}$, assume that it satisfies the hypothesis h_{ℓ_o}. By the third part of the lemma, for each $x \in X$ there is a neighborhood U and a constant K such that $g_K|U$ is 2-acyclic. We can assume that $\ell_o \leq K$ hence $\mathcal{G}_K|U$ is 2-acyclic. To prove that $\mathcal{R}_k|U$ is formally integrable it is enough to check that $\mathcal{R}_{m+1}|U \to \mathcal{R}_m|U \to 0$ is exact for $k \leq m \leq K$. This amounts to checking condition 1) of the theorem and replacing condition 2) by $H^2_m \mathcal{G}|U = 0$ for $k \leq m \leq K-1$. If, moreover, g_{m_o} is 1-acyclic for some m_o then U can be replaced by X and K depends only on n, m_o and *rank* E.

The δ-Poincaré lemma also has implications on the cohomology of $(8.6)_\ell$. Assume that $\mathcal{R}_k$ is formally integrable. Diagram $(8.7)_\ell$ can be completed with bottom arrows terminating in zeros such that the columns are exact. If, further, $\mathcal{R}_k$ satisfies the hypothesis h_{ℓ_o}, there is for every $x \in X$ a neighborhood U and a constant K_o such that the first line of $(8.7)_\ell$ restricted to U is exact for $\ell \geq K_o + n$, hence the cohomologies of $(8.6)_{\ell-1}$ and $(8.6)_\ell$ restricted to U are isomorphic. If moreover g_{m_o} is 1-acyclic for some m_o then U can be replaced by X and K_o depends only on n, m_o and *rank* E. These isomorphic cohomologies will be referred to as the *stable* cohomology of the first complex (or simply of $\mathcal{R}_k$) and the range $\ell \geq K_o + n$ as the *stable* range.

Let $\mathcal{R}_k$ be a formally integrable linear differential equation. Using diagram (8.7) we can define, as in Section 7, a second linear complex associated to the equation $\mathcal{R}_k$

$$(8.10)_\ell \qquad 0 \longrightarrow \mathcal{S} \xrightarrow{j_\ell} \mathcal{C}^0_\ell \mathcal{R}_k \xrightarrow{D'} \mathcal{C}^1_\ell \mathcal{R}_k \xrightarrow{D'} \cdots \xrightarrow{D'} \mathcal{C}^n_\ell \mathcal{R}_k \longrightarrow 0$$

where $\mathcal{C}^0_\ell \mathcal{R}_k = \mathcal{R}_\ell$ and, for $p \geq 0$,

$$\mathcal{C}^{p+1}_\ell \mathcal{R}_k = \underline{\Lambda^{p+1} T^*} \otimes \mathcal{R}_\ell \Big/ \zeta_{p+1.\ell} \circ \underline{\delta}_{p,\ell+1} (\underline{\Lambda^p T^*} \otimes \mathcal{G}_{\ell+1})$$

A diagram chase shows that the complex $(8.10)_\ell$ is exact at $\mathcal{C}^p_\ell \mathcal{R}_k$ whenever $(8.6)_{p+\ell+1}$ is exact at $\underline{\Lambda^p T^*} \otimes \mathcal{R}_{\ell+1}$. Moreover, one also proves that the cohomology of $(8.10)_\ell$, for $\ell \geq K_o - 1$, is isomorphic to the stable cohomology of the first complex.

9. *Linear differential operators*

Let $\mathcal{E}$ and $\mathcal{F}$ be two sheaves of $\mathcal{O}$-modules on X. A linear differential operator of order $\leq k$ on $\mathcal{E}$ with values in $\mathcal{F}$ is an **R**-linear sheaf map $D : \mathcal{E} \to \mathcal{F}$ which can be extended to a left $\mathcal{O}$-linear sheaf map $\phi : \mathcal{J}_k \otimes \mathcal{E} \to \mathcal{F}$ in the sense that $D = \phi \circ j_k$. Since $\mathcal{J}_k \otimes \mathcal{E}$ is left generated by its right holonomic elements, it follows that the extension ϕ, when it exists, is unique. The ℓth prolongation of D is the differential operator

$$D_\ell = j_\ell \circ D : \mathcal{E} \to \mathcal{J}_\ell \otimes \mathcal{F}$$

which is of order $\leq k + \ell$ since $p_\ell \phi \circ j_{k+\ell} = D_\ell$ (cf. diagram 8.4).

Let D be a differential operator of order $\leq k$ and call $\mathcal{R}_k = \ker \phi$ the linear differential equation (of order k) associated to D. If $\mathcal{S}$ is the sheaf of solutions of $\mathcal{R}_k$ then $0 \to \mathcal{S} \to \mathcal{E} \xrightarrow{D} \mathcal{F}$ is exact. Moreover the exact commutative diagram

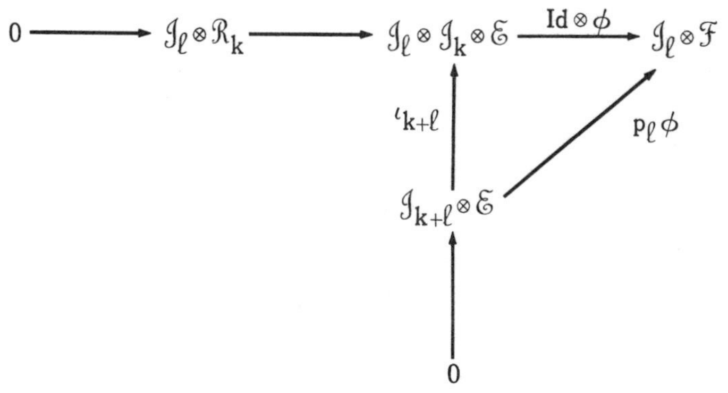

shows that $\mathcal{R}_{k+\ell} = \ker p_\ell \phi$, hence the equation (of order $k+\ell$) associated to D_ℓ is the ℓth prolongation of $\mathcal{R}_k$. Diagram (8.5) gives a trivial proof that $\mathcal{R}_{(k+\ell)+h} = \mathcal{R}_{k+(\ell+h)}$. Conversely, given a differential equation $\mathcal{R}_k \subset \mathcal{J}_k \otimes \mathcal{E}$ there exists a differential operator D of order $\leq k$ whose associated equation (of order k) is $\mathcal{R}_k$: for example take $D = \phi \circ j_k$ where ϕ: $\mathcal{J}_k \otimes \mathcal{E} \to \mathcal{J}_k \otimes \mathcal{E}/\mathcal{R}_k$ is the quotient map. The operators D and D′ in (8.6) and (8.10) are linear of order 1. The operator j_k is of order k. An $\mathcal{O}$-linear sheaf map is an operator of order 0.

CHAPTER II

LINEAR LIE EQUATIONS

10. *Brackets in* $\mathcal{J}_k T$

The Lie bracket of germs of vertical vector fields (elements of $\underline{T_V}$) is a left $\mathcal{O}$-bilinear operation, hence defines a left $\mathcal{O}$-linear map

$$[\ ,\ ]\colon \Lambda^2_{\mathcal{O}} \underline{T_V} \to \underline{T_V}$$

(exterior product with respect to the left structure of $\underline{T_V}$). Since the Lie derivative $\mathcal{L}(\xi)$, $\xi \in \underline{TX^2}$, satisfies $\mathcal{L}(\xi)(\mathcal{J}^{k+1}) \subset \mathcal{J}^k$, it follows that the above bracket factors to a left $\mathcal{O}$-linear map

$$[\ ,\ ]\colon \Lambda^2_{\mathcal{O}} \mathcal{J}_k T \to \mathcal{J}_{k-1} T$$

(recall that $\mathcal{J}_k T = \underline{T_V}/\mathcal{J}^{k+1}\underline{T_V} \simeq \mathcal{J}_k \otimes_{\mathcal{O}} \underline{T}$). This latter bracket is transformed by Π_2 (cf. Section 4) into a sheaf map which is the extension, to germs, of the bundle map

$$[\ ,\ ]\colon \Lambda^2 J_k T \to J_{k-1} T$$

defined by $[j_k\xi(x), j_k\eta(x)] = j_{k-1}([\xi,\eta])(x)$ where ξ, η are vector fields on X. We can similarly define a right $\mathcal{O}$-linear bracket $\Lambda^2_{\mathcal{O}} \mathcal{J}'_k T \to \mathcal{J}'_{k-1} T$ (exterior product with respect to the right structure) quotient of the Lie bracket of horizontal vector fields. It does not yield anything new since r_k transforms one bracket into the other. These brackets have the inconvenience of lowering the order of jets since vertical (or horizontal) vector fields are not compatible with the *diagonal structure* of X^2 in the sense that the filtration $(\mathcal{J}^k)_{k \geq 1}$ of $\mathcal{O}_{X^2}$ is not invariant under $\mathcal{L}(\xi)$, ξ vertical or horizontal. We want to define a second bracket on $\mathcal{J}_k \otimes_{\mathcal{O}} \underline{T}$ which preserves the order of jets, hence defines a Lie algebra structure on $\mathcal{J}_k \otimes_{\mathcal{O}} \underline{T}$. This will be achieved by choosing a suitable Lie algebra subsheaf Θ of $\underline{TX}^2$ such that:

(i) Θ is compatible with the *diagonal structure*, i.e., $\mathcal{L}(\theta)\mathcal{J} \subset \mathcal{J}$ for $\theta \in \Theta$ and

(ii) $\Theta/(\Theta \cap \mathcal{J}^{k+1}\underline{TX}^2)$ is isomorphic with $\mathcal{J}_k \otimes_{\mathcal{O}} \underline{T}$ where $\Theta \cap \mathcal{J}^{k+1}\underline{TX}^2$ is the subsheaf of Θ of germs vanishing to order $k+1$ on Δ.

A good candidate for (i) is the sheaf Θ_Δ of germs of diagonal vector fields on X^2 (tangent to the diagonal) since $\theta \in \Theta_\Delta$ if and only if $\mathcal{L}(\theta)\mathcal{J} \subset \mathcal{J}$. But Θ_Δ does not fulfill condition (ii) and furthermore it presents the technical inconvenience that $\Theta_\Delta \cap \mathcal{J}^{k+1}\underline{TX}^2 \supsetneq \mathcal{J}^{k+1}\Theta_\Delta$ (in R^2, $(y-x)\frac{\partial}{\partial x} \in \Theta_\Delta \cap \mathcal{J}\underline{TR}^2$ does not belong to $\mathcal{J}\Theta_\Delta$).

Let $\mathcal{F}(X^2)$ be the ring of C^∞-functions on X^2 and $\mathfrak{X}(X^2)$ the $\mathcal{F}(X^2)$-module of vector fields. Each $\xi \in \mathfrak{X}(X^2)$ decomposes uniquely into $\xi = \xi_V + \xi_H$ where $\xi_V = V \circ \xi$ is the vertical component and $\xi_H = H \circ \xi$ is the horizontal component, hence $\mathfrak{X}(X^2) = \mathfrak{X}(X^2)_V \oplus \mathfrak{X}(X^2)_H$ where $\mathfrak{X}(X^2)_V$ is the $\mathcal{F}(X^2)$-submodule of vertical vector fields (annihilated by π_{1*}) and similarly for $\mathfrak{X}(X^2)_H$. We now define a *diagonal* $\mathcal{F}(X^2)$-module structure on $\mathfrak{X}(X^2)$ by the relation

$$f \,\Delta\, \xi = f\xi_V + g\xi_H$$

where $g(x,y) = f(x,x)$. $\mathfrak{X}(X^2)_V$ is a diagonal $\mathcal{F}(X^2)$-submodule of $\mathfrak{X}(X^2)$ and the induced diagonal structure coincides with the natural one since $\xi_H = 0$ for any $\xi \in \mathfrak{X}(X^2)_V$. $\mathfrak{X}(X^2)_H$ is also a diagonal submodule and the induced diagonal structure reduces to the natural left $\mathcal{F}(X)$-module structure via the map $\Delta^*\colon f \in \mathcal{F}(X^2) \mapsto f \circ \Delta \in \mathcal{F}(X)$ where $\Delta\colon X \to X^2$ is the diagonal injection. More explicitly, denoting by $i\colon \mathcal{F}(X) \to \mathcal{F}(X^2)$ the map $f \mapsto f \circ \pi_1$, we have $f \,\Delta\, \xi = i(\Delta^* f)\xi$ for any $f \in \mathcal{F}(X^2)$ and $\xi \in \mathfrak{X}(X^2)_H$. It follows that the direct sum decomposition $\mathfrak{X}(X^2) = \mathfrak{X}(X^2)_V \oplus \mathfrak{X}(X^2)_H$ also holds for the diagonal structure. Let $\check{\mathfrak{X}}(X^2)$ be the set of π_1-projectable vector fields on X^2. $\check{\mathcal{X}}(X^2)$ is a submodule of $\mathcal{X}(X^2)$ for the diagonal $\mathcal{F}(X^2)$-module structure though it is not a submodule for the natural structure. $\mathcal{X}(X^2)_V$ is a diagonal submodule of $\check{\mathcal{X}}(X^2)$. Let $\tilde{\mathcal{X}}(X^2)$ be the set of π_1-projectable diagonal (tangent to Δ) vector fields. $\tilde{\mathcal{X}}(X^2)$ is a diagonal submodule of $\check{\mathcal{X}}(X^2)$ hence a diagonal submodule of $\mathcal{X}(X^2)$ but it is not a natural submodule. Remark that an element $\xi \in \tilde{\mathcal{X}}(X^2)$ is entirely determined by its vertical component ξ_V since $\xi_H = i \circ \pi_{2*}(\xi_V|_\Delta)$ where $i: \mathcal{X}(X) \to \mathcal{X}(X^2)_{\underset{\sim}{H}}$ is the horizontal lifting of vector fields. The vertical projection $\xi \in \tilde{\mathcal{X}}(X^2) \mapsto \xi_V \in \mathcal{X}(X^2)_V$ is a diagonal $\mathcal{F}(X^2)$-linear isomorphism. Let $\mathcal{X}(X^2)_{PH}$ be the set of horizontal π_1-projectable vector fields. $\mathcal{X}(X^2)_{PH}$ is a diagonal submodule of $\check{\mathcal{X}}(X^2)$ hence a diagonal submodule of $\mathcal{X}(X^2)$ that is not a natural submodule. We define similarly $\mathcal{X}(X^2)_{PV}$. $\check{\mathcal{X}}(X^2)$ is homogeneous with respect to the direct sum decomposition of $\mathcal{X}(X^2)$ and the induced decomposition is given by $\check{\mathcal{X}}(X^2) = \mathcal{X}(X^2)_V \oplus \mathcal{X}(X^2)_{PH}$ (diagonal structure). The elements of $\tilde{\mathcal{X}}(X^2)$ will be called *admissible* vector fields.

In the sequel it will be understood that all the sheaves on X^2 are actually restricted to Δ. Let

$$\underline{TX^2},\ \Theta_V = \underline{T_V}\ \check{\Theta}, \tilde{\Theta}, \Theta_{PH} \quad \text{and} \quad \Theta_{PV}$$

be the sheaves of germs corresponding to

$$\mathcal{X}(X^2),\ \mathcal{X}(X^2)_V,\ \check{\mathcal{X}}(X^2),\ \tilde{\mathcal{X}}(X^2),\ \mathcal{X}(X^2)_{PH} \quad \text{and} \quad \mathcal{X}(X^2)_{PV}$$

respectively. $\underline{TX^2}$ carries a natural $\mathcal{O}_{X^2}$-module structure and a diagonal $\mathcal{O}_{X^2}$-module structure. If $f \in \mathcal{O}_{X^2}$ and $\xi \in \underline{TX^2}$ then $f \,\Delta\, \xi = f\xi_V + i(g)\xi_H$ where $g = \Delta^* f$. This gives rise, via i and j, to a diagonal left $\mathcal{O}$-module structure which coincides with the natural one and a diagonal right $\mathcal{O}$-module structure $\xi \,\Delta\, g = j(g)\xi_V + i(g)\xi_H$ with $g \in \mathcal{O}$. The sheaf $\check{\Theta}$ is a diagonal submodule of $\underline{TX^2}$ and $\tilde{\Theta}, \Theta_V, \Theta_{PH}$ are diagonal submodules of $\check{\Theta}$. The induced diagonal structure on Θ_V coincides with the natural structure and the induced diagonal structure on Θ_{PH} reduces, via Δ^*, to the diagonal left $\mathcal{O}$-module structure. The diagonal left, diagonal right and natural left $\mathcal{O}$-module structures all coincide on Θ_{PH} and the map $i\colon \underline{T} \to \Theta_{PH}$ (horizontal lifting of vector fields) is a left $\mathcal{O}$-linear isomorphism. The decomposition $\underline{TX}^2 = \underline{T_V} \oplus \underline{T_H}$ holds for the natural and diagonal $\mathcal{O}_{X^2}$-module structures, $\breve{\Theta}$ is homogeneous and the induced direct sum decomposition is given by $\breve{\Theta} = \Theta_V \oplus \Theta_{PH}$ (diagonal structure). The sheaf Θ_{PV} is only a diagonal right $\mathcal{O}$-submodule of $\breve{\Theta}$ and the induced right structure coincides with the natural right $\mathcal{O}$-module structure. The map $j\colon \underline{T} \to \Theta_{PV}$ (vertical lifting of vector fields) is a right $\mathcal{O}$-linear isomorphism. It follows that

$$\tilde{j} = j + i\colon \ \underline{T} \to \breve{\Theta}$$

is a diagonal right linear injection whose image is contained in $\tilde{\Theta}$.

If $\xi \in \underline{TX}^2$ and $f \in \mathcal{J}$ then $f\,\Delta\,\xi$ is vertical and vanishes on Δ. Furthermore, if ξ is π_1-projectable and vanishes on Δ then ξ is vertical. It follows that

$$\breve{\Theta} \cap \mathcal{J}^{k+1}\underline{TX}^2 = \tilde{\Theta} \cap \mathcal{J}^{k+1}\underline{TX}^2 = \Theta_V \cap \mathcal{J}^{k+1}\underline{TX}^2 = \mathcal{J}^{k+1}\Delta\breve{\Theta} =$$
$$\mathcal{J}^{k+1}\Delta\tilde{\Theta} = \mathcal{J}^{k+1}\Delta\Theta_V = \mathcal{J}^{k+1}\Theta_V\,.$$

Define the quotient sheaf

$$\breve{\mathcal{J}}_k T = \breve{\Theta}/\mathcal{J}^{k+1}\Delta\breve{\Theta}\,.$$

This sheaf carries a diagonal $\mathcal{J}_k$-module structure and diagonal left, right $\mathcal{O}$-module structures in such a way that it becomes a module over the left, right $\mathcal{O}$-algebra $\mathcal{J}_k$. The diagonal left structure coincides with the natural left structure. The natural right $\mathcal{O}$-module structure is not defined. The quotient sheaves $\tilde{\mathcal{J}}_k T = \tilde{\Theta}/\mathcal{J}^{k+1}\Delta\tilde{\Theta}$ and $\mathcal{J}_k T = \Theta_V/\mathcal{J}^{k+1}\Delta\Theta_V \simeq \mathcal{J}_k \otimes_{\mathcal{O}} \underline{T}$ are diagonal $\mathcal{J}_k$-submodules of $\breve{\mathcal{J}}_k T$ and the induced diagonal structure on $\mathcal{J}_k T$ coincides with the natural one. Since $\mathcal{J}\,\Delta\Theta_{PH} = 0$ it follows that $\Theta_{PH}/\mathcal{J}^{k+1}\Delta\Theta_{PH} = \Theta_{PH} \simeq \underline{T}$ is a diagonal $\mathcal{J}_k$-submodule of $\breve{\mathcal{J}}_k T$. The induced diagonal $\mathcal{J}_k$-module structure reduces, via $\beta_k\colon \mathcal{J}_k \to \mathcal{O}$, to the diagonal left $\mathcal{O}$-module structure and the latter coincides with the natural left and diagonal right $\mathcal{O}$-module structures. The direct sum decomposition of $\check{\Theta}$ yields the direct sum decomposition

$$\check{\mathcal{J}}_k T = \mathcal{J}_k T \oplus \underline{T}$$

for the diagonal $\mathcal{J}_k$-module structure. We also have a decomposition $\check{\Theta} = \tilde{\Theta} \oplus \Theta_{PH}$ (diagonal structure) which factors to

$$\check{\mathcal{J}}_k T = \tilde{\mathcal{J}}_k T \oplus \underline{T} .$$

Furthermore $\check{\Theta} = \Theta_V + \tilde{\Theta}$, hence

$$\check{\mathcal{J}}_k T = \mathcal{J}_k T + \tilde{\mathcal{J}}_k T$$

(diagonal structure, the sum not being direct). The vertical projection $\xi \mapsto \xi_V$ of $\check{\Theta}$ will be denoted by

$$\varepsilon : \check{\Theta} \to \Theta_V .$$

It is a diagonal $\mathcal{O}_{X^2}$-linear retraction onto Θ_V and its restriction $\varepsilon : \tilde{\Theta} \to \Theta_V$ is an isomorphism whose inverse is defined by $\xi \mapsto \xi + i \circ \pi_{2*}(\xi|_\Delta)$, $\xi \in \theta_V$. Passage to the quotient yields the $\mathcal{J}_k$-linear retraction

$$\varepsilon_k : \check{\mathcal{J}}_k T \to \mathcal{J}_k T$$

which is of course the vertical projection $A \mapsto A_V$ (upon the first factor) in $\check{\mathcal{J}}_k T = \mathcal{J}_k T \oplus \underline{T}$. The restriction $\varepsilon_k : \tilde{\mathcal{J}}_k T \to \mathcal{J}_k T$ is a $\mathcal{J}_k$-linear isomorphism. If $A \in \check{\mathcal{J}}_k T$ is represented by $\xi \in \check{\Theta}$ then the horizontal projection A_H (upon the second factor) is equal to $\pi_1 * \xi$. The map $\tilde{j}$ composed with the quotient map yields the diagonal right $\mathcal{O}$-linear injection

$$\tilde{j}_k = j_k + i_k : \underline{T} \to \check{\mathcal{J}}_k T$$

whose image is in $\tilde{\mathcal{J}}_k T$. Moreover, $j_k = \varepsilon_k \circ \tilde{j}_k$ and $\varepsilon_h \circ \tilde{\rho}_h = \rho_h \circ \varepsilon_k$ where ρ_h, $\tilde{\rho}_h$ and $\check{\rho}_h$ are the standard projections of k-jets onto h-jets.

The left, right and the $\mathcal{J}_0 = \mathcal{O}$-module structures all coincide on $\check{\mathcal{J}}_0 T$ and the map $\xi \in \check{\Theta} \mapsto (\pi_{2*}(\xi|_\Delta), \pi_{1*}\xi) \in \underline{T} \oplus \underline{T}$ factors to an $\mathcal{O}$-linear isomorphism $\check{\mathcal{J}}_0 T \to \underline{T} \oplus \underline{T}$ which transforms $\tilde{\mathcal{J}}_0 T$ into the diagonal of $\underline{T} \oplus \underline{T}$. Projecting the image on either of the factors, we can identify $\tilde{\mathcal{J}}_0 T$ with $\underline{T}$. For later use we remark that the identification $\tilde{\mathcal{J}}_0 T \simeq \underline{T}$ can also be obtained by

a) factoring the map $\xi \in \tilde{\Theta} \mapsto \Delta_*^{-1}(\xi|_\Delta) \in \underline{T}$ or

b) factoring $\pi_{1*}: \tilde{\Theta} \to \underline{T}$ or

c) taking the inverse of the $\mathcal{O}$-linear isomorphism $\tilde{j}_0: \underline{T} \to \tilde{\mathcal{J}}_0 T$.

Recall that $\mathcal{J}_0 T$ identifies with $\underline{T}$ via the quotient of $\xi \in \Theta_V \mapsto \pi_{2*}(\xi|_\Delta) \in \underline{T}$, this quotient being the inverse of $j_0: \underline{T} \to \mathcal{J}_0 T$. We remark finally that all the previous quotient sheaves $\check{\mathcal{J}}_k T$, $\tilde{\mathcal{J}}_k T$ and $\mathcal{J}_k T$ are locally free of finite rank for the left and right diagonal $\mathcal{O}$-module structures and, excepting $\check{\mathcal{J}}_k T$, also for the $\mathcal{J}_k$-module structure. The sheaf $\check{\mathcal{J}}_k T$ is locally finitely generated over $\mathcal{J}_k$. The isomorphism $\varepsilon_k^{-1} \circ \psi_k: \mathcal{J}_k \otimes_{\mathcal{O}} \underline{T} \to \tilde{\mathcal{J}}_k T$ is the quotient of the map $f \otimes \xi \in \mathcal{O}_{X^2} \otimes_{\mathcal{O}} \underline{T} \mapsto f \Delta \tilde{j}\xi \in \tilde{\Theta}$.

$\underline{TX}^2$, with the bracket of vector fields, is an R-Lie algebra sheaf and $\check{\Theta}, \tilde{\Theta}, \Theta_V, \Theta_{PH}$ are Lie subalgebras. If $\xi \in \underline{TX}^2$ and $f \in \mathcal{J}^{k+1}$ then $\mathcal{L}(\xi)f \in \mathcal{J}^k$ hence the induced bracket $\Lambda_R^2 \check{\Theta} \to \check{\Theta}$ on $\check{\Theta}$ factors to an R-linear map

$$[\ ,\ ]: \Lambda_R^2 \check{\mathcal{J}}_k T \to \check{\mathcal{J}}_{k-1} T$$

with the loss of one unit in the order of jets. In particular, this bracket vanishes on $\check{\mathcal{J}}_0 T$. If $f, g \in \mathcal{O}$ and $A, B \in \check{\mathcal{J}}_k T$ then $[f\Delta A, g\Delta B] = (fg) \Delta [A, B] + f(A_H g) \Delta B - g(B_H f) \Delta A$ where $A_H \in \underline{T}$ is the horizontal component of A, $A_H g$ is the Lie derivative of g by A_H and $f \Delta A$ is diagonal left $\mathcal{O}$-multiplication (which coincides with the natural left $\mathcal{O}$-multiplication). Moreover we have the following commutation formulas: $[\tilde{j}_k \theta, \tilde{j}_k \theta'] = \tilde{j}_{k-1}([\theta, \theta'])$ and $[\check{\rho}_h A, \check{\rho}_h B] = \check{\rho}_{h-1}[A, B]$ where $\theta, \theta' \in \underline{T}$ and $A, B \in \check{\mathcal{J}}_k T$. We shall now examine the restriction of $[\ ,\ ]$ to the various subsheaves.

The restriction of $[\ ,\ ]$ to the horizontal component $\underline{T}$ of $\check{\mathcal{J}}_k T$ is also obtained by factoring the bracket $\Lambda^2_R \Theta_{PH} \to \Theta_{PH}$. Since $\mathcal{J}\,\Delta\Theta_{PH} = 0$, it is clear that $0 = [\xi, f\Delta\eta] \in \mathcal{J}^{k+1}\Delta\Theta_{PH}$ for any $\xi, \eta \in \Theta_{PH}$ and $f \in \mathcal{J}^{k+1}$ hence the restriction $[\ ,\ ]|_{\underline{T}}$ can be redefined to take its values in the horizontal component $\underline{T}$ of $\check{\mathcal{J}}_k T$ (instead of $\check{\mathcal{J}}_{k-1} T$). This restriction obviously coincides with the usual Lie bracket in $\underline{T}$.

The restriction of $[\ ,\ ]$ to the vertical component $\mathcal{J}_k T$ of $\check{\mathcal{J}}_k T$ is also obtained by factoring the bracket $\Lambda^2_R \Theta_V \to \Theta_V$ which is left $\mathcal{O}$-bilinear hence reduces to $\Lambda^2_{\mathcal{O}} \Theta_V \to \Theta_V$. This restriction $[\ ,\ ]|_{\mathcal{J}_k T}$ coincides with the bracket defined in the beginning of this section. If $\theta, \theta' \in \underline{T}$ then $[j_k\theta, j_k\theta'] = j_{k-1}([\theta, \theta'])$.

The restriction of $[\ ,\ ]$ to $\tilde{\mathcal{J}}_k T$ is also obtained by factoring the bracket $\Lambda^2_R \tilde{\Theta} \to \tilde{\Theta}$. The relation

$$\mathcal{L}(\xi)(f \,\Delta\, \eta) = (\mathcal{L}(\xi)f) \,\Delta\, \eta + f \,\Delta\, \mathcal{L}(\xi)\eta + (i(g) - f) \,\Delta\, \varepsilon^{-1}[\xi_V, \eta_H]$$

for $\xi, \eta \in \tilde{\Theta}$, $f \in \mathcal{O}_{X^2}$ and $g(x) = f(x, x)$, together with $\mathcal{L}(\xi)\,\mathcal{J}^k \subset \mathcal{J}^k$ shows that $\mathcal{L}(\xi)(\mathcal{J}^{k+1}\Delta\tilde{\Theta}) \subset \mathcal{J}^{k+1}\Delta\tilde{\Theta}$ hence $\Lambda^2_R \tilde{\Theta} \to \tilde{\Theta}$ factors actually to an R-linear map

$$[\![\ ,\ ]\!] : \Lambda^2_R \tilde{\mathcal{J}}_k T \to \tilde{\mathcal{J}}_k T$$

and we obviously have $[A, B] = \tilde{\rho}_{k-1}[\![A, B]\!]$ for any $A, B \in \tilde{\mathcal{J}}_k T$. The restriction $[\ ,\ ]|_{\tilde{\mathcal{J}}_k T}$ is thus redefined by $[\![\ ,\ ]\!]$ to take its values in $\tilde{\mathcal{J}}_k T$. The sheaf $\tilde{\mathcal{J}}_k T$ together with $[\![\ ,\ ]\!]$ is an R-Lie algebra sheaf, the injection $\tilde{j}_k$ and the projections $\tilde{\rho}_h$ are Lie algebra morphisms. In particular, $\tilde{j}_0$ identifies the Lie algebras $\underline{T}$ and $\tilde{\mathcal{J}}_0 T$. If $f, g \in \mathcal{O}$ and $A, B \in \tilde{\mathcal{J}}_k T$ then

$$[\![f \,\Delta\, A, g \,\Delta\, B]\!] = (fg) \,\Delta\, [\![A, B]\!] + f(\tilde{\rho}_0 A)g \,\Delta\, B - g(\tilde{\rho}_0 B)f \,\Delta\, A\,.$$

The bracket $[\![\ ,\ ]\!]$ on $\tilde{\mathcal{J}}_k T$ transports via ε_k to a bracket, still noted by $[\![\ ,\ ]\!]$, on $\mathcal{J}_k T \simeq \mathcal{J}_k \otimes_{\mathcal{O}} \underline{T}$. Let $A, B \in \mathcal{J}_k T$ be represented by $\xi, \eta \in \Theta_V$ and write $\xi'_H = (\varepsilon^{-1}\xi)_H$, $\eta'_H = (\varepsilon_\eta^{-1})_H$. The field ξ'_H is the unique π_1-projectable horizontal (germ of) vector field such that

$\xi'_H|_\Delta = (r_*\xi)|_\Delta$ (i.e., $\xi|_\Delta$ thought of as horizontal). Then

$$[\![A, B]\!] = [\xi, \eta] + [\xi, \eta'_H] + [\xi'_H, \eta] \ \mathrm{mod}\ \mathcal{J}^{k+1}\Theta_V$$

hence $[\![\ ,\ ]\!]$ is equal to $[\ ,\ ]$ plus the correcting term $[\xi, \eta'_H] + [\xi'_H, \eta]\ \mathrm{mod}\ \mathcal{J}^{k+1}\Theta_V$. If A and B are holonomic, i.e., $A = j_k\theta$ and $B = j_k\theta'$ with $\theta, \theta' \in \underline{T}$, then $[\![A, B]\!] = j_k([\theta, \theta'])$ (since $j_k = \varepsilon_k \circ \tilde{j}_k$) and $\rho_{k-1}[\![A, B]\!] = [A, B]$. The last relation does not hold for arbitrary $A, B \in \mathcal{J}_k T$. If $A, B \in \mathcal{J}\Theta_V/\mathcal{J}^{k+1}\Theta_V = \mathcal{J}_k^0 T \simeq (\mathcal{J}/\mathcal{J}^{k+1}) \otimes_{\mathcal{O}} \underline{T}$ then $[A, B]$ is defined $\mathrm{mod}\ \mathcal{J}^{k+1}\Theta_V$ and $[A, B] = [\![A, B]\!] \in \mathcal{J}_k^0 T$. Since $\mathcal{J}_k \otimes_{\mathcal{O}} \underline{T}$ is left $\mathcal{O}$-generated by its holonomic elements then $[\![\ ,\ ]\!]$ is the unique Lie algebra bracket on $\mathcal{J}_k \otimes_{\mathcal{O}} \underline{T}$ which satisfies the following properties:

1) $j_k: \underline{T} \to \mathcal{J}_k \otimes_{\mathcal{O}} \underline{T}$ is a Lie algebra morphism and
2) $[\![A, fB]\!] = f[\![A, B]\!] + [(\rho_0 A)f]B$ where $f \in \mathcal{O}$ and $\rho_0: \mathcal{J}_k \otimes_{\mathcal{O}} \underline{T} \to \mathcal{J}_0 \otimes_{\mathcal{O}} \underline{T} \simeq \underline{T}$.

Observe that 1) and 2) can be replaced by the unique condition

(10.1) $$[\![f j_k\mu, g j_k\eta]\!] = fg j_k[\mu, \eta] + f[\mathcal{L}(\mu)g] j_k\eta - g[\mathcal{L}(\eta)f] j_k\mu$$

where $f, g \in \mathcal{O}$ and $\mu, \eta \in \underline{T}$. Property 1) implies that on $\mathcal{J}_0 \otimes_{\mathcal{O}} \underline{T} \simeq \underline{T}$ the bracket $[\![\ ,\ ]\!]$ is equal to the Lie bracket of vector fields. A definition of $[\![\ ,\ ]\!]$ based on the properties 1) and 2) was given by Ngô Van Quê [10(a)].

The Lie bracket of a vertical vector field with a π_1-projectable diagonal vector field is again vertical hence the restriction of $[\ ,\ ]$ to $\tilde{\mathcal{J}}_k T \otimes_{\mathbb{R}} \mathcal{J}_k T$ yields an R-linear map $\tilde{\mathcal{J}}_k T \otimes_{\mathbb{R}} \mathcal{J}_k T \to \mathcal{J}_{k-1} T$ which is also obtained by factoring the map $\tilde{\Theta} \otimes_{\mathbb{R}} \Theta_V \to \Theta_V$. Since diagonal vector fields preserve $\mathcal{J}^k$ it follows that $[\xi, f\eta] \in \mathcal{J}^k\Theta_V$ for any $\xi \in \tilde{\Theta}$, $f \in \mathcal{J}^k$ and $\eta \in \Theta_V$ hence the preceding map actually factors to

$$[\![\ ,\ ]\!]: \tilde{\mathcal{J}}_k T \otimes_{\mathbb{R}} \mathcal{J}_{k-1} T \to \mathcal{J}_{k-1} T$$

and $[A, B] = [\![A, \rho_{k-1} B]\!]$ for any $A \otimes B \in \tilde{\mathcal{J}}_k T \otimes_{\mathbb{R}} \mathcal{J}_k T$. The restriction of $[\ ,\ ]$ is thus redefined by $[\![\ ,\ ]\!]$ and the latter satisfies the following conditions:

a) $\llbracket f\,\Delta\,\tilde{A}, gB \rrbracket = fg\llbracket \tilde{A}, B\rrbracket + [f(\tilde{\rho}_0\tilde{A})g]B$, $f, g \in \mathcal{O}$, $\tilde{A} \in \tilde{\mathcal{J}}_k T$ and $B \in \mathcal{J}_{k-1}T$.

b) $\llbracket \tilde{A}, B\rrbracket \in \mathcal{J}^0_{k-1}T$ for any $\tilde{A} \in \tilde{\mathcal{J}}_k T$ and $B \in \mathcal{J}^0_{k-1}T$.

c) $\llbracket \llbracket \tilde{A}, \tilde{B}\rrbracket, C\rrbracket = \llbracket \tilde{A}, \llbracket \tilde{B}, C\rrbracket\rrbracket - \llbracket \tilde{B}, \llbracket \tilde{A}, C\rrbracket\rrbracket$, $\tilde{A}, \tilde{B} \in \tilde{\mathcal{J}}_k T$ and $C \in \mathcal{J}_{k-1}T$.

d) $\llbracket \tilde{A}, [B, C]\rrbracket = [\llbracket \tilde{A}, B\rrbracket, C] + [B, \llbracket \tilde{A}, C\rrbracket]$, $\tilde{A} \in \tilde{\mathcal{J}}_k T$ and $B, C \in \mathcal{J}_{k-1}T$.

d′) If $B, C \in \mathcal{J}^0_{k-1}T$ then each term in d) belongs to $\mathcal{J}^0_{k-1}T$ and $[\ ,\ ] = \llbracket\ ,\ \rrbracket$.

This bracket transports via ε_{k-1} to an R-linear map

$$\llbracket\ ,\ \rrbracket : \mathcal{J}_k T \otimes_R \mathcal{J}_{k-1}T \to \mathcal{J}_{k-1}T$$

and satisfies analogous properties (drop $\sim$ and Δ). If $A \in \mathcal{J}_k T$ is represented by $\xi \in \Theta_V$ and $B \in \mathcal{J}_{k-1}T$ is represented by $\eta \in \Theta_V$ then

$$\llbracket A, B\rrbracket = [\xi, \eta] + [\xi'_H, \eta] \ \mathit{mod}\ \mathcal{J}^k\Theta_V$$

(compare with $\llbracket A, B\rrbracket$). This implies the property

e) $\llbracket A, B\rrbracket = [A, B]$ for any $A \in \mathcal{J}^0_k T$ and $B \in \mathcal{J}_{k-1}T$.

For any $\xi \in \tilde{\Theta}$, the Lie derivative $\mathcal{L}(\xi)$ operates as a (germ of) derivation on $\mathcal{O}_{X^2}$ and preserves the filtration $(\mathcal{J}^k)$. Hence the Lie algebra representation

$$\mathcal{L} : \tilde{\Theta} \to \mathcal{D}er_R\, \mathcal{O}_{X^2}$$

factors to a left $\mathcal{O}$-linear faithful Lie algebra representation

$$\mathcal{L}_1 : \tilde{\mathcal{J}}_k T \to \mathcal{D}er_R\, \mathcal{J}_k$$

which transports, via ε_k, to a left $\mathcal{O}$-linear faithful Lie algebra representation

$$\mathcal{L}_1 : \mathcal{J}_k T \simeq \mathcal{J}_k \otimes_{\mathcal{O}} \underline{T} \to \mathcal{D}er_R\, \mathcal{J}_k$$

($\mathcal{J}_k T$ with $[\![\ ,\ ]\!]$). If $f \in \mathcal{O}$ and $A \in \mathcal{J}_k T$ then $\mathcal{L}_1(A)(i_k f) = i_k(\mathcal{L}(\rho_0 A)f)$. The sheaf $\mathcal{J}_k T$ (or $\tilde{\mathcal{J}}_k T$) is thus represented as a sheaf of derivation of $\mathcal{O}_{\Delta(k)}$.

Let $\mathcal{D}er_{\mathbf{R}}\, \mathcal{J}_k T$ be the sheaf of germs of derivations of the $\mathbf{R}$-Lie algebra ($\mathcal{J}_k T$, $[\![\ ,\ ]\!]$). The bracket $[\![\ ,\ ]\!]$ defines an adjoint representation

$$\mathcal{L}_2\colon\ \mathcal{J}_k T \to \mathcal{D}er_{\mathbf{R}}\, \mathcal{J}_k T \quad (\text{resp. } \tilde{\mathcal{J}}_k T \to \mathcal{D}er_{\mathbf{R}}\, \tilde{\mathcal{J}}_k T)$$

and the bracket $[\![\ ,\ ]\!]$ defines an adjoint representation

$$\mathcal{L}_3\colon\ \mathcal{J}_k T \to \mathcal{H}om_{\mathbf{R}}\, \mathcal{J}_{k-1} T \quad (\text{resp. } \tilde{\mathcal{J}}_k T \to \mathcal{H}om_{\mathbf{R}}\, \mathcal{J}_{k-1} T)$$

which restricted to the subalgebra $\mathcal{J}_k^0 T$ gives a representation

$$\mathcal{J}_k T \to \mathcal{D}er_{\mathbf{R}}\, \mathcal{J}_{k-1}^0 T \quad (\text{resp. } \tilde{\mathcal{J}}_k T \to \mathcal{D}er_{\mathbf{R}}\, \mathcal{J}_{k-1}^0 T)$$

in view of the properties (c), (d) and (d′). We use the notation $\mathcal{L}_i$ for all these derivations since they are always induced by Lie derivatives. If T is any tensor bundle on X then $\mathcal{J}_k T$ (resp. $\tilde{\mathcal{J}}_k T$) operates on $\underline{T}$ via $\rho_0\colon \mathcal{J}_k T \to \mathcal{J}_0 T \simeq \underline{T}$ (resp. $\tilde{\rho}_0\colon \tilde{\mathcal{J}}_k T \to \tilde{\mathcal{J}}_0 T \simeq \underline{T}$) and the standard Lie derivative of tensor fields, hence a Lie algebra representation

$$\mathcal{L}_4\colon\ \mathcal{J}_k T \to \mathcal{H}om_{\mathbf{R}}\, \underline{T} \quad (\text{resp. } \tilde{\mathcal{J}}_k T \to \mathcal{H}om_{\mathbf{R}}\, \underline{T})\ .$$

We can now extend these representations to (germs of) tensor fields with coefficients in jet sheaves. For example, the representation $\mathcal{L}_2$ extends to a Lie algebra representation

$$\mathcal{L}_2\colon\ \mathcal{J}_k T \to \mathcal{H}om_{\mathbf{R}}\, \underline{\Lambda^p T^*} \otimes_{\mathcal{O}} \mathcal{J}_k T$$

$$(\text{alternatively } \mathcal{J}_k \otimes_{\mathcal{O}} \underline{T} \to \mathcal{H}om_{\mathbf{R}}\, \underline{\Lambda^p T^*} \otimes_{\mathcal{O}} \mathcal{J}_k \otimes_{\mathcal{O}} \underline{T})$$

defined by $\mathcal{L}_2(A)(\omega \otimes B) = [\mathcal{L}_4(A)\omega] \otimes B + \omega \otimes \mathcal{L}_2(A)B$. This is well defined in view of formula (2). In the same way $\mathcal{L}_3$ induces the representation

$$\mathcal{L}_3\colon\ \mathcal{J}_k T \to \mathcal{H}om_{\mathbf{R}}\, \underline{\Lambda^p T^*} \otimes_{\mathcal{O}} \mathcal{J}_{k-1} T\ .$$

Replacing $\mathcal{J}_k T$ by $\tilde{\mathcal{J}}_k T$ we also obtain the representations

$$\tilde{\mathcal{L}}_2 : \tilde{\mathcal{J}}_k T \to \mathcal{H}om_{\mathbf{R}}\, \underline{\Lambda^p T^*} \otimes_{\mathcal{O}} \tilde{\mathcal{J}}_k T$$

and

$$\tilde{\mathcal{L}}_3 : \tilde{\mathcal{J}}_k T \to \mathcal{H}om_{\mathbf{R}}\, \underline{\Lambda^p T^*} \otimes_{\mathcal{O}} \mathcal{J}_{k-1} T$$

which are equivalent to $\mathcal{L}_2$ and $\mathcal{L}_3$ via ε_k. If we identify $\underline{\Lambda^p T^*} \otimes_{\mathcal{O}} \mathcal{J}_k T$ with $\underline{F}/\mathcal{J}^{k+1}\underline{F}$ where $F = (\Lambda^p T_H^*) \otimes T_V$ is a vector bundle on X^2 (cf. end of Section 4) and choose a representative $\xi \in \underline{T_V}$ for $A \in \mathcal{J}_k T$ then $\mathcal{L}_3(A)$ operating on $\underline{\Lambda^p T^*} \otimes_{\mathcal{O}} \mathcal{J}_{k-1} T$ is the quotient of the Lie derivative $\mathcal{L}(\xi + \xi'_H)$ operating on (germs of) horizontal exterior differential p-forms with values in the bundle of vertical vectors (i.e., section of F). Similarly, if $\xi \in \tilde{\Theta}$ represents $A \in \tilde{\mathcal{J}}_k T$ then $\tilde{\mathcal{L}}_2(A)$ is the quotient of the Lie derivative $\mathcal{L}(\xi)$ operating on (germs of) horizontal exterior differential p-forms u with coefficients in $\tilde{\Theta}$, i.e., u is horizontal with values in TX^2 and $u(i\theta_1, \ldots, i\theta_p) \in \tilde{\Theta}$ for any $\theta_i \in \underline{T}$ or, in coordinates, $u = \Sigma\, dx^{\wedge\beta} \otimes \xi_\beta$ with $\xi_\beta \in \tilde{\Theta}$ (cf. Section 14).

11. *Coordinates*

Let (x, y) be a system of local coordinates in X^2 as in Section 5. Then

$$\frac{\partial}{\partial x} = \left(\frac{\partial}{\partial x_1}, \ldots, \frac{\partial}{\partial x_n}\right)$$

is an $\mathcal{O}_{X^2}$-basis for (germs of) horizontal vector fields,

$$\frac{\partial}{\partial y} = \left(\frac{\partial}{\partial y_1}, \ldots, \frac{\partial}{\partial y_n}\right)$$

is an $\mathcal{O}_{X^2}$-basis for vertical vector fields and

$$\frac{\partial}{\partial x} + \frac{\partial}{\partial y} = \left(\frac{\partial}{\partial x_1} + \frac{\partial}{\partial y_1}, \ldots, \frac{\partial}{\partial x_n} + \frac{\partial}{\partial y_n}\right)$$

is a diagonal $\mathcal{O}_{X^2}$-basis for π_1-projectable diagonal vector fields. Using matrix notation $F\frac{\partial}{\partial x} = \Sigma f_i \frac{\partial}{\partial x_i}$ where $F = (f_1, \ldots, f_n)$, the f_i are local functions on X^2 and $\frac{\partial}{\partial x}$ is written as a column matrix, the vertical fields read $F(x,y)\frac{\partial}{\partial y}$, the π_1-projectable diagonal fields read

$$F\,\Delta\left(\frac{\partial}{\partial x} + \frac{\partial}{\partial y}\right) = F(x,y)\frac{\partial}{\partial y} + F(x,x)\frac{\partial}{\partial x}\,,$$

ε reads

$$F(x,y)\frac{\partial}{\partial y} + F(x,x)\frac{\partial}{\partial x} \mapsto F(x,y)\frac{\partial}{\partial y}$$

and ε^{-1} reads

$$F(x,y)\frac{\partial}{\partial y} \mapsto F(x,y)\frac{\partial}{\partial y} + F(x,x)\frac{\partial}{\partial x}$$

(the notation $F(x,y)$ and $F(x,x)$ indicates the variables upon which the function depends so that $F(x,x)$ denotes $(x,y) \to F(x,x)$). An element $A \in \mathcal{J}_k T$ is the class *mod* $\mathcal{J}^{k+1}\underline{T_V}$ of a unique (germ of) polynomial vertical vector field

$$\sum_{\substack{|\alpha| \leq k \\ 1 \leq i \leq n}} a_{\alpha,i}(x)\frac{(y-x)^\alpha}{\alpha!}\frac{\partial}{\partial y_i} = \sum_{|\alpha| \leq k} a_\alpha(x)\frac{(y-x)^\alpha}{\alpha!}\frac{\partial}{\partial y}$$

and an element $\tilde{A} \in \tilde{\mathcal{J}}_k T$ is the class *mod* $\mathcal{J}^{k+1}\Delta\tilde{\Theta}$ of a unique (germ of) polynomial π_1-projectable diagonal vector field

$$\sum_{|\alpha| \leq k} \tilde{a}_\alpha(x)\frac{(y-x)^\alpha}{\alpha!}\,\Delta\left(\frac{\partial}{\partial x} + \frac{\partial}{\partial y}\right) = \sum_{|\alpha| \leq k} \tilde{a}_\alpha(x)\frac{(y-x)^\alpha}{\alpha!}\frac{\partial}{\partial y} + a_0(x)\frac{\partial}{\partial x}\,.$$

The first bracket in $\mathcal{J}_k T$ reads

$$[A,B] = \sum_{\substack{|\alpha+\beta| \leq k-1 \\ 1 \leq i,\, j \leq n}} (a_{\alpha,j}\, b_{\beta+1_j,i} - a_{\alpha+1_j,i}\, b_{\beta,j})\frac{(y-x)^{\alpha+\beta}}{\alpha!\,\beta!}\frac{\partial}{\partial y_i}\,.$$

The second bracket in $\mathcal{J}_k T$ reads

$$\llbracket A,B\rrbracket = \sum_i \left\{ \sum_{\substack{|\alpha+\beta|\le k\\ |\alpha|>0,\, j}} a_{\alpha,j}\, b_{\beta+1_j,i}\frac{(y-x)^{\alpha+\beta}}{\alpha!\,\beta!} - \right.$$

$$\sum_{\substack{|\alpha+\beta|\le k\\ |\beta|>0,\, j}} a_{\alpha+1_j,i}\, b_{\beta,j}\frac{(y-x)^{\alpha+\beta}}{\alpha!\,\beta!} +$$

$$\left.\sum_{\substack{|\alpha|\le k\\ j}} \left(a_{0,j}\frac{\partial b_{\alpha,i}}{\partial x_j} - b_{0,j}\frac{\partial a_{\alpha,i}}{\partial x_j}\right)\frac{(y-x)^{\alpha}}{\alpha!}\right\}\frac{\partial}{\partial y_i}$$

and in $\tilde{\mathcal{J}}_k T$ reads

$$\llbracket \tilde{A},\tilde{B}\rrbracket = \sum_i \{idem\}\,\Delta\left(\frac{\partial}{\partial x_i}+\frac{\partial}{\partial y_i}\right).$$

The third bracket $\mathcal{J}_k T \otimes_R \mathcal{J}_{k-1} T \to \mathcal{J}_{k-1} T$ reads

$$\llbracket A,B\rrbracket = \sum_i \left\{ \sum_{\substack{|\alpha+\beta|\le k-1\\ |\alpha|>0,\, j}} a_{\alpha,j}\, b_{\beta+1_j,i}\frac{(y-x)^{\alpha+\beta}}{\alpha!\,\beta!} - \right.$$

$$\sum_{\substack{|\alpha+\beta|\le k-1\\ j}} a_{\alpha+1_j,i}\, b_{\beta,j}\frac{(y-x)^{\alpha+\beta}}{\alpha!\,\beta!} +$$

$$\left.\sum_{\substack{|\beta|\le k-1\\ j}} a_{0,j}\frac{\partial b_{\beta,i}}{\partial x_j}\frac{(y-x)^{\beta}}{\beta!}\right\}\frac{\partial}{\partial y_i}$$

and similarly for $\llbracket \tilde{A},B\rrbracket$.

12. *Lie equations*

In this section we introduce differential equations which are invariant under the bracket operations. Such equations arise in the study of pseudogroups.

Recall (cf. end of Section 6) that $\mathcal{J}_k \otimes_{\mathcal{O}} \underline{T}$ identifies with $\mathcal{J}_k T = \underline{T_V}/\mathcal{J}^{k+1}\underline{T_V}$ via the quotient of the map

$$f \otimes \theta \in \mathcal{O}_{X^2} \otimes_{\mathcal{O}} \underline{T} \mapsto f(j\theta) \in \underline{T_V}$$

and more generally, $\underline{\Lambda^p T^*} \otimes_{\mathcal{O}} \mathcal{J}_k \otimes_{\mathcal{O}} \underline{T}$ identifies with $\underline{F}^p/\mathcal{J}^{k+1}\underline{F}^p$, $F^p = \Lambda^p T_H^* \otimes_R T_V$, via the quotient of

$$\omega \otimes f \otimes \xi \in \underline{\Lambda^p T^*} \otimes_{\mathcal{O}} \mathcal{O}_{X^2} \otimes_{\mathcal{O}} \underline{T} \mapsto f(\pi_1^*\omega) \otimes j\xi \in \underline{F}^p .$$

$\underline{F}^p$ is the sheaf of germs of horizontal exterior differential p-forms with values in T_V. The operator

$$D: \underline{\Lambda^p T^*} \otimes \mathcal{J}_k \otimes \underline{T} \to \underline{\Lambda^{p+1} T^*} \otimes \mathcal{J}_{k-1} \otimes \underline{T}$$

is the quotient of $d_H: \underline{F}^p \to \underline{F}^{p+1}$ where d_H is exterior differentiation with respect to the variable x, namely $(d_H u)(v_1, \ldots, v_{p+1}) = (du_y)(Hv_1, \ldots, Hv_{p+1})$ where $v_i \in T_{(x,y)}X^2$, $u_y = u|_{X \times \{y\}}$ and $H: TX^2 \to T_H$ is the horizontal projection. The differential du_y is defined since $T_V|_{X \times \{y\}}$ is a trivial bundle with fibre T_yX. In particular

$$D: \mathcal{J}_k \otimes \underline{T} \to \underline{T}^* \otimes \mathcal{J}_{k-1} \otimes \underline{T}$$

is the quotient of $d_H: \underline{T_V} \to \underline{T_H^*} \otimes \underline{T_V}$. Let $A \in \underline{T}^* \otimes \mathcal{J}_{k-1} \otimes \underline{T}$ be represented by the (germ of) horizontal differential 1-form u with values in T_V, $u \in \underline{T_H^*} \otimes \underline{T_V}$, and let $\theta \in \underline{T}$. Then $\langle\theta, A\rangle = \langle i\theta, u\rangle \bmod \mathcal{J}^k\underline{T_V}$ where $i: \underline{T} \to \underline{T_H}$ is the horizontal lifting of vector fields (we assume of course that all elements are in stalks over the same base point). In particular, if $\xi \in \underline{T_V}$ represents the element $A \in \mathcal{J}_k \otimes \underline{T}$ then $DA \in \underline{T}^* \otimes \mathcal{J}_{k-1} \otimes \underline{T}$ is represented by $d_H\xi \in \underline{T_H^*} \otimes \underline{T_V}$ and $\langle i\theta, d_H\xi\rangle = \mathcal{L}(i\theta)\xi = [i\theta, \xi]$ hence we obtain the relation $\langle\theta, DA\rangle = [i\theta, \xi] \bmod \mathcal{J}^k\underline{T_V}$ for any $\theta \in \underline{T}$ and $A \in \mathcal{J}_k \otimes \underline{T}$.

LEMMA (Malgrange [9(c)]). *Let* $\mathcal{R}_k \subset J_k \otimes T$ *be a linear differential equation and assume that* $\mathcal{R}_{k+1} \to \mathcal{R}_k \to 0$ *is exact. The following conditions are equivalent:*

1) $[\![\mathcal{R}_k, \mathcal{R}_k]\!] \subset \mathcal{R}_k$
2) $[\mathcal{R}_{k+1}, \mathcal{R}_{k+1}] \subset \mathcal{R}_k$
3) $[\![\mathcal{R}_{k+1}, \mathcal{R}_k]\!] \subset \mathcal{R}_k$

Proof: 1) $\Rightarrow$ 2). Let $A, B \in \mathcal{R}_{k+1}$ be represented by $\xi, \eta \in \underline{T_V}$ i.e., $A = \xi \bmod J^{k+2}\underline{T_V}$ and $B = \eta \bmod J^{k+2}\underline{T_V}$. Then $A' = \rho_k A$ and $B' = \rho_k B$ are also represented by $\xi, \eta \bmod J^{k+1}\underline{T_V}$. Condition 1) implies that

$$[\xi, \eta] + [\xi, \eta'_H] + [\xi'_H, \eta] \bmod J^{k+1}\underline{T_V} = [\![A', B']\!] \in \mathcal{R}_k .$$

But $DA, DB \in \underline{T}^* \otimes \mathcal{R}_k$ and $\eta'_H = i(\eta'')$, $\xi'_H = i(\xi'')$ with $\eta'', \xi'' \in \underline{T}$ so that $[\xi, \eta'_H] \bmod J^{k+1}\underline{T_V} = -\langle \eta'', DA \rangle \in \mathcal{R}_k$ and $[\xi'_H, \eta] \bmod J^{k+1}\underline{T_V} = \langle \xi'', DB \rangle \in \mathcal{R}_k$, hence

$$[A, B] = [\xi, \eta] \bmod J^{k+1}\underline{T_V} \in \mathcal{R}_k .$$

2) $\Rightarrow$ 3). Let $A \in \mathcal{R}_{k+1}$ and $B' \in \mathcal{R}_k$. Since $\mathcal{R}_{k+1} \to \mathcal{R}_k \to 0$ is exact then $B' = \rho_k B$ with $B \in \mathcal{R}_{k+1}$. Let ξ, η represent respectively A, B. By assumption, $[A, B] = [\xi, \eta] \bmod J^{k+1}\underline{T_V} \in \mathcal{R}_k$, $\xi'_H = i(\xi'')$ with $\xi'' \in \underline{T}$ and $[\xi'_H, \eta] \bmod J^{k+1}\underline{T_V} = \langle \xi'', DB \rangle \in \mathcal{R}_k$, hence

$$[\xi, \eta] + [\xi'_H, \eta] \bmod J^{k+1}\underline{T_V} = [\![A, B']\!] \in \mathcal{R}_k .$$

3) $\Rightarrow$ 1) is proved similarly using surjectivity. It is interesting to remark that (for obvious reasons) the implications 1) $\Rightarrow$ 2), 1) $\Rightarrow$ 3) and 3) $\Rightarrow$ 2) do not need the surjectivity assumption. This will be used in the next proof.

DEFINITION. *A linear differential equation* $\mathcal{R}_k \subset J_k \otimes \underline{T}$ *is called a Lie equation if* $[\![\mathcal{R}_k, \mathcal{R}_k]\!] \subset \mathcal{R}_k$, *i.e.,* $\mathcal{R}_k$ *is a Lie algebra subsheaf of* $J_k \otimes \underline{T}$ *for the bracket* $[\![\ ,\ ]\!]$.

THEOREM. *If* $\mathcal{R}_k$ *is a Lie equation then* $\mathcal{R}_{k+\ell}$, $\ell \geq 0$, *is also a Lie equation.*

Proof: By induction it is enough to prove that $\mathcal{R}_{k+1}$ is a Lie equation. Let $A, B \in \mathcal{R}_{k+1}$ be represented by $\xi, \eta \in \underline{T_V}$ modulo $\mathcal{J}^{k+2}\underline{T_V}$. It is clear that $\rho_k([\![A, B]\!]) = [\![\rho_k A, \rho_k B]\!] \in \mathcal{R}_k$ and that $D([\![A, B]\!]) \in \underline{T}^* \otimes \mathcal{R}_k$ is equivalent to $<\theta, D([\![A, B]\!])> \in \mathcal{R}_k$ for any $\theta \in \underline{T}$. The latter is proved as follows:

$$[\![A, B]\!] = [\xi, \eta] + [\xi'_H, \eta] + [\xi, \eta'_H] \ mod\ \mathcal{J}^{k+2}\underline{T_V}$$

and

$$<\theta, D([\![A, B]\!])> = [i\theta, [\xi, \eta] + [\xi'_H, \eta] + [\xi, \eta'_H]] \ mod\ \mathcal{J}^{k+1}\underline{T_V} .$$

But

$$[i\theta, [\xi, \eta]] = [[i\theta, \xi], \eta] + [\xi, [i\theta, \eta]] = [<i\theta, d_H\xi>, \eta] + [\xi, <i\theta, d_H\eta>] ,$$

$$[i\theta, [\xi'_H, \eta]] = [[i\theta, \xi'_H], \eta] + [\xi'_H, [i\theta, \eta]] = <i[\theta, \pi_1 * \xi'_H], d_H\eta> + [\xi'_H, <i\theta, d_H\eta>] ,$$

$$[i\theta, [\xi, \eta'_H]] = [[i\theta, \xi], \eta'_H] + [[\eta'_H, i\theta], \xi] = [<i\theta, d_H\xi>, \eta'_H] + <i[\pi_1 * \eta'_H, \theta], d_H\xi> .$$

Set $\xi'' = [\theta, \pi_1 * \xi'_H]$ and $\eta'' = [\pi_1 * \eta'_H, \theta]$. $\xi'', \eta'' \in \underline{T}$ and since ξ, η represent elements of $\mathcal{R}_{k+1}$ it follows that $<i\theta, d_H\eta> = \zeta$, $<i\theta, d_H\xi> = \mu$, $<i\xi'', d_H\eta> = \zeta'$ and $<i\eta'', d_H\xi> = \mu'$ represent elements of $\mathcal{R}_k$. Let $A' = \zeta \ mod\ \mathcal{J}^{k+1}\underline{T_V}$ and $B' = \mu \ mod\ \mathcal{J}^{k+1}\underline{T_V}$. Then $A', B' \in \mathcal{R}_k$ and the lemma implies that

$$<\theta, D([\![A, B]\!])> = ([\xi, \zeta] + [\xi'_H, \zeta]) - ([\eta, \mu] + [\eta'_H, \mu]) + \zeta' + \mu' \ mod\ \mathcal{J}^{k+1}\underline{T_V} = [\![A, A']\!] - [\![B, B']\!] + (\zeta' + \mu' \ mod\ \mathcal{J}^{k+1}\underline{T_V}) \in \mathcal{R}_k.$$

CHAPTER III

DERIVATIONS AND BRACKETS

13. *Derivations of scalar differential forms*

In this section we shall recall some basic facts about derivations of scalar differential forms and make explicit the notations. For a detailed account we refer the reader to [3] and [4]. Let M, N be modules over the same ring and $u: \Lambda^r M \to M$, $v: \Lambda^s M \to N$ skew-symmetric forms of degree r and s respectively. Define $i(u)v = u \barwedge v: \Lambda^t M \to N$, $t = r + s - 1$ by

$$u \barwedge v(x_1, \dots, x_t) = \frac{1}{r!(s-1)!} \sum_{\sigma} \varepsilon(\sigma) v(u(x_{\sigma(1)}, \dots, x_{\sigma(r)}), x_{\sigma(r+1)}, \dots, x_{\sigma(t)})$$

where divisibility assumptions are made on the ring (which in fact is irrelevant, cf. [3], [4]). If u and v are decomposable, namely if $u = \alpha \otimes \xi \in (\Lambda^r M)^* \otimes M$, $v = \beta \otimes \eta \in (\Lambda^s M)^* \otimes N$, then $u \barwedge v = (\alpha \wedge i(\xi)\beta) \otimes \eta = (u \barwedge \beta) \otimes \eta$ where $i(\xi)$ is the usual interior product.

Let X be a differentiable manifold, $\mathcal{F}(X)$ the C^∞ functions on X, $T = TX$ the tangent bundle and $\mathfrak{X}(X)$ the Lie algebra of vector fields which is also an $\mathcal{F}(X)$-module. Denote by $A^r(X)$ the $\mathcal{F}(X)$-module of scalar exterior differential r-forms on X and by $B^r(X)$ the $\mathcal{F}(X)$-module of exterior differential r-forms on X with values in T (vector valued forms on X). $A^r(X) = [\Lambda^r \mathfrak{X}(X)]^*$, $B^r(X) = \mathrm{Hom}_{\mathcal{F}(X)}(\Lambda^r \mathfrak{X}(X), \mathfrak{X}(X))$, $B^0(X) = \mathfrak{X}(X)$ and $A^0(X) = \mathcal{F}(X)$. $A(X) = \bigoplus A^r(X)$ is a graded $\mathcal{F}(X)$-algebra and $B(X) = \bigoplus B^r(X)$ is a graded left module over the graded algebra $A(X)$. In particular $B(X)$ is a graded $\mathcal{F}(X)$-module and $A(X)$ is a graded $\mathbf{R}$-algebra. A derivation of degree p of the graded $\mathbf{R}$-algebra $A(X)$ is an $\mathbf{R}$-linear map $D: A(X) \to A(X)$ which satisfies the two conditions:

i) $D(A^r(X)) \subset A^{r+p}(X)$ (D is homogeneous of degree p)

ii) $D(\omega \wedge \mu) = (D\omega) \wedge \mu + (-1)^{pr} \omega \wedge D\mu$ where $\omega \in A^r(X)$.

Set $\text{Der } A(X) = \bigoplus_{\mathbb{R}} \text{Der}^p A(X)$ where $\text{Der}^p A(X)$ is the R-vector space of derivations of degree p and define a bracket on this space by extending the following formula given for homogeneous elements

$$[D_1, D_2] = D_1 \circ D_2 - (-1)^{p_1 p_2} D_2 \circ D_1$$

where D_i is of degree p_i. Der A(X) together with [,] is a graded R-Lie algebra:

1) [,] is R-bilinear,

2) $[\text{Der}^p, \text{Der}^q] \subset \text{Der}^{p+q}$ i.e., $[\ ,\ ]: \text{Der } A(X) \otimes_{\mathbb{R}} \text{Der } A(X) \to \text{Der } A(X)$ is homogeneous of degree zero,

3) $[D_1, D_2] = (-1)^{p_1 p_2 + 1} [D_2, D_1]$, D_i homogeneous of degree p_i,

4) $(-1)^{p_1 p_3} [D_1, [D_2, D_3]] + (-1)^{p_1 p_2} [D_2, [D_3, D_1]] + (-1)^{p_2 p_3} [D_3, [D_1, D_2]] = 0$ (the Jacobi identity).

Furthermore, Der A(X) carries a natural graded left module structure over the graded algebra A(X), in particular it is a graded module over $\mathcal{F}(X)$. A derivation D is a local operator in the sense that $D\omega|_U$ only depends upon $\omega|_U$ where U is any open set. This implies the existence of a natural map $\text{Der } A(X) \to \text{Der } A(U)$ which is semi-linear with respect to the algebra morphism $A(X) \to A(U)$. We next distinguish some special derivations:

1) The exterior differentiation $d \in \text{Der}^1$, the Lie derivative along a vector field ξ, $\mathcal{L}(\xi) \in \text{Der}^0$, and the interior product $i(\xi) \in \text{Der}^{-1}$. We also remark that $\text{Der}_p = 0$ for $p < -1$ and $p > n = \mathit{dim}\, X$.

2) Any $u \in B^r(X)$ defines an *interior* derivation $i(u) \in \text{Der}^{r-1}$, $i(u)\omega = u \barwedge \omega$. The map $i: B(X) \to \text{Der } A(X)$ is a homogeneous left A(X)-linear injective map of degree -1 whose image is the graded Lie subalgebra and A(X)-submodule of those derivations which anni-

hilate $\mathcal{F}(X)$ or equivalently, which are $\mathcal{F}(X)$-linear. The Lie algebra structure of the image transports to $B(X)$ and is defined for homogeneous elements by $[u, v] = u \overline{\wedge} v - (-1)^{rs} v \overline{\wedge} u$ where $\deg u = r + 1$ and $\deg v = s + 1$. This bracket is bilinear over $\mathcal{F}(X)$ hence $B(X)$ and its image in $\mathrm{Der}\, A(X)$ are Lie algebras over $\mathcal{F}(X)$.

3) Any $u \in B^r(X)$ defines a *Lie derivation* $\mathcal{L}(u) \in \mathrm{Der}^r$ by the formula $\mathcal{L}(u) = [i(u), d]$. The map $\mathcal{L}\colon B(X) \to \mathrm{Der}\, A(X)$ is a homogeneous $\mathbb{R}$-linear injective map of degree 0 whose image is the graded Lie subalgebra of those derivations which commute with d, i.e., $[D, d] = 0$. The Lie algebra structure of the image transports to $B(X)$ and defines a bracket $[u, v]$ such that $\mathcal{L}([u, v]) = [\mathcal{L}(u), \mathcal{L}(v)]$. This bracket is only $\mathbb{R}$-bilinear and will be referred to as the Nijenhuis bracket.

4) $\mathrm{Der}\, A(X) = \mathcal{L}(B(X)) \oplus_{\mathbb{R}} i(B(X))$
$\mathrm{Der}^p\, A(X) = \mathcal{L}(B^p(X)) \oplus_{\mathbb{R}} i(B^{p+1}(X))$.

In particular all derivations of degree -1 are interior and all derivations of degree $n = \dim X$ are Lie. The direct sum decomposition also implies that for any derivation $D \in \mathrm{Der}\, A(U)$ and any point $x \in U$ there is a derivation $D' \in \mathrm{Der}\, A(X)$ and a neighborhood U' of x in U such that $D'|_{U'} = D|_{U'}$.

5) $\mathcal{L}(\mathrm{Id}) = d$, $[u, \mathrm{Id}] = 0$ and $i(\mathrm{Id}) = \oplus\, r\mathrm{Id}_r$ where Id is the identity map of $\mathfrak{X}(X)$ and Id_r is the identify map of $A^r(X)$. If $u \in B^0(X) = \mathfrak{X}(X)$ then $\mathcal{L}(u)$ is the usual Lie derivative by the vector field u and $[u, v]$ is the usual Lie derivative of the vector form v by the vector field u.

6) Assume that u and v are decomposable, i.e.,

$$u = \alpha \otimes \xi \in A^r(X) \otimes_{\mathcal{F}(X)} \mathfrak{X}(X) \quad \text{and} \quad v = \beta \otimes \eta \in A^s(X) \otimes_{\mathcal{F}(X)} \mathfrak{X}(X) .$$

Then

$$\mathcal{L}(u)\omega = \alpha \wedge \mathcal{L}(\xi)\omega + (-1)^r d\alpha \wedge i(\xi)\omega, \qquad i(u)\omega = \alpha \wedge i(\xi)\omega$$

and

$[u, v] = (\mathcal{L}(u)\beta)\otimes\eta - (-1)^{rs}(\mathcal{L}(v)\alpha)\otimes\xi + (\alpha\wedge\beta)\otimes[\xi, \eta]$ (Nijenhuis bracket) .

7) The map Der A(X) $\to$ Der A(U) is a morphism of graded Lie algebras which commutes with $\mathcal{L}$, i and B(X) $\to$ B(U). In particular this map preserves the direct sum decomposition given in 4).

Consider now the sheaf $\Lambda\underline{T}^*\otimes_{\mathcal{O}}\underline{T}$ of germs of exterior differential forms with values in T and the sheaf $\Lambda\underline{T}^*$ of germs of scalar forms. For any $x \in X$ the stalk $(\Lambda\underline{T}^*)_x$ is a graded $\mathcal{O}_x$-algebra and $(\Lambda\underline{T}^*\otimes_{\mathcal{O}}\underline{T})_x$ is a graded module over the graded algebra $(\Lambda\underline{T}^*)_x$. A derivation of degree p of the R-algebra $(\Lambda\underline{T}^*)_x$ is an R-linear map D which satisfies conditions i) and ii) where forms are replaced by germs of forms. All the previous considerations for Der A(X) transcribe without change for Der $(\Lambda\underline{T}^*)_x$ when forms are replaced by germs. In particular the decomposition Der $(\Lambda\underline{T}^*)_x = \mathcal{L}(\Lambda\underline{T}^*\otimes_{\mathcal{O}}\underline{T})_x \oplus_{R} i(\underline{T}^*\otimes_{\mathcal{O}}\underline{T})_x$ holds. Since derivations of A(X) are local operators it follows that any $D \in$ Der A(X) induces a derivation $D_x \in$ Der $(\Lambda\underline{T}^*)_x$. The previous remark shows that conversely any derivation of $(\Lambda\underline{T}^*)_x$ is induced by a derivation of A(X) and the latter is locally unique in the sense that if D and D′ induce the same derivation of $(\Lambda\underline{T}^*)_x$ there is a neighborhood U of x such that $D|_U = D'|_U$. We infer that $\mathcal{D}er\, \Lambda\underline{T}^* = \bigcup_x \text{Der}\,(\Lambda\underline{T}^*)_x$ is the sheaf of germs of derivations of A(X). Furthermore, $\mathcal{D}er\, \Lambda\underline{T}^*$ is a sheaf of graded left $\Lambda\underline{T}^*$-modules and a sheaf of graded Lie algebras when equipped with the Nijenhuis bracket. The remark 7) implies that all computations at the germ level can first be performed locally with representatives and then factored to germs.

14. *The bracket* $[\![\ ,\ ]\!]$

In this section we define the sheaves of germs of vector valued exterior differential forms for which the previous section will be applied. It will be understood that all the sheaves on X^2 will be restricted to Δ hence will be considered as sheaves on X. We start by making a few simple remarks which will be useful in subsequent proofs. Let ξ be a vector field on X^2. Then ξ vanishes to order $k+1$ on the diagonal (as a section of TX^2) if and only if the components of ξ, relative to a local basis of TX^2, vanish to order $k+1$ on the diagonal. Let u be an exterior differential r-form on X^2 with values in TX^2. The form u is a section of the bundle $F^r = (\Lambda^r T^*X^2) \otimes TX^2$ and as such, it vanishes to order $k+1$ on Δ if and only if for any vector fields ξ_i on X^2 the vector field $u(\xi_1, \ldots, \xi_r)$ vanishes to order $k+1$ on Δ. Passing to germs, we infer that the element $u \in \underline{F}^r = (\Lambda^r \underline{T^*X^2}) \otimes_{\mathcal{O}_{X^2}} \underline{TX}^2$ belongs to $\mathcal{J}^{k+1}\underline{F}^r$ if and only if $u(\xi_1, \ldots, \xi_r) \in \mathcal{J}^{k+1}\underline{TX}^2$ for any germs of vector fields ξ_i.

In Section 10 we introduced the diagonal $\mathcal{F}(X^2)$-module structure of $\mathfrak{X}(X^2)$. We now define a diagonal $\mathcal{F}(X^2)$-module structure on $B(X^2)$ by the relation

$$(f\Delta u)(\xi_1, \ldots, \xi_r) = f\Delta(u(\xi_1, \ldots, \xi_r))$$

where $u \in B^r(X^2)$. The diagonal structure commutes with the natural one. Similarly (or taking germs) the sheaf

$$\underline{F} = \Lambda \underline{T^*X^2} \otimes_{\mathcal{O}_{X^2}} \underline{TX}^2$$

carries a diagonal $\mathcal{O}_{X^2}$-module structure which commutes with the natural one and is the extension to $\underline{F}$ of the diagonal structure of $\underline{TX}^2$.

Recall that a form $u \in B^r(X^2)$ is horizontal if $u(\xi_1, \ldots, \xi_r) = u(H\xi_1, \ldots, H\xi_r)$. The set of horizontal forms is an $\mathcal{F}(X^2)$-submodule of $B(X^2)$ for the natural and diagonal structures. Correspondingly, the sheaf of germs of horizontal forms is an $\mathcal{O}_{X^2}$-submodule of $\underline{F}$ for both natural and diagonal structures. This submodule identifies with $\Lambda \underline{T}^* \otimes_{\mathcal{O}} \underline{TX}^2$. Similar remarks hold for vertical forms.

We say that a horizontal r-form u has values in $\check{\mathfrak{X}}(X^2)$ if $u(\xi_1, \ldots, \xi_r) \in \check{\mathfrak{X}}(X^2)$ for any projectable horizontal vector fields ξ_i. Let $\check{\Phi}^r$ denote the sheaf of germs of horizontal r-forms with values in $\check{\mathfrak{X}}(X^2)$. An element $u \in \underline{F}^r$ belongs to $\check{\Phi}^r$ if and only if

1) $u(\xi_1, \dots, \xi_r) = u(H\xi_1, \dots, H\xi_r)$ for any $\xi_i \in \underline{TX}^2$ (u is horizontal)

2) $u(\xi_1, \dots, \xi_r) \in \check{\Theta}$ for any $\xi_i \in \Theta_{PH}$ (ξ_i projectable horizontal).

The sheaf $\check{\Phi}^r$ is not stable under the natural $\mathcal{O}_{X^2}$-module structure of $\underline{F}^r$. However $\check{\Phi}^r$ is a diagonal $\mathcal{O}_{X^2}$-submodule of $\underline{F}^r$. Next remark that if a horizontal forms u takes values in $\check{\mathfrak{X}}(X^2)$ and vanishes on Δ then u takes vertical values, i.e., values in T_V. It follows that

$$\check{\Phi}^r \cap \mathcal{J}^{k+1}\underline{F}^r = \mathcal{J}^{k+1}\Delta\check{\Phi}^r .$$

The quotient sheaf $\check{\Phi}^r/\mathcal{J}^{k+1}\Delta\check{\Phi}^r$ carries an induced diagonal module structure over the left, right $\mathcal{O}$-algebra $\mathcal{J}_k$. The direct sum $\check{\Phi} = \bigoplus \check{\Phi}^r$ is a diagonal homogeneous submodule of $\underline{F}$ and

$$\check{\Phi}/\mathcal{J}^{k+1}\Delta\check{\Phi} = \bigoplus \check{\Phi}^r/\mathcal{J}^{k+1}\Delta\check{\Phi}^r .$$

The sheaf $\wedge\underline{T}^* \otimes_{\mathcal{O}} \check{\mathcal{J}}_k T$ also carries a diagonal $\mathcal{J}_k$-module structure inherited from the corresponding diagonal structure of $\check{\mathcal{J}}_k T$. The mapping $\wedge\underline{T}^* \otimes_{\mathcal{O}} \check{\Theta} \to \check{\Phi}$ defined by $\omega \otimes \theta \mapsto (\pi_1^* \omega) \otimes \theta$ is a diagonal $\mathcal{O}_{X^2}$-linear isomorphism which is homogeneous of degree zero. This isomorphism factors to a $\mathcal{J}_k$-linear homogeneous isomorphism

$$\wedge\underline{T}^* \otimes_{\mathcal{O}} \check{\mathcal{J}}_k T \to \check{\Phi}/\mathcal{J}^{k+1}\Delta\check{\Phi}$$

of degree zero.

Let $\tilde{\Phi}^r$ be the sheaf of germs of horizontal r-forms with values in $\tilde{\mathfrak{X}}(X^2)$. An element $u \in \underline{F}^r$ belongs to $\tilde{\Phi}^r$ if and only if u is horizontal and $u(\xi_1, \dots, \xi_r) \in \tilde{\Theta}$ for any $\xi_i \in \Theta_{PH}$. The sheaf $\tilde{\Phi}^r$ is an $\mathcal{O}_{X^2}$-submodule of $\check{\Phi}^r$ (diagonal structure) and $\tilde{\Phi}^r/\mathcal{J}^{k+1}\Delta\tilde{\Phi}^r$ is a $\mathcal{J}_k$-submodule of $\check{\Phi}^r/\mathcal{J}^{k+1}\Delta\check{\Phi}^r$ since

$$\tilde{\Phi}^r \cap \mathcal{J}^{k+1}\underline{F}^r = \mathcal{J}^{k+1}\Delta\tilde{\Phi}^r = \mathcal{J}^{k+1}\Delta\check{\Phi}^r .$$

The direct sum $\tilde{\Phi} = \bigoplus \tilde{\Phi}^r$ is a homogeneous $\mathcal{O}_{X^2}$-submodule of $\check{\Phi}$ and

$$\tilde{\Phi}/\mathcal{J}^{k+1}\Delta\tilde{\Phi} = \bigoplus \tilde{\Phi}^r/\mathcal{J}^{k+1}\Delta\tilde{\Phi}^r$$

is a homogeneous $\mathcal{J}_k$-submodule of $\check{\Phi}/\mathcal{J}^{k+1}\Delta\check{\Phi}$. The sheaf $\Lambda\underline{T}^* \otimes_{\mathcal{O}} \tilde{\Theta}$ identifies with $\tilde{\Phi}$ and therefore, $\Lambda\underline{T}^* \otimes_{\mathcal{O}} \tilde{\mathcal{J}}_k T$ identifies with $\tilde{\Phi}/\mathcal{J}^{k+1}\Delta\tilde{\Phi}$. Since $\tilde{\mathcal{J}}_0 T \simeq \underline{T}$, it follows that $\Lambda\underline{T}^* \otimes_{\mathcal{O}} \tilde{\mathcal{J}}_0 T \simeq \Lambda\underline{T}^* \otimes \underline{T}$. Any horizontal r-form u with values in $\tilde{\mathcal{X}}(X^2)$ induces at each point $p \in \Delta$ and r-form on $(T_H)_p$ with values in $T_p\Delta$. Denote by $\Delta: X \to X^2$ the diagonal injection. The map $\Delta^*: \tilde{\Phi} \to \Lambda\underline{T}^* \otimes_{\mathcal{O}} \underline{T}$ factors to an $\mathcal{O}$-linear isomorphism

$$\tilde{\Phi}/\mathcal{J}\Delta\tilde{\Phi} \simeq \Lambda\underline{T}^* \otimes_{\mathcal{O}} \tilde{\mathcal{J}}_0 T \to \Lambda\underline{T}^* \otimes_{\mathcal{O}} \underline{T}$$

which coincides with the previous identification.

Let Φ_V^r be the sheaf of germs of horizontal r-forms with values in $\mathcal{X}(X^2)_V$. Remark that a horizontal form with values in $\mathcal{X}(X^2)_V$ is simply a horizontal form with values in the bundle T_V. The form $u \in \underline{F}^r$ belongs to Φ_V^r if and only if u is horizontal and $u(\xi_1, \ldots, \xi_r) \in \Theta_V = \underline{T_V}$ for any $\xi_i \in \Theta_{PH}$. The sheaf Φ_V^r is an $\mathcal{O}_{X^2}$-submodule of $\check{\Phi}^r$ (diagonal structure) and it is clearly an $\mathcal{O}_{X^2}$-submodule of $\underline{F}^r$ for the natural structure. The induced diagonal and natural structures on Φ_V^r coincide. Moreover $\Phi_V^r/\mathcal{J}^{k+1}\Phi_V^r$ is a $\mathcal{J}_k$-submodule of $\check{\Phi}^r/\mathcal{J}^{k+1}\Delta\check{\Phi}^r$ since

$$\mathcal{J}^{k+1}\Delta\check{\Phi}^r = \mathcal{J}^{k+1}\Delta\tilde{\Phi}^r = \mathcal{J}^{k+1}\Delta\Phi_V^r = \mathcal{J}^{k+1}\Phi_V^r = \Phi_V^r \cap \mathcal{J}^{k+1}\underline{F}^r .$$

The direct sum $\Phi_V = \bigoplus \Phi_V^r$ is a homogeneous $\mathcal{O}_{X^2}$-submodule of $\check{\Phi}$ whose diagonal structure coincides with the natural one and $\Phi_V/\mathcal{J}^{k+1}\Phi_V = \bigoplus \Phi_V^r/\mathcal{J}^{k+1}\Phi_V^r$ is a homogeneous $\mathcal{J}_k$-submodule of $\check{\Phi}/\mathcal{J}^{k+1}\Delta\check{\Phi}$. Moreover, $\Lambda\underline{T}^* \otimes_{\mathcal{O}} \Theta_V$ identifies with Φ_V and therefore

$$\Lambda\underline{T}^* \otimes_{\mathcal{O}} \mathcal{J}_k T \simeq \Lambda\underline{T}^* \otimes_{\mathcal{O}} \mathcal{J}_k \otimes_{\mathcal{O}} \underline{T}$$

identifies with $\Phi_V/\mathcal{J}^{k+1}\Phi_V$.

Let Φ_{PH}^r be the sheaf of germs of horizontal r-forms with values in $\mathcal{X}(X^2)_{PH}$. A horizontal form with values in $\mathcal{X}(X^2)_{PH}$ is a horizontal form u with values in the bundle T_H such that $u(\xi_1, \ldots, \xi_r)$ is projectable horizontal for any projectable horizontal ξ_i. It follows that $u \in \underline{F}^r$ belongs to Φ_{PH}^r if and only if it is horizontal and $u(\xi_1, \ldots, \xi_r) \in \Theta_{PH}$

for any $\xi_i \in \Theta_{PH}$. Φ^r_{PH} is a diagonal $\mathcal{O}_{X^2}$-submodule of $\check{\Phi}^r$ and $\mathcal{J}\Delta\Phi^r_{PH} = 0$ hence $\Phi^r_{PH} = \Phi^r_{PH}/\mathcal{J}^{k+1}\Delta\Phi^r_{PH}$ is a $\mathcal{J}_k$-submodule of $\check{\Phi}^r/\mathcal{J}^{k+1}\Delta\check{\Phi}^r$. $\Phi_{PH} = \bigoplus \Phi^r_{PH}$ is a homogeneous $\mathcal{O}_{X^2}$-submodule of $\check{\Phi}$ and a homogeneous $\mathcal{J}_k$-submodule of $\check{\Phi}/\mathcal{J}^{k+1}\Delta\check{\Phi}$. Moreover the induced left and right diagonal $\mathcal{O}$-module structures coincide and the $\mathcal{J}_k$-module structure reduces to the $\mathcal{O}$-module structure via β_k (the map induced by $f \in \mathcal{O}_{X^2} \mapsto f \circ \Delta \in \mathcal{O}$). By means of π_1^*, the sheaf Φ_{pH} identifies with

$$\Lambda\underline{T}^* \otimes_{\mathcal{O}} \Theta_{PH} \simeq \Lambda\underline{T}^* \otimes_{\mathcal{O}} \underline{T}\,.$$

All the previous quotient sheaves are locally free of finite rank for the left and right $\mathcal{O}$-module structures and (except the quotients of $\check{\Phi}$ and Φ_{PH}) also for the $\mathcal{J}_k$-module structure. The quotients of $\check{\Phi}$ and Φ_{PH} are locally finitely generated over $\mathcal{J}_k$.

Any differential form u with values in TX^2 decomposes uniquely into vertical and horizontal components: $u = u_V + u_H$ where $u_V = V \circ u$ takes values in T_V and $u_H = H \circ u$ takes values in T_H. The direct sum decomposition

$$B(X^2) = B(X^2)_V \oplus B(X^2)_H$$

holds for the natural as well as the diagonal $\mathcal{F}(X^2)$-module structure. Passing to germs we obtain the direct sum decomposition with respect to natural and diagonal $\mathcal{O}_{X^2}$-module structures

$$\underline{F} = \left[\Lambda\underline{T^*X^2} \otimes_{\mathcal{O}_{X^2}} \underline{T_V}\right] \oplus \left[\Lambda\underline{T^*X^2} \otimes_{\mathcal{O}_{X^2}} \underline{T_H}\right]$$

which of course is also a consequence of $\underline{TX^2} = \underline{T_V} \oplus \underline{T_H}$. The diagonal submodule $\check{\Phi}$ is homogeneous with respect to this decomposition. The induced direct sum decomposition is $\check{\Phi} = \Phi_V \oplus \Phi_{PH}$ (diagonal structure). If $u \in \check{\Phi}$, its horizontal component u_H is horizontal with projectable horizontal values and its vertical component u_V is of course horizontal with vertical values. Taking quotients yields the direct sum decomposition for the diagonal $\mathcal{J}_k$-module structure

$$\Lambda\underline{T}^* \otimes_{\mathcal{O}} \check{\mathcal{J}}_k T = (\Lambda\underline{T}^* \otimes_{\mathcal{O}} \mathcal{J}_k T) \oplus (\Lambda\underline{T}^* \otimes_{\mathcal{O}} \underline{T})$$

which is compatible with the direct sum $\check{\mathcal{J}}_k T = \mathcal{J}_k T \oplus_{\mathcal{O}} \underline{T}$ (cf. Section 10).

The vertical projection $u \mapsto u_V$ restricted to $\check{\Phi}$ will be denoted by

$$\varepsilon \colon \check{\Phi} \to \Phi_V \ .$$

It is a diagonal $\mathcal{O}_{X^2}$-linear retraction onto Φ_V whose restriction to $\check{\Phi}^0 = \check{\Theta}$ is the map $\varepsilon\colon \check{\Theta} \to \Theta_V$ defined in Section 10. The restricted map

$$\varepsilon \colon \tilde{\Phi} \to \Phi_V$$

is an isomorphism. For any $u \in \check{\Phi}^r$ we have

$$(\varepsilon u)(\xi_1, \ldots, \xi_r) = \varepsilon(u(\xi_1, \ldots, \xi_r)) \ .$$

However, if $u \in \Phi_V^r$ then the relation

$$(\varepsilon^{-1} u)(\xi_1, \ldots, \xi_r) = \varepsilon^{-1}(u(\xi_1, \ldots, \xi_r))$$

is only true for projectable horizontal germs of vector fields ξ_i where the first ε^{-1} is the map $\Phi_V^r \to \tilde{\Phi}^r$ and the second is the map $\Theta_V \to \tilde{\Theta}$. Passage to the quotients yields the $\mathcal{J}_k$-linear retraction

$$\varepsilon_k \colon \Lambda\underline{T}^* \otimes_{\mathcal{O}} \check{\mathcal{J}}_k T \to \Lambda\underline{T}^* \otimes_{\mathcal{O}} \mathcal{J}_k T$$

which is of course the projection upon the vertical factor $u \mapsto u_V$. Moreover, $\varepsilon_k = \mathrm{Id} \otimes \varepsilon_k$ and the restriction

$$\mathrm{Id} \otimes \varepsilon_k \colon \Lambda\underline{T}^* \otimes_{\mathcal{O}} \tilde{\mathcal{J}}_k T \to \Lambda\underline{T}^* \otimes_{\mathcal{O}} \mathcal{J}_k T$$

is an isomorphism whose inverse is $\mathrm{Id} \otimes \varepsilon_k^{-1}$.

LEMMA 1. *The sheaves* $\check{\Phi}$, $\tilde{\Phi}$, Φ_V *and* Φ_{PH} *are closed under the Nijenhuis bracket. The map* $\pi_1^* \colon \Lambda\underline{T}^* \otimes_{\mathcal{O}} \underline{T} \to \Phi_{PH}$ *is an isomorphism of graded Lie algebras.*

Proof: Let $u \in \check{\Phi}^r$ and $v \in \check{\Phi}^s$. We can assume that u and v are decomposable of the form $u = (\pi_1^* a) \otimes \xi$, $v = (\pi_1^* \beta) \otimes \eta$ with $a \in \Lambda^r \underline{T}^*$, $\beta \in \Lambda^s \underline{T}^*$ and $\xi, \eta \in \check{\Theta}$. Then

$$[u, v] = [\mathfrak{L}(u)(\pi_1^*\beta)] \otimes \eta - (-1)^{rs}[\mathfrak{L}(v)(\pi_1^* a)] \otimes \xi + \pi_1^*(a \wedge \beta) \otimes [\xi, \eta] ,$$

$$\mathfrak{L}(u)(\pi_1^*\beta) = (\pi_1^* a) \wedge [\mathfrak{L}(\xi)(\pi_1^*\beta)] + (-1)^r (\pi_1^* da) \wedge [i(\xi)(\pi_1^*\beta)] =$$

$$\pi_1^*(a \wedge \mathfrak{L}(\pi_{1*}\xi)\beta + (-1)^r da \wedge i(\pi_{1*}\xi)\beta)$$

and similarly for $\mathfrak{L}(v)(\pi_1^* a)$. These relations show that the covariant part in each term of $[u, v]$ is of the form $\pi_1^*\omega$ with $\omega \in \Lambda^{r+s}\underline{T}^*$, hence $[u, v]$ is horizontal. Moreover, if ξ_i is projectable horizontal then

$$[u, v](\xi_1, \ldots, \xi_{r+s}) \in \check{\Theta}$$

since each covariant part applied to $(\xi_1, \ldots, \xi_{r+s})$ gives an element in $i(\mathcal{O})$, $[\xi, \eta] \in \check{\Theta}$ and $\check{\Theta}$ is stable under the natural left $\mathcal{O}$-multiplication. The same argument applies to $\tilde{\Phi}$, Φ_V and Φ_{PH}. The last assertion is obvious.

LEMMA 2. *If* $u \in \check{\Phi}$ *and* $v \in \mathfrak{J}^{k+1}\Delta\check{\Phi}$ *then* $[u, v] \in \mathfrak{J}^k\Delta\check{\Phi}$.

Proof: We can assume further that β is of the form $\beta = dx_{i_1} \wedge \ldots \wedge dx_{i_s}$. Then $v \in \mathfrak{J}^{k+1}\Delta\check{\Phi}$ implies that

$$\eta = v\left(\partial/\partial x_{i_1}, \ldots, \partial/\partial x_{i_s}\right) \in \mathfrak{J}^{k+1}\Delta\check{\Theta}$$

which in turn implies that

$$[\mathfrak{L}(u)(\pi_1^*\beta)] \otimes \eta \in \mathfrak{J}^{k+1}\Delta\check{\Phi} .$$

Moreover, $[\xi, \eta] \in \mathfrak{J}^k\Delta\check{\Theta}$ implies that $\pi_1^*(a \wedge \beta) \otimes [\xi, \eta] \in \mathfrak{J}^k\Delta\check{\Phi}$. Finally $[\mathfrak{L}(v)(\pi_1^* a)] \otimes \xi = 0$ since

$$v \in \mathfrak{J}^{k+1}\Delta\check{\Phi} = \mathfrak{J}^{k+1}\Phi_V \subset \Phi_V$$

and therefore $\mathfrak{L}(v)(\pi_1^* a) = 0$.

LEMMA 3. *If* $u \in \tilde{\Phi}$ *and* $v \in \mathfrak{J}^{k+1}\Delta\tilde{\Phi}$ *then* $[u, v] \in \mathfrak{J}^{k+1}\Delta\tilde{\Phi}$.

Proof: Observe that $[\xi, \eta] \in \mathcal{J}^{k+1}\Delta\tilde{\Theta}$ (cf. Section 10).

The second lemma implies that the Nijenhuis bracket on $\breve{\Phi}$ factors to a bracket (homogeneous R-linear map of degree zero):

$$[\Lambda\underline{T}^* \otimes_{\mathcal{O}} \breve{\mathcal{J}}_k T] \otimes_{\mathbf{R}} [\Lambda\underline{T}^* \otimes_{\mathcal{O}} \breve{\mathcal{J}}_k T] \to \Lambda\underline{T}^* \otimes_{\mathcal{O}} \breve{\mathcal{J}}_{k-1} T$$

which lowers the order of jets and satisfies properties analogous to those of a graded Lie algebra bracket (with obvious modification). The third lemma implies that the Nijenhuis bracket on $\tilde{\Phi}$ factors to a bracket

$$[\Lambda\underline{T}^* \otimes_{\mathcal{O}} \tilde{\mathcal{J}}_k T] \otimes_{\mathbf{R}} [\Lambda\underline{T}^* \otimes_{\mathcal{O}} \tilde{\mathcal{J}}_k T] \to \Lambda\underline{T}^* \otimes_{\mathcal{O}} \tilde{\mathcal{J}}_k T$$

and defines on $\Lambda\underline{T}^* \otimes_{\mathcal{O}} \tilde{\mathcal{J}}_k T$ a structure of graded Lie algebra.

In Section 6 we defined the operator

$$D: \Lambda^r\underline{T}^* \otimes_{\mathcal{O}} \mathcal{J}_k T \to \Lambda^{r+1}\underline{T}^* \otimes_{\mathcal{O}} \mathcal{J}_{k-1} T$$

which transports via ε to

$$\tilde{D}: \Lambda^r\underline{T}^* \otimes_{\mathcal{O}} \breve{\mathcal{J}}_k T \to \Lambda^{r+1}\underline{T}^* \otimes_{\mathcal{O}} \tilde{\mathcal{J}}_{k-1} T ,$$

where $\tilde{D} = \varepsilon_{k-1}^{-1} \circ D \circ \varepsilon_k$. Define

$$[\![u, v]\!] = [u, v] + (-1)^{r+1}\,\tilde{D}u \barwedge \breve{\rho}_{k-1}v + (-1)^{rs+s}\,\tilde{D}v \barwedge \breve{\rho}_{k-1}u$$

where $u \in \Lambda^r\underline{T}^* \otimes_{\mathcal{O}} \breve{\mathcal{J}}_k T$, $v \in \Lambda^s\underline{T}^* \otimes_{\mathcal{O}} \breve{\mathcal{J}}_k T$,

$$\breve{\rho}_{k-1}: \Lambda\underline{T}^* \otimes_{\mathcal{O}} \breve{\mathcal{J}}_k T \to \Lambda\underline{T}^* \otimes_{\mathcal{O}} \breve{\mathcal{J}}_{k-1} T$$

is the natural projection and

$$\tilde{D}u \barwedge \breve{\rho}_{k-1}v = (\tilde{\rho}_0\tilde{D}u) \barwedge \breve{\rho}_{k-1}v$$

recalling that $\tilde{\mathcal{J}}_0 T \simeq \underline{T}$. The bracket $[\![\ ,\ ]\!]$ defines a homogeneous R-linear map of degree zero

$$[\Lambda\underline{T}^* \otimes_{\mathcal{O}} \breve{\mathcal{J}}_k T] \otimes_{\mathbf{R}} [\Lambda\underline{T}^* \otimes_{\mathcal{O}} \breve{\mathcal{J}}_k T] \to \Lambda\underline{T}^* \otimes_{\mathcal{O}} \breve{\mathcal{J}}_{k-1} T$$

which lowers the order of jets. If u and v are homogeneous of degree zero, i.e., $u, v \in \check{\mathcal{J}}_k T$ then $[\![u, v]\!] = [u, v]$ is the bracket in $\check{\mathcal{J}}_k T$ as defined in Section 10.

Assume now that u, v take values in $\tilde{\mathcal{J}}_k T$. Then $[u, v] \in \Lambda \underline{T}^* \otimes_{\mathcal{O}} \tilde{\mathcal{J}}_k T$ and the last two terms of $[\![u, v]\!]$ can be redefined by deleting $\check{\rho}_{k-1}$. More precisely

$$[\![u, v]\!] = [u, v] + (-1)^{r+1} \tilde{D}u \barwedge v + (-1)^{rs+s} \tilde{D}v \barwedge u$$

defines a homogeneous $\mathbf{R}$-linear map of degree zero

$$[\Lambda \underline{T}^* \otimes_{\mathcal{O}} \tilde{\mathcal{J}}_k T] \otimes_{\mathbf{R}} [\Lambda \underline{T}^* \otimes_{\mathcal{O}} \tilde{\mathcal{J}}_k] \to \Lambda \underline{T}^* \otimes_{\mathcal{O}} \tilde{\mathcal{J}}_k T$$

If u, v are homogeneous of degree zero then $[\![u, v]\!]$ is the bracket of the Lie algebra $\tilde{\mathcal{J}}_k T$ as defined in Section 10. If $k = 0$ then $u, v \in \Lambda \underline{T}^* \otimes \underline{T}$ and $[\![u, v]\!] = [u, v]$ is the Nijenhuis bracket of vector forms on X.

Assume now that $u, v \in \Lambda \underline{T}^* \otimes_{\mathcal{O}} \underline{T} =$ horizontal component of $\Lambda \underline{T}^* \otimes_{\mathcal{O}} \check{\mathcal{J}}_k$. Then $[\![u, v]\!] = [u, v]$ since $\tilde{D}u = \tilde{D}v = 0$ and Lemma 1 implies that $\Lambda \underline{T}^* \otimes_{\mathcal{O}} \underline{T}$ with the bracket $[\![\ ,\ ]\!]$ is a graded Lie algebra isomorphic to the graded Lie algebra $\Lambda \underline{T}^* \otimes_{\mathcal{O}} \underline{T}$ of vector forms on X.

Let $u \in \Lambda \underline{T}^* \otimes_{\mathcal{O}} \tilde{\mathcal{J}}_k T \subset \Lambda \underline{T}^* \otimes_{\mathcal{O}} \check{\mathcal{J}}_k T$. Then $\tilde{\rho}_0 u \in \Lambda \underline{T}^* \otimes \underline{T}$ is equal to $u_H \in \Lambda \underline{T}^* \otimes \underline{T}$ where u_H is the horizontal component of u in the direct sum decomposition of $\Lambda \underline{T}^* \otimes_{\mathcal{O}} \check{\mathcal{J}}_k T$, hence

$$\tilde{D}u \barwedge \check{\rho}_{k-1} v = (\tilde{D}u)_H \barwedge \check{\rho}_{k-1} v \,.$$

If $\mathcal{U} \in \check{\Phi}^r$ and $\mathcal{V} \in \check{\Phi}^s$ are representatives of $u, v \in \Lambda \underline{T}^* \otimes_{\mathcal{O}} \check{\mathcal{J}}_k T$ then

$$[\![u, v]\!] = [\mathcal{U}, \mathcal{V}] + (-1)^{r+1} \tilde{d}_H \mathcal{U} \barwedge \mathcal{V} + (-1)^{rs+s} \tilde{d}_H \mathcal{V} \barwedge \mathcal{U} \mod \mathcal{J}^k \Delta \check{\Phi}$$

where

$$\tilde{d}_H = \varepsilon^{-1} \circ d_H \circ \varepsilon : \ \check{\Phi} \to \tilde{\Phi} \,.$$

The bracket $[\![\ ,\ ]\!]$ on $\Lambda \underline{T}^* \otimes_{\mathcal{O}} \check{\mathcal{J}}_k T$ obviously satisfies the relation

$$[\![u, v]\!] = (-1)^{rs+1} [\![v, u]\!]$$

for homogeneous elements of degrees r and s. However, this bracket does not satisfy a Jacobi identity (in the general sense). Take for example $X^2 = \mathbf{R}^2$ and $u = xy\frac{\partial}{\partial y} + x^2\frac{\partial}{\partial x}$, $v = \frac{\partial}{\partial x}$, $w = xdx \otimes \frac{\partial}{\partial x}$ $(mod\ \mathcal{J}^{k+1}\Delta\check{\Phi})$. Then the graded Jacobi relation for u, v and w is equal to $xdx \otimes \frac{\partial}{\partial x}$ $(mod\ \mathcal{J}^{k-1}\Delta\check{\Phi})$.

We will now prove that $[\![\ ,\ ]\!]$ defines a graded Lie algebra structure on $\Lambda\underline{T}^* \otimes_{\mathcal{O}} \tilde{\mathcal{J}}_k T$. For $k = 0$ this is obvious since $[\![\ ,\ ]\!] = [\ ,\ ]$. For $k > 0$ the Jacobi identity can be checked by a straightforward though excessively long computation. To avoid this, we shall give a second definition of $[\![\ ,\ ]\!]$ and prove the Jacobi identity for the latter.

15. *The adjoint representation*

Let Ψ^ℓ be the subsheaf of $\Lambda^\ell \underline{T^*X^2}$ of germs of scalar horizontal ℓ-forms ω on X^2 such that $\omega(\xi_1, \ldots, \xi_\ell) \in i(\mathcal{F}(X))$ for any projectable horizontal vector fields ξ_i $(i: f \in \mathcal{F}(X) \mapsto f \circ \pi_1 \in \mathcal{F}(X^2))$. An element $\omega \in \Lambda^\ell \underline{T^*X}^2$ belongs to Ψ^ℓ if and only if ω is horizontal and $\omega(\xi_1, \ldots, \xi_\ell) \in i(\mathcal{O})$ for any $\xi_i \in \Theta_{PH}$. The sheaf Ψ^ℓ is a left $\mathcal{O}$-submodule of $\Lambda^\ell \underline{T^*X}^2$ and $\Psi = \bigoplus \Psi^\ell$ is a homogeneous left $\mathcal{O}$-subalgebra of $\Lambda \underline{T^*X}^2$ which is invariant by d. The map $\pi_1^*: \Lambda\underline{T}^* \to \Psi$ is an $\mathcal{O}$-linear isomorphism of algebras whose inverse is given by $\Delta^*: \Psi \to \Lambda\underline{T}^*$ where $\Delta: X \to X^2$ is the diagonal inclusion. For any $\mathfrak{U} \in \check{\Phi}$, Ψ is invariant by $\mathcal{L}(\mathfrak{U})$ since it is invariant by d and $i(\mathfrak{U})\omega = i(\mathfrak{U}_H)\omega$ where $\mathfrak{U}_H \in \Phi_{PH}$ is the horizontal component of $\mathfrak{U}$. The previous argument shows that $\mathcal{L}(\mathfrak{U})\omega = \mathcal{L}(\mathfrak{U}_H)\omega$. Moreover, if $\omega = \pi_1^*\alpha$ and $\mathfrak{U}_H = \pi_1^*\mathfrak{V}$ with $\alpha \in \Lambda\underline{T}^*$ and $\mathfrak{V} \in \Lambda\underline{T}^* \otimes \underline{T}$, then $i(\mathfrak{U}_H)\omega = \pi_1^*[i(\mathfrak{V})\alpha]$ hence $\mathcal{L}(\mathfrak{U})\omega = \pi_1^*[\mathcal{L}(\mathfrak{V})\alpha]$ where $\mathcal{L}(\mathfrak{V})\alpha$ is the usual Lie derivative of α by $\mathfrak{V}$ on the manifold X. The map

$$\check{\Phi} \to \mathcal{D}er\,\Psi,\ \mathfrak{U} \mapsto \mathcal{L}(\mathfrak{U}),$$

is $\mathbf{R}$-linear and homogeneous of degree zero. It transports via π_1^* to $\check{\Phi} \to \mathcal{D}er\,\Lambda\underline{T}^*$ and the latter factors to

$$\mathcal{L}:\ \Lambda\underline{T}^* \otimes_{\mathcal{O}} \check{\mathcal{J}}_k T \to \mathcal{D}er\,\Lambda\underline{T}^*,$$

for any $k \geq 0$, since $\mathfrak{U} \in \mathcal{I}\Delta\check{\Phi}$ implies that $\mathfrak{U} \in \Phi_V$ hence $\mathfrak{U}_H = 0$ and consequently $\mathcal{L}(\mathfrak{U})\omega = 0$. Furthermore, for any $u \in \Lambda\underline{T}^* \otimes_{\mathcal{O}} \check{\mathcal{J}}_k T$ and $\alpha \in \Lambda\underline{T}^*$ we have $\mathcal{L}(u)\alpha = \mathcal{L}(u_H)\alpha$ where $u_H \in \Lambda\underline{T}^* \otimes \underline{T}$ is the horizontal component of u in the direct sum decomposition

$$\Lambda\underline{T}^* \otimes_{\mathcal{O}} \check{\mathcal{J}}_k T = (\Lambda\underline{T}^* \otimes_{\mathcal{O}} \mathcal{J}_k T) \oplus (\Lambda\underline{T}^* \otimes_{\mathcal{O}} \underline{T})$$

and $\mathcal{L}(u_H)\alpha$ is the usual Lie derivative of α by the vector form u_H on X. We infer that $\mathcal{L}$ maps onto the subalgebra of Lie derivations in $\mathcal{D}er\ \Lambda\underline{T}^*$. A simple computation shows that $[u,v]_H = [u_H, v_H]$ hence $\mathcal{L}$ transforms the bracket [,] into the derivations bracket for $k \geq 1$. In particular, for any $k \geq 0$, the restriction

$$\mathcal{L}:\ \Lambda\underline{T}^* \otimes_{\mathcal{O}} \tilde{\mathcal{J}}_k T \to \mathcal{D}er\ \Lambda\underline{T}^*$$

is a Lie algebra representation with respect to the Nijenhuis bracket and the derivations bracket, whose image is the subalgebra of Lie derivations. It is injective for $k = 0$.

Define the map

$$ad:\ \Lambda\underline{T}^* \otimes_{\mathcal{O}} \check{\mathcal{J}}_k T \to \mathcal{D}er\ \Lambda\underline{T}^*$$

by

$$ad(u) = \mathcal{L}(u) + (-1)^{r+1}\, i(\tilde{D}u)$$

where $u \in \Lambda^r\underline{T}^* \otimes_{\mathcal{O}} \check{\mathcal{J}}_k T$ and $i(\tilde{D}u)\alpha = i(\tilde{\rho}_0 \tilde{D}u)\alpha = i((\tilde{D}u)_H)\alpha$. If $u = \alpha \otimes \xi \in \Lambda^r\underline{T}^* \otimes_{\mathcal{O}} \check{\mathcal{J}}_k T$ and $v = \beta \otimes \eta \in \Lambda^s\underline{T}^* \otimes_{\mathcal{O}} \check{\mathcal{J}}_k T$ then

$$[\![u,v]\!] = ad(u)\beta \otimes \check{\rho}_{k-1}\eta - (-1)^{rs}\, ad(v)\alpha \otimes \check{\rho}_{k-1}\xi + \alpha\wedge\beta \otimes [\xi,\eta] \in \Lambda^{r+s}\underline{T}^* \otimes_{\mathcal{O}} \check{\mathcal{J}}_{k-1} T$$

where [,] is the bracket in $\check{\mathcal{J}}_k T$ with values in $\check{\mathcal{J}}_{k-1} T$ (cf. Section 10). If u and v take values in $\underset{\sim}{\mathcal{J}}_k T$ then

$$[\![u,v]\!] = ad(u)\beta \otimes \eta - (-1)^{rs}\, ad(v)\alpha \otimes \xi + (\alpha\wedge\beta) \otimes [\![\xi,\eta]\!] \in \Lambda^{r+s}\underline{T}^* \otimes_{\mathcal{O}} \tilde{\mathcal{J}}_k T .$$

A simple computation shows that

$$ad(\alpha\wedge u) = \alpha\wedge ad(u)$$

for any $\alpha \in \Lambda\underline{T}^*$, $u \in \Lambda\underline{T}^* \otimes_{\mathcal{O}} \tilde{\mathcal{J}}_k T$ and $k \geq 1$, hence

$$ad\colon\ \Lambda\underline{T}^* \otimes_{\mathcal{O}} \tilde{\mathcal{J}}_k T \to \mathcal{D}er\, \Lambda\underline{T}^*$$

is left $\Lambda\underline{T}^*$-linear ($\mathcal{D}er\, \Lambda\underline{T}^*$ is a left $\Lambda\underline{T}^*$-module).(*) Let $\mathcal{U} \in \breve{\Phi}^r$ be a representative of u and take $\omega = \pi_1^*\alpha \in \Psi$, $\alpha \in \Lambda\underline{T}^*$. Then

$$ad(u)\alpha = \Delta^*(\mathcal{L}(\mathcal{U})\omega + (-1)^{r+1}(\tilde{d}_H\mathcal{U})\overline{\wedge}\omega)\,.$$

We shall now give an alternate description of ad. The sheaf $\Lambda^\ell \underline{T^*_V}$ identifies with the subsheaf of $\Lambda^\ell \underline{T^*X^2}$ of germs of scalar vertical differential ℓ-forms on X^2 namely the forms ν such that $\nu(\xi_1,\ldots,\xi_\ell) = \nu(V\xi_1,\ldots,V\xi_\ell)$ for any vector fields ξ_i on X^2. $\Lambda^\ell \underline{T^*_V}$ is an $\mathcal{O}_{X^2}$-submodule of $\Lambda^\ell \underline{T^*X^2}$, the map $j\colon \Lambda^\ell\underline{T}^* \to \Lambda^\ell\underline{T^*_V}$, induced by π_2^*, is right $\mathcal{O}$-linear and factors to an $\mathcal{O}$-linear isomorphism

$$j_0\colon\ \Lambda^\ell\underline{T}^* \to \Lambda^\ell\underline{T^*_V}/\mathcal{J}\Lambda^\ell\underline{T^*_V} = \mathcal{J}_0\Lambda^\ell T^*$$

(the left, right and $\mathcal{J}_0 = \mathcal{O}$-module structures all agree on the quotient). j_0^{-1} is the quotient of the map $\Delta^*\colon \Lambda^\ell\underline{T^*_V} \to \Lambda^\ell\underline{T}^*$. More generally,

$$j_0\colon\ \Lambda\underline{T}^* \to \Lambda\underline{T^*_V}/\mathcal{J}\Lambda\underline{T^*_V} = \mathcal{J}_0\Lambda T^*$$

is an isomorphism of graded $\mathcal{O}$-algebras.

Let $\xi \in \tilde{\Theta}$ and $\nu \in \Lambda^\ell\underline{T^*_V}$. Define

$$L(\xi)\nu = V^*[\mathcal{L}(\xi)\nu]$$

or, more explicitly

$$[L(\xi)\nu]\,(\xi_1,\ldots,\xi_\ell) = [\mathcal{L}(\xi)\nu]\,(V\xi_1,\ldots,V\xi_\ell)$$

where $\mathcal{L}(\xi)$ is the Lie derivative along ξ. $L(\xi)\nu \in \Lambda^\ell\underline{T^*_V}$ and if

(*) (Added in proof) In particular, $ad\colon \tilde{\mathcal{J}}_k T \to \mathcal{D}er\, \Lambda\underline{T}^*$ is the left $\mathcal{O}$-linear sheaf map (cf. Section 9) which corresponds to the first order differential operator $\mathcal{L}\colon \theta \in \underline{T} \mapsto \mathcal{L}(\theta) \in \mathcal{D}er\, \Lambda\underline{T}^*$. The relation $\mathcal{L}([\theta,\theta']) = [\mathcal{L}(\theta), \mathcal{L}(\theta')]$ and the fact

$\nu \in \mathcal{I}\Lambda^{\ell}T^*_{\underline{V}}$ then $\mathcal{L}(\xi)\nu$ vanishes on the diagonal (since ξ is diagonal) hence $L(\xi)\nu \in \mathcal{I}\Lambda^{\ell}T^*_{\underline{V}}$. Moreover, $L(\xi) \in \mathcal{D}er^0\, \Lambda T^*_{\underline{V}}$ and L is left $\mathcal{O}$-linear, i.e., $L(f\xi) = fL(\xi)$ where $f \in \mathcal{O}$, $f\xi$ is left multification and $fL(\xi) = i(f)L(\xi)$ ($\mathcal{D}er\, \Lambda T^*_{\underline{V}}$ is a left $\Lambda T^*_{\underline{V}}$-module). The first assertion is obvious and the second follows from the formula

$$\mathcal{L}(f\xi) = f\mathcal{L}(\xi) + df\wedge i(\xi)$$

and the fact that df is horizontal. Using the relation

$$\mathcal{L}(\xi) <\xi_1\wedge\ldots\wedge\xi_\ell, \nu> = <\mathcal{L}(\xi)(\xi_1\wedge\ldots\wedge\xi_\ell), \nu> + <\xi_1\wedge\ldots\wedge\xi_\ell, \mathcal{L}(\xi)\nu>$$

and the fact that $[\Theta_V, \tilde{\Theta}] \subset \Theta_V = \underline{T_V}$ one proves that

$$L([\xi,\xi']) = [L(\xi), L(\xi')]$$

for any $\xi, \xi' \in \tilde{\Theta}$ (the second bracket in $\mathcal{D}er^0\Lambda T^*_{\underline{V}}$). Finally, if $\xi \in \mathcal{I}^{k+1}\Delta\tilde{\Theta}, k \geq 1$, then $L(\xi)\nu \in \mathcal{I}\Lambda T^*_{\underline{V}}$ for any $\nu \in \Lambda T^*_{\underline{V}}$ (in fact $L(\xi)\nu \in \mathcal{I}^k\Lambda T^*_{\underline{V}}$). The previous discussion shows that $L(\xi)$ factors to a derivation of $\Lambda \underline{T}^*$ and the map

$$\xi \in \tilde{\Theta} \to L(\xi) \in \mathcal{D}er^0\, \Lambda\underline{T}^*$$

factors to

$$L\colon\ \tilde{\mathcal{J}}_k T \to \mathcal{D}er^0\, \Lambda\underline{T}^*$$

for any $k \geq 1$. If $\xi \in \tilde{\mathcal{J}}_k T$ is represented by $\xi \in \tilde{\Theta}$ and $\nu = \pi_2^*a$ with $a \in \Lambda\underline{T}^*$ then

$$L(\xi)a = \Delta^*(V^*[\mathcal{L}(\xi)\nu]) .$$

The map L is a left $\mathcal{O}$-linear Lie algebra morphism ($\tilde{\mathcal{J}}_k T$ with **[,]** and $\mathcal{D}er^0\Lambda\underline{T}^*$ with the commutator). We now define

$$L\colon\ \Lambda\underline{T}^* \otimes_{\mathcal{O}} \tilde{\mathcal{J}}_k T \to \mathcal{D}er\, \Lambda\underline{T}^*$$

by the relation

$$L(a\otimes\xi) = a\wedge L(\xi) .$$

---

that $\tilde{\mathcal{J}}_k T$ is left $\mathcal{O}$-generated by $\tilde{j}_k\underline{T}$ yields a simpler proof of the theorem of Section 15 without resorting to the representation L. However, the latter is useful in Section 22.

This definition is compatible with the tensor product since $L(\xi)$ is left $\mathcal{O}$-linear with respect to ξ and $\mathcal{D}er\,\Lambda\underline{T}^*$ is a graded left module over the graded $\mathcal{O}$-algebra $\Lambda\underline{T}^*$ ($\tilde{L}$ is the standard left $\Lambda\underline{T}^*$-linear extension of the left $\mathcal{O}$-linear map $L: \tilde{\mathcal{J}}_k T \to \mathcal{D}er^0\,\Lambda\underline{T}^*$). If $u \in \Lambda^r\underline{T}^* \otimes_{\mathcal{O}} \tilde{\mathcal{J}}_k T$ then $L(u) \in \mathcal{D}er^r\,\Lambda\underline{T}^*$, i.e., L is homogeneous of degree zero. It should be remarked that:

a) the usual expression

$$\alpha \wedge L(\xi) + (-1)^r d\alpha \wedge i(\xi), \ r = deg\ \alpha ,$$

does not factor to the tensor product. This is to be expected since $L(\xi)$ is not a Lie derivative. Its component of type i is not trivial (see Lemma 3 and the i component of ad).

b) If $u \in \Lambda^r\underline{T}^* \otimes_{\mathcal{O}} \tilde{\mathcal{J}}_k T$, $r \geq 1$, and $\mathcal{U} \in \tilde{\Phi}^r$ is a representative of u then

$$L(u)\alpha \neq \Delta^*(V^*[\mathcal{L}(\mathcal{U})\nu])$$

with $\nu = \pi_2^*\alpha$ and $\alpha \in \Lambda\underline{T}^*$. In fact $V^*[\mathcal{L}(\mathcal{U})\nu] = 0$.

The extended map L is left $\Lambda\underline{T}^*$-linear and homogeneous of degree zero. For $k \geq 1$, $u = \alpha \otimes \xi \in \Lambda^r\underline{T}^* \otimes_{\mathcal{O}} \tilde{\mathcal{J}}_k T$ and $v = \beta \otimes \eta \in \Lambda^s\underline{T}^* \otimes_{\mathcal{O}} \tilde{\mathcal{J}}_k T$ define the bracket

$$\{u, v\} = (L(u)\beta) \otimes \eta - (-1)^{rs}(L(v)\alpha) \otimes \xi + (\alpha \wedge \beta) \otimes [\![\xi, \eta]\!] .$$

This bracket defines a homogeneous $\mathbf{R}$-linear map of degree zero

$$[\Lambda\underline{T}^* \otimes_{\mathcal{O}} \tilde{\mathcal{J}}_k T] \otimes_{\mathbf{R}} [\Lambda\underline{T}^* \otimes_{\mathcal{O}} \tilde{\mathcal{J}}_k T] \to \Lambda\underline{T}^* \otimes_{\mathcal{O}} \tilde{\mathcal{J}}_k T .$$

LEMMA 1. $\Lambda\underline{T}^* \otimes_{\mathcal{O}} \tilde{\mathcal{J}}_k T$ *together with* $\{\ ,\ \}$ *is a graded Lie algebra* ($k \geq 1$).

The relation

$$\{u, v\} = (-1)^{rs+1}\{v, u\}$$

is obvious for u and v homogeneous of degrees r and s. The Jacobi identity is easily verified for decomposable elements.

LEMMA 2. $L(\{u,v\}) = [L(u), L(v)]$.

This relation is easily verified for decomposable elements. The lemma implies that L is a graded Lie algebra morphism.

LEMMA 3. $L(u) = ad(u)$.

Proof: Since L and ad are left $\Lambda\underline{T}^*$-linear it is enough to prove that $L(\xi) = ad(\xi)$ for $\xi \in \tilde{\mathcal{J}}_k T$, hence $L(\xi)\alpha = ad(\xi)\alpha$ for any $\alpha \in \Lambda\underline{T}^*$. Take a representative $\xi \in \tilde{\Theta}$ and let ξ_H and ξ_V be its horizontal and vertical components. We can further assume that $\alpha \in \Lambda^\ell \underline{T}^*$ is of the form $\alpha = F\, dG_1 \wedge \ldots \wedge dG_\ell$ with $F, G_k \in \mathcal{O}$ hence $\omega = \pi_1^*\alpha \in \Psi^\ell$ will be given by $\omega = f\, dg_1 \wedge \ldots \wedge dg_\ell$ where $f = F \circ \pi_1$ and $g_k = G_k \circ \pi_1$ belong to $i(\mathcal{O})$. Then

$$ad(\xi)\alpha = \Delta^*(\mathcal{L}(\xi)\omega - (\tilde{d}_H\xi)\bar{\wedge}\omega)$$

and

$$\mathcal{L}(\xi)\omega - (\tilde{d}_H\xi)\bar{\wedge}\omega = \mathcal{L}(\xi_H)\omega - \sum_{i=1}^{n} dx_i \wedge \left[\left(\varepsilon^{-1}\frac{\partial\xi_V}{\partial x_i}\right)\bar{\wedge}\omega\right] =$$

$$(\xi_H f)dg_1 \wedge \ldots \wedge dg_\ell + f\sum_{k=1}^{\ell} dg_1 \wedge \ldots \wedge d(\xi_H g_k) \wedge \ldots \wedge dg_\ell -$$

$$f\sum_{i=1}^{n} dx_i \wedge \sum_{k=1}^{\ell} (-1)^{k+1} \left\langle \left(\varepsilon^{-1}\frac{\partial\xi_V}{\partial x_i}\right)_H, dg_k \right\rangle dg_1 \wedge \ldots \wedge \widehat{dg_k} \wedge \ldots \wedge dg_\ell$$

hence

$$ad(\xi)\alpha = [(\xi_H f)\circ\Delta]\, dG_1 \wedge \ldots \wedge dG_\ell + F\sum_{k=1}^{\ell} dG_1 \wedge \ldots \wedge d[(\xi_H g_k)\circ\Delta] \wedge \ldots \wedge dG_\ell -$$

$$F\sum_{i=1}^{n} dx_i \wedge \sum_{k=1}^{\ell} (-1)^{k+1} \left(\left\langle \left(\varepsilon^{-1}\frac{\partial\xi_V}{\partial x_i}\right)_H, dg_k \right\rangle \circ \Delta\right) dG_1 \wedge \ldots \wedge \widehat{dG_k} \wedge \ldots \wedge dG_\ell .$$

To compute $L(\xi)\alpha$, let $\nu = \pi_2^*\alpha = \phi\, d\gamma_1 \wedge \ldots \wedge d\gamma_\ell \in \Lambda^\ell \underline{T}_V^*$ where $\phi = F \circ \pi_2$ and $\gamma_k = G_k \circ \pi_2$ belong to $j(\mathcal{O})$. Then

$$V^*[\mathcal{L}(\xi)\nu] = (\xi_V\phi)\, d\gamma_1 \wedge \ldots \wedge d\gamma_\ell + \phi\sum_{k=1}^{\ell} d\gamma_1 \wedge \ldots \wedge d_V(\xi_V\gamma_k) \wedge \ldots \wedge d\gamma_\ell .$$

Write $\xi_V = \eta + \zeta$ where $\eta = j(\pi_{1*}\xi)$. The field η is projectable vertical, $\eta = r_*\xi_H$ with $r: (x, y) \mapsto (y, x)$ the reflection with respect to the diagonal and ζ is vertical and vanishes on Δ. It follows that

$$\eta\phi|_\Delta = \xi_H f|_\Delta, \eta\gamma_k|_\Delta = \xi_H g_k|_\Delta \quad \text{and} \quad \zeta\phi|_\Delta = 0$$

and therefore

$$V^*[\mathfrak{L}(\xi)\nu] = (\eta\phi)d\gamma_1\wedge\ldots\wedge d\gamma_\ell + (\zeta\phi)d\gamma_1\wedge\ldots\wedge d\gamma_\ell + \phi\sum_{k=1}^{\ell} d\gamma_1\wedge\ldots\wedge d(\eta\gamma_k)\wedge\ldots\wedge d\gamma_\ell$$
$$\phi\sum_{i=1}^{n} dy_i\wedge\sum_{k=1}^{\ell}(-1)^{k+1}\frac{\partial}{\partial y_i}(\zeta\gamma_k)\,d\gamma_1\wedge\ldots\wedge\widehat{d\gamma_k}\wedge\ldots\wedge d\gamma_\ell$$

and

$$L(\xi)\alpha = \Delta^*[V^*[\mathfrak{L}(\xi)\nu]) = [(\xi_H f)\circ\Delta]\,dG_1\wedge\ldots\wedge dG_\ell +$$
$$F\sum_{k=1}^{\ell} dG_1\wedge\ldots\wedge d\,[(\xi_H g_k)\circ\Delta]\wedge\ldots\wedge dG_\ell +$$
$$F\sum_{i=1}^{n} dx_i\wedge\sum_{k=1}^{\ell}(-1)^{k+1}\left(\frac{\partial\zeta\gamma_k}{\partial y_i}\circ\Delta\right)dG_1\wedge\ldots\wedge\widehat{dG_k}\wedge\ldots\wedge dG_\ell\,.$$

It remains to prove that

$$<\left(\varepsilon^{-1}\frac{\partial\xi_V}{\partial x_i}\right)_H, dg_k>\circ\Delta = -\frac{\partial\zeta\gamma_k}{\partial y_i}\circ\Delta$$

or, equivalently, that

$$\left(\varepsilon^{-1}\frac{\partial\xi_V}{\partial x_i}\right)_H g_k|_\Delta = -\frac{\partial}{\partial y_i}(\zeta\gamma_k)|_\Delta\,.$$

Trivializing TX^2 and observing that ζ vanishes on Δ we get

$$\frac{\partial}{\partial y_i}(\zeta\gamma_k)|_\Delta = \frac{\partial\zeta}{\partial y_i}\gamma_k|_\Delta\,.$$

On the other hand, since $\frac{\partial\eta}{\partial x_i} = 0$, then

$$\frac{\partial \xi_V}{\partial x_i} = \frac{\partial \zeta}{\partial x_i}$$

and, since ζ vanishes on Δ, we also obtain

$$0 = \left[\left(\frac{\partial}{\partial x_i} + \frac{\partial}{\partial y_i}\right)\zeta\right]_\Delta = \left(\frac{\partial}{\partial x_i}\zeta\right)_\Delta + \left(\frac{\partial}{\partial y_i}\zeta\right)_\Delta .$$

It follows that

$$\frac{\partial}{\partial y_i}(\zeta\gamma_k)|_\Delta = \frac{\partial \zeta}{\partial y_i}\gamma_k|_\Delta = -\frac{\partial \zeta}{\partial x_i}\gamma_k|_\Delta = -\frac{\partial \xi_V}{\partial x_i}\gamma_k|_\Delta = -\left(\varepsilon^{-1}\frac{\partial \xi_V}{\partial x_i}\right)_H g_k|_\Delta .$$

LEMMA 4. $[\![u, v]\!] = \{u, v\}$ *and* $ad([\![u, v]\!]) = [ad\,u, ad\,v]$ *for any* $u, v \in \Lambda\underline{T}^* \otimes_{\mathcal{O}} \tilde{\mathcal{J}}_k T$.

THEOREM. $\Lambda\underline{T}^* \otimes_{\mathcal{O}} \tilde{\mathcal{J}}_k T$ *together with* $[\![\ ,\ ]\!]$ *is a graded Lie algebra. If* $k > 0$ *then the mapping*

$$ad = ad_k : \Lambda\underline{T}^* \otimes_{\mathcal{O}} \tilde{\mathcal{J}}_k T \to \mathcal{D}er\, \Lambda\underline{T}^*$$

is a Lie algebra representation whose image is the subalgebra $\bigoplus_{r \geq 0} \mathcal{D}er^r\, \Lambda\underline{T}^*$. *The representation* ad *is left* $\Lambda\underline{T}^*$*-linear and homogeneous of degree zero. The relations*

$$ad_k = ad_1 \circ \tilde{\rho}_1, \quad ker\, ad_1 = \underline{\delta}(\Lambda\underline{T}^* \otimes_{\mathcal{O}} S^2\underline{T}^* \otimes_{\mathcal{O}} \underline{T})$$

hold and, in particular,

$$ad_1 : \tilde{\mathcal{J}}_1 T \to \mathcal{D}er^0\, \Lambda\underline{T}^*$$

is an isomorphism. If $k = 0$ *then*

$$ad = \mathcal{L} : \Lambda\underline{T}^* \otimes_{\mathcal{O}} \underline{T} \to \mathcal{D}er\, \Lambda\underline{T}^* .$$

For a more general statement see Theorems 2, 3 and Lemma 5 of Section 29.

Proof: The Lie algebra structure assertions are consequences of Lemma 4. Let $k > 0$. The following diagram is commutative and exact

$$\begin{array}{ccccc} \Lambda^r\underline{T}^*\otimes_{\mathcal{O}} S^1\underline{T}^*\otimes_{\mathcal{O}} \underline{T} & \xrightarrow{\underline{\delta}} & \Lambda^{r+1}\underline{T}^*\otimes_{\mathcal{O}} \underline{T} & \longrightarrow & 0 \\ \downarrow & & \| & & \\ \Lambda^r\underline{T}^*\otimes_{\mathcal{O}} \tilde{\mathcal{J}}_1 T & \xrightarrow{\tilde{D}} & \Lambda^{r+1}\underline{T}^*\otimes_{\mathcal{O}} \tilde{\mathcal{J}}_0 T & \longrightarrow & 0 \end{array} .$$

Since $\mathit{ad}_k(u) = \mathcal{L}(\tilde{\rho}_0 u) + (-1)^{r+1} i(\tilde{\rho}_0 \tilde{D} u)$ then $\mathit{ad}_k = \mathit{ad}_1 \circ \tilde{\rho}_1$ is obvious. Since $\tilde{\rho}_0$ is surjective we infer that the derivation $\mathit{ad}_1(u)$ can have any Lie component. If $u \in \Lambda^r\underline{T}^*\otimes_{\mathcal{O}} S^1\underline{T}^*\otimes_{\mathcal{O}} \underline{T}$ then $\mathit{ad}_1(u) = (-1)^{r+1} i(\underline{\delta} u)$ hence, by the surjectivity of $\underline{\delta}$, $\mathit{ad}_1(u)$ can be any interior derivation of degree ≥ 0. This proves the assertion on $\mathit{im}\ \mathit{ad}_k$ since $\tilde{\rho}_1$ is surjective. If $\mathit{ad}_1(u) = 0$ then $\tilde{\rho}_0 u = 0$ ($\mathcal{L}$ is injective), i.e., $u \in \Lambda^r\underline{T}^*\otimes_{\mathcal{O}} S^1\underline{T}^*\otimes_{\mathcal{O}}$ and $i(\underline{\delta} u) = 0$ hence $\underline{\delta} u = 0$ (i is injective). The assertion on $\mathit{ker}\ \mathit{ad}_1$ follows from the exactness of the $\underline{\delta}$-complex.

Remark: We can extend L to a map $L: \Lambda\underline{T}^*\otimes_{\mathcal{O}} \check{\mathcal{J}}_k T \to \mathcal{D}er\, \Lambda\underline{T}^*$ as follows. We first set $L(\xi)\alpha = \Delta^*(V^*[\mathcal{L}(\xi)\pi_2^*\alpha])$ with $\alpha \in \Lambda\underline{T}^*$, $\xi \in \check{\mathcal{J}}_k T$ (represented by $\xi \in \check{\Theta}$) and $k \geq 1$. Then $L: \check{\mathcal{J}}_k T \to \mathcal{D}er^0\, \Lambda\underline{T}^*$ is again left $\mathcal{O}$-linear, hence we can define the extended map by $L(\alpha\otimes\xi) = \alpha \wedge L(\xi)$ with $\alpha\otimes\xi \in \Lambda\underline{T}^*\otimes_{\mathcal{O}} \check{\mathcal{J}}_k T$. The map L satisfies all the properties of the previously defined map $L: \Lambda\underline{T}^*\otimes_{\mathcal{O}} \tilde{\mathcal{J}}_k T \to \mathcal{D}er\, \Lambda\underline{T}^*$ (with the obvious modifications of lowering the order of jets in brackets) but $L(u) \neq \mathit{ad}(u)$ when $u \notin \Lambda\underline{T}^*\otimes_{\mathcal{O}} \tilde{\mathcal{J}}_k T$. The extended map L will not be used in the sequel since the useful representation is ad.

16. *The fundamental identities*

The main purpose of this section is to show that the operators D and $\tilde{D}$ are adjoint derivations to the previous brackets. The intrinsic proofs of the relations to follow are straightforward, though lengthy computations and we shall only indicate the main steps. It turns out that the *coordinate* proofs, though less aesthetic, are much shorter.

Let ξ be a vertical vector field on X^2 and η a horizontal vector field. Since T_V is horizontally trivial (i.e., $T_V|_{X\times\{y\}}$ is trivial) the restriction of ξ to $X\times\{y\}$ is a function with values in the vector space T_yX and $\eta|_{X\times\{y\}}$ is a vector field tangent to $X\times\{y\}$. We denote by $\eta\bullet\xi$ the Lie derivative of the function ξ along the vector field η. Then $\eta\bullet\xi$ is a vertical vector field on X^2 which does not coincide in general with the bracket $[\eta,\xi] = \mathfrak{L}(\eta)\xi$ (Lie derivative of the vector field ξ) since the latter need not even be vertical. However if η is projectable horizontal then $\eta\bullet\xi = [\eta,\xi]$.

Let u be a horizontal exterior differential r-form on X^2 with values in T_V. We denote by u_y the induced r-form $u|_{X\times\{y\}}$ on the submanifold $X\times\{y\}$. The form u_y takes values in the vector space T_yX. In Section 6 we defined the horizontal exterior differentiation by

$$(d_H u)_{(x,y)} = H^*[d(u_y)]_{(x,y)}$$

or, equivalently,

$$d_H u(v_1,\ldots,v_{r+1}) = d(u_y)(Hv_1,\ldots,Hv_{r+1})$$

where $v_i \in T_{(x,y)}X^2$ and H is the horizontal projection. Let ξ_i be vector fields on X^2 and ξ_{iH} their horizontal components. A simple computation shows that

$$d_H u(\xi_1,\ldots,\xi_{r+1}) = \sum_{i=1}^{r+1} (-1)^{i+1}\, \xi_{iH}\bullet u(\xi_1,\ldots,\hat{\xi}_i,\ldots,\xi_{r+1}) +$$

$$\sum_{i<j} (-1)^{i+j}\, u([\xi_{iH},\xi_{jH}],\xi_1,\ldots,\hat{\xi}_i,\ldots,\hat{\xi}_j,\ldots,\xi_{r+1}) .$$

In particular, if the ξ_i are projectable horizontal, the previous formula reduces to the usual one, namely

$$d_H u(\xi_1, \ldots, \xi_{r+1}) = \sum_{i=1}^{r+1} (-1)^{i+1} \mathfrak{L}(\xi_i) u(\xi_1, \ldots, \hat{\xi}_i, \ldots, \xi_{r+1}) +$$

$$\sum_{i<j} (-1)^{i+j} u([\xi_i, \xi_j], \xi_1, \ldots, \hat{\xi}_i, \ldots, \hat{\xi}_j, \ldots, \xi_{r+1}) .$$

In order to preserve standard notation we denote by χ the vector valued 1-form on X^2 defined by the horizontal projection $H: TX^2 \to T_H$ and by Ω the vector valued 1-form defined by the vertical projection $V: TX^2 \to T_V$ ($\chi = H$ and $\Omega = V$ with a slight abuse). χ is a horizontal 1-form with values in $\mathfrak{X}(X^2)_{PH}$ since $<\xi, \chi> = \xi$ is projectable horizontal for any projectable horizontal ξ hence χ defines a section of Φ^1_{PH}. Similarly, Ω is a vertical 1-form with values in projectable vertical fields since $<\xi, \Omega> = \xi$ is π_2-projectable vertical for any π_2-projectable vertical vector field ξ. If Id is the identity of $\mathfrak{X}(X)$ then $\chi = \pi_1^* \mathrm{Id}$ and $\Omega = \pi_2^* \mathrm{Id}$.

Let $\tilde{u}$ be a horizontal r-form with values in $\tilde{\mathfrak{X}}(X^2)$, i.e., $\tilde{u}(\xi_1, \ldots, \xi_r)$ is diagonal and π_1-projectable for any projectable horizontal vector fields ξ_i. Let $u = \tilde{u} \barwedge \Omega = V \circ \tilde{u}$ be the corresponding horizontal form with values in T_V. The mapping $\tilde{u} \mapsto u$ is bijective and linear for the diagonal $\mathfrak{F}(X^2)$-module structure. On the germs level it induces the map ε.

LEMMA 1. *If* ω *is a scalar horizontal* r-*form on* X^2 *then* $d\omega = d_H\omega + i(\Omega)d\omega$. *If* ω *is a scalar vertical* r-*form on* X^2 *then* $d\omega = d_V\omega + i(\chi)d\omega$.

The proof is a simple verification left to the reader. We observe that Lemma 1 is a corollary of general results for almost-product structures where the operator d splits canonically into two components. In fact $d = \mathfrak{L}(\chi) + \mathfrak{L}(\Omega)$ and, in our (integrable) case, the partial operators d_H and d_V are simply the restrictions of $\mathfrak{L}(\chi)$ to horizontal forms and $\mathfrak{L}(\Omega)$ to vertical forms respectively.

LEMMA 2. $d_H u = - [\Omega, u]$.

Proof: We have to prove that $\mathcal{L}(d_H u) = -\mathcal{L}([\Omega, u])$. Since these are Lie derivations, it is enough to show that $\mathcal{L}(d_H u)f = \mathcal{L}([\Omega, u])f$ for any function f on X^2. But

$$\mathcal{L}([\Omega, u])f = i(\Omega)d(u \barwedge df) + (-1)^{r+1}\, i(u)d(\Omega \barwedge df) - d(u \barwedge df)$$

and

$$\mathcal{L}(d_H u)f = i(d_H u)df = df \circ d_H u \,.$$

Furthermore,

$$\Omega \barwedge df = df \circ \Omega = d_V f$$

and Lemma 1 applied to $\omega = u \barwedge df = df \circ u$ gives

$$\mathcal{L}([\Omega, u])f = -\, d_H(df \circ u) + (-1)^{r+1}\, i(u)d(d_V f) \,.$$

It remains to prove that

$$d_H(df \circ u) + (-1)^r i(u)d(d_V f) - (df) \circ d_H u = 0 \,.$$

The first and last terms are clearly horizontal and $d_V d_V f = 0$ implies that the second term is also horizontal (use second part of Lemma 1). It is therefore enough to check the above relation when applied to projectable horizontal vector fields. This is left to the reader.

LEMMA 3. $d_H u = -\, [\Omega, \tilde{u}]$.

Proof: We shall prove that $[\Omega, \tilde{u}] = [\Omega, u]$. Since

$$\mathcal{L}([\Omega, \tilde{u}])f = i(\Omega)d(df \circ \tilde{u}) + (-1)^{r+1}\, i(\tilde{u})d(df \circ \Omega) - d(df \circ u)$$

it is enough to prove that

$$i(\Omega)d(df \circ \tilde{u}) + (-1)^{r+1}\, i(\tilde{u})d(df \circ \Omega) = i(\Omega)d(df \circ u) + (-1)^{r+1}\, i(u)d(df \circ \Omega) \,.$$

Write $\tilde{u} = u + v$ where $v = \tilde{u}_H$ is the horizontal component of $\tilde{u}$. The form v is horizontal with values in $\mathcal{X}(X^2)_{PH}$. Replacing $\tilde{u}$ by $u + v$, it all amounts to proving that

$$i(\Omega)d(df\circ v) + (-1)^{r+1}\, i(v)d(df\circ\Omega) = 0\,.$$

A simple computation shows that the left expression is a scalar horizontal form (each term separately is not horizontal) and the relation will follow by applying it to projectable horizontal vector fields.

LEMMA 4. $d_H u = [\chi, u] = [\chi, \tilde{u}]$.

Proof: $\chi = \mathrm{Id} - \Omega$ and $[\chi, u] = [\mathrm{Id}, u] - [\Omega, u] = -[\Omega, u]$ where Id is the identity of $\mathcal{X}(X^2)$.

THEOREM 1. *Let* $\tilde{u} \in \Lambda\underline{T}^* \otimes_{\mathcal{O}} \tilde{\mathcal{J}}_k T$ *and* $u = (\varepsilon_k \tilde{u}) \in \Lambda\underline{T}^* \otimes_{\mathcal{O}} \mathcal{J}_k T$. *Denote by* $\tilde{\mathfrak{U}}$ *(resp.* $\mathfrak{U} = \varepsilon\tilde{\mathfrak{U}}$*) a representative of* $\tilde{u}$ *(resp.* u*) in* $\tilde{\Phi}$ *(resp.* Φ_V*). Then*

$$Du = -[\Omega, \mathfrak{U}] \bmod \mathcal{J}^k\Phi_V = -[\Omega, \tilde{\mathfrak{U}}] \bmod \mathcal{J}^k\Phi_V =$$

$$[\chi, \mathfrak{U}] \bmod \mathcal{J}^k\Phi_V = [\chi, \tilde{\mathfrak{U}}] \bmod \mathcal{J}^k\Phi_V\,.$$

The form χ is a section of $\check{\Phi}^1$ $(\Phi_{PH} \subset \check{\Phi})$ and we denote by χ_k the induced section in

$$\underline{T}^* \otimes_{\mathcal{O}} \check{\mathcal{J}}_k T \simeq \check{\Phi}^1 / \mathcal{J}^{k+1}\Delta\check{\Phi}^1\,.$$

On $\Lambda\underline{T}^* \otimes_{\mathcal{O}} \check{\mathcal{J}}_k T$ we have two brackets [,] and **[,]** which map into $\Lambda\underline{T}^* \otimes_{\mathcal{O}} \check{\mathcal{J}}_{k-1} T$.

THEOREM 2. *For any* $\tilde{u} \in \Lambda\underline{T}^* \otimes_{\mathcal{O}} \tilde{\mathcal{J}}_k T$ *and* $u = (\varepsilon_k \tilde{u}) \in \Lambda\underline{T}^* \otimes_{\mathcal{O}} \mathcal{J}_k T$ *we have the relations*

$$Du = [\chi_k, u] = [\chi_k, \tilde{u}] \in \Lambda\underline{T}^* \otimes_{\mathcal{O}} \mathcal{J}_{k-1} T$$

and

$$\tilde{D}u = \tilde{D}\tilde{u} = [\![\chi_k, u]\!] = [\![\chi_k, \tilde{u}]\!] \in \Lambda\underline{T}^* \otimes_{\mathcal{O}} \tilde{\mathcal{J}}_{k-1} T\,.$$

Proof: The first formula is a restatement of the previous theorem. The second formula can be proved as follows. The relation

$$\tilde{D}u = \varepsilon_{k-1}^{-1} Du = Du + (\tilde{D}u)_H$$

gives the vertical and horizontal components of $\tilde{D}u$ in the direct sum decomposition

$$\Lambda\underline{T}^* \otimes_{\mathcal{O}} \check{\mathcal{J}}_{k-1}T = [\Lambda\underline{T}^* \otimes_{\mathcal{O}} \mathcal{J}_{k-1}T] \oplus [\Lambda\underline{T}^* \otimes \underline{T}] .$$

Furthermore,

$$[\![\chi_k, u]\!] = [\chi_k, u] + \tilde{D}\chi_k \barwedge \rho_{k-1} u + \tilde{D}u \barwedge \chi_{k-1} ,$$

$[\chi_k, u] = Du$, $\tilde{D}\chi_k \barwedge \rho_{k-1} u = 0$ since $\varepsilon_k \chi_k = 0$ and

$$\tilde{D}u \barwedge \chi_{k-1} = (\tilde{D}u)_H \barwedge Id = (\tilde{D}u)_H$$

since $\chi_{k-1} = (\chi_{k-1})_H = Id \in \underline{T}^* \otimes \underline{T}$ in the direct sum decomposition

$$\underline{T}^* \otimes_{\mathcal{O}} \check{\mathcal{J}}_{k-1}T = [\underline{T}^* \otimes_{\mathcal{O}} \mathcal{J}_{k-1}T] \oplus [\underline{T}^* \otimes_{\mathcal{O}} T] .$$

Hence $\tilde{D}u = [\![\chi_k, u]\!]$ and we obviously have $\tilde{D}u = \tilde{D}\tilde{u}$ and $[\![\chi_k, u]\!] = [\![\chi_k, \tilde{u}]\!]$.

LEMMA 5. *For any horizontal form* $u \in \Lambda\underline{T}^* \otimes_{\mathcal{O}} \check{\mathcal{J}}_k T$ (*i.e.*, $u = u_H$), *we have* $[\chi_k, u] = [\![\chi_k, u]\!] = 0$.

Proof: Let $\mathfrak{U} \in \Phi_{PH}$ be a representative of u. Then $\mathfrak{U} = \pi_1^* \mathcal{V}$ and $\chi = \pi_1^* Id$ where $\mathcal{V}, Id \in \Lambda\underline{T}^* \otimes \underline{T}$ are vector forms on X. We infer that (cf. the first lemma of Section 14) $[\chi, \mathfrak{U}] = \pi_1^*[Id, \mathcal{V}] = 0$ and therefore $[\![\chi_k, u]\!] = [\chi_k, u] = 0$ since $\tilde{D}\chi_k = \tilde{D}u = 0$.

Remark: The last lemma yields a trivial proof of Lemma 3. In fact

$$-[\Omega, \tilde{u}] = [\chi, \tilde{u}] = [\chi, u] + [\chi, \tilde{u}_H] = [\chi, u] = [\Omega, u] .$$

THEOREM 3. *For any* $u \in \Lambda\underline{T}^* \otimes_{\mathcal{O}} \check{\mathcal{J}}_k T$ *we have* $\tilde{D}u = [\![\chi_k, u]\!]$.

Proof: Let $u = u_V + u_H$. Then $\tilde{D}u = \tilde{D}u_V$ and $[\![\chi_k, u]\!] = [\![\chi_k, u_V]\!]$ since $[\![\chi_k, u_H]\!] = 0$.

LEMMA 6. *For any* $u \in \Lambda^r \underline{T}^* \otimes_{\mathcal{O}} \tilde{\mathcal{J}}_k T$, $v \in \Lambda \underline{T}^* \otimes_{\mathcal{O}} \tilde{\mathcal{J}}_k T$ *and* $k \geq 2$ *we have the relation*

$$\tilde{D}[\![u, v]\!] = [\![\tilde{D}u, \tilde{\rho}_{k-1} v]\!] + (-1)^r [\![\tilde{\rho}_{k-1} u, \tilde{D}v]\!] .$$

Proof: We can assume that $v \in \Lambda^s \underline{T}^* \otimes_{\mathcal{O}} \tilde{\mathcal{J}}_k T$. Since

$$\varepsilon_k \colon \Lambda \underline{T}^* \otimes_{\mathcal{O}} \tilde{\mathcal{J}}_k T \to \Lambda \underline{T}^* \otimes_{\mathcal{O}} \mathcal{J}_k T$$

is an isomorphism it is enough to prove that

$$\varepsilon_{k-1} \tilde{D}[\![u, v]\!] = \varepsilon_{k-1} [\![\tilde{D}u, \tilde{\rho}_{k-1} v]\!] + (-1)^r \varepsilon_{k-1} [\![\tilde{\rho}_{k-1} u, \tilde{D}v]\!] .$$

Let $\mathfrak{U} \in \tilde{\Phi}^r$ and $\mathfrak{V} \in \tilde{\Phi}^s$ be representatives of u and v. For simplicity of notation we shall with $\varepsilon^{-1}\mathfrak{W} = \mathfrak{W}^{\sim} \in \tilde{\Phi}$ for any $\mathfrak{W} \in \Phi_V$. Let $\mathfrak{U} = \mathfrak{U}_V + \mathfrak{U}_H$ and $\mathfrak{V} = \mathfrak{V}_V + \mathfrak{V}_H$ be the decomposition of $\mathfrak{U}$ and $\mathfrak{V}$ into vertical and horizontal parts. Then $\mathfrak{U}_V = \varepsilon \mathfrak{U}$ and $\mathfrak{V}_V = \varepsilon \mathfrak{V}$ belong to Φ_V and $\mathfrak{U}_H, \mathfrak{V}_H \in \Phi_{PH}$. Furthermore

$$[\![u, v]\!] = [\mathfrak{U}, \mathfrak{V}] + (-1)^{r+1} (d_H \mathfrak{U}_V)^{\sim} \barwedge \mathfrak{V} + (-1)^{rs+s} (d_H \mathfrak{V}_V)^{\sim} \barwedge \mathfrak{U} \mod \mathcal{J}^{k+1} \Delta \tilde{\Phi} .$$

We observe next that the first three terms in

$$[\mathfrak{U}, \mathfrak{V}] = [\mathfrak{U}_V, \mathfrak{V}_V] + [\mathfrak{U}_V, \mathfrak{V}_H] + [\mathfrak{U}_H, \mathfrak{V}_V] + [\mathfrak{U}_H, \mathfrak{V}_H]$$

belong to Φ_V and the last term belongs to Φ_{PH}. It follows that

$$\varepsilon [\mathfrak{U}, \mathfrak{V}] = [\mathfrak{U}_V, \mathfrak{V}_V] + [\mathfrak{U}_V, \mathfrak{V}_H] + [\mathfrak{U}_H, \mathfrak{V}_V] .$$

Similarly,

$$(d_H \mathfrak{U}_V)^{\sim} \barwedge \mathfrak{V} = (d_H \mathfrak{U}_V)^{\sim} \barwedge \mathfrak{V}_V + (d_H \mathfrak{U}_V)^{\sim} \barwedge \mathfrak{V}_H ,$$

the first term belongs to Φ_V and the second to Φ_{PH} hence

$$\varepsilon ((d_H \mathfrak{U}_V)^{\sim} \barwedge \mathfrak{V}) = (d_H \mathfrak{U}_V)^{\sim} \barwedge \mathfrak{V}_V .$$

A corresponding result holds for

$$(d_H \mathfrak{V}_V)^{\sim} \barwedge \mathfrak{U} .$$

We infer that

$$\varepsilon_k \llbracket u, v \rrbracket = [\mathcal{U}_V, \mathcal{V}_V] + [\mathcal{U}_V, \mathcal{V}_H] + [\mathcal{U}_H, \mathcal{V}_V] + (-1)^{r+1}(d_H \mathcal{U}_V)^{\sim} \barwedge \mathcal{V}_V +$$
$$(-1)^{rs+s}(d_H \mathcal{V}_V)^{\sim} \barwedge \mathcal{U}_V \ \mathit{mod}\ \mathcal{J}^{k+1}\Phi_V .$$

By Lemma 4,

$$\varepsilon_k \llbracket u, v \rrbracket = \llbracket u, v \rrbracket_V = [\mathcal{U}_V, \mathcal{V}_V] + [\mathcal{U}_V, \mathcal{V}_H] + [\mathcal{U}_H, \mathcal{V}_V] +$$
$$(-1)^{r+1}[\chi, \mathcal{U}_V]^{\sim} \barwedge \mathcal{V}_V + (-1)^{rs+s}[\chi, \mathcal{V}_V]^{\sim} \barwedge \mathcal{U}_V \ \mathit{mod}\ \mathcal{J}^{k+1}\Phi_V ,$$

$$\varepsilon_{k-1} \tilde{D} \llbracket u, v \rrbracket = D(\llbracket u, v \rrbracket_V) = [\chi, [\mathcal{U}_V, \mathcal{V}_V]] + [\chi, [\mathcal{U}_V, \mathcal{V}_H]] + [\chi, [\mathcal{U}_H, \mathcal{V}_V]] +$$
$$(-1)^{r+1}[\chi, [\chi, \mathcal{U}_V]^{\sim} \barwedge \mathcal{V}_V] + (-1)^{rs+s}[\chi, [\chi, \mathcal{V}_V]^{\sim} \barwedge \mathcal{U}_V] \ \mathit{mod}\ \mathcal{J}^{k}\Phi_V$$

and

$$\varepsilon_{k-1} \llbracket \tilde{D}u, \tilde{\rho}_{k-1} v \rrbracket = \llbracket \tilde{D}u, \tilde{\rho}_{k-1} v \rrbracket_V = [d_H \mathcal{U}_V, \mathcal{V}_V] + [d_H \mathcal{U}_V, \mathcal{V}_H] +$$
$$[(d_H \mathcal{U}_V)_{\tilde{H}}, \mathcal{V}_V] + (-1)^{rs}(d_H \mathcal{V}_V)^{\sim} \barwedge d_H \mathcal{U}_V \ \mathit{mod}\ \mathcal{J}^{k}\Phi_V =$$
$$[[\chi, \mathcal{U}_V], \mathcal{V}_V] + [[\chi, \mathcal{U}_V], \mathcal{V}_H] + [[\chi, \mathcal{U}_V]_{\tilde{H}}, \mathcal{V}_V] +$$
$$(-1)^{rs}[\chi, \mathcal{V}_V]^{\sim} \barwedge [\chi, \mathcal{U}_V] \ \mathit{mod}\ \mathcal{J}^{k}\Phi_V$$

since $d_H d_H \mathcal{U}_V = 0$. Similarly

$$\varepsilon_{k-1} \llbracket \tilde{\rho}_{k-1} u, \tilde{D} v \rrbracket = [\mathcal{U}_V, [\chi, \mathcal{V}_V]] + [\mathcal{U}_H, [\chi, \mathcal{V}_V]] + [\mathcal{U}_V, [\chi, \mathcal{V}_V]_{\tilde{H}}] +$$
$$(-1)^{r+1}[\chi, \mathcal{U}_V]^{\sim} \barwedge [\chi, \mathcal{V}_V] \ \mathit{mod}\ \mathcal{J}^{k}\Phi_V .$$

The Jacobi identity in the graded algebra $\check{\Phi}$ (with the Nijenhuis bracket) and the relations

$$[\mathcal{V}_H, \chi] = [\chi, \mathcal{U}_H] = 0 \quad \text{(cf. Lemma 5)}$$

imply that

$$[\chi, [\mathcal{U}_V, \mathcal{V}_V]] = [[\chi, \mathcal{U}_V], \mathcal{V}_V] + (-1)^r [\mathcal{U}_V, [\chi, \mathcal{V}_V]] ,$$

$$[\chi, [\mathcal{U}_V, \mathcal{V}_H]] = [[\chi, \mathcal{U}_V], \mathcal{V}_H]$$

and

$$[\chi, [\mathcal{U}_H, \mathcal{V}_V]] = (-1)^r [\mathcal{U}_H, [\chi, \mathcal{V}_V]] .$$

It remains to prove that

$$(-1)^{r+1} [\chi, [\chi, \mathcal{U}_V]^\sim \barwedge \mathcal{V}_V] + (-1)^{rs+s} [\chi, [\chi, \mathcal{V}_V]^\sim \barwedge \mathcal{U}_V] =$$
$$[[\chi, \mathcal{U}_V]^\sim_H, \mathcal{V}_V] + (-1)^{rs} [\chi, \mathcal{V}_V]^\sim \barwedge [\chi, \mathcal{U}_V] +$$
$$(-1)^r [\mathcal{U}_V, [\chi, \mathcal{V}_V]^\sim_H] - [\chi, \mathcal{U}_V]^\sim \barwedge [\chi, \mathcal{V}_V] .$$

The right hand side in each $\barwedge$ operation is a horizontal form hence we can replace everywhere $[\chi, \mathcal{U}_V]^\sim$ and $[\chi, \mathcal{V}_V]^\sim$ by their horizontal components. Moreover we can assume that $\mathcal{U} = \omega \otimes \xi$ and $\mathcal{V} = \mu \otimes \eta$ where $\omega \in \Psi^r$, $\mu \in \Psi^s$ (cf. beginning of Section 15), $\xi, \eta \in \tilde{\Theta}$ and $d\omega = d\mu = 0$. From the formulas

$$[\chi, \mathcal{U}_V] = d_H(\omega \otimes \xi_V) = (-1)^r \omega \wedge dx_i \otimes \frac{\partial \xi_V}{\partial x_i} ,$$

$$[\chi, \mathcal{U}_V]^\sim_H = (-1)^r \omega \wedge dx_i \otimes \left(\frac{\partial \xi_V}{\partial x_i}\right)^\sim_H$$

and

$$[\chi, \mathcal{U}_V]^\sim_H \barwedge \mathcal{V}_V = (-1)^r \omega \wedge dx_i \wedge \left[i\left(\frac{\partial \xi_V}{\partial x_i}\right)^\sim_H \mu \right] \otimes \eta_V$$

we infer that

$$\text{(I)}\quad (-1)^{r+1} [\chi, [\chi, \mathcal{U}_V]^\sim_H \barwedge \mathcal{V}_V] = (-1)^r \omega \wedge dx_i \wedge \left[di\left(\frac{\partial \xi_V}{\partial x_i}\right)^\sim_H \mu \right] \otimes \eta_V +$$
$$(-1)^{r+s+1} \omega \wedge dx_i \wedge \left[i\left(\frac{\partial \xi_V}{\partial x_i}\right)^\sim_H \mu \right] \wedge dx_j \otimes \frac{\partial \eta_V}{\partial x_j}$$

$$\text{(II)}\quad (-1)^{rs+s} [\chi, [\chi, \mathcal{V}_V]^\sim_H \barwedge \mathcal{U}_V] = (-1)^{rs+s+1} \mu \wedge dx_i \wedge \left(di\left(\frac{\partial \eta_V}{\partial x_i}\right)^\sim_H \omega \right) \otimes \xi_V +$$
$$(-1)^{rs+r+s} \mu \wedge dx_i \wedge \left[i\left(\frac{\partial \eta_V}{\partial x_i}\right)^\sim_H \omega \right] \wedge dx_j \otimes \frac{\partial \xi_V}{\partial x_j}$$

$$\text{(III)} \qquad [[\chi, \mathfrak{U}_V]_H^\sim, \mathfrak{V}_V] = (-1)^r \omega \wedge dx_i \wedge \left[\mathfrak{L}\left(\frac{\partial \xi_V}{\partial x_i}\right)_H^\sim \mu\right] \otimes \eta_V +$$

$$(-1)^r \omega \wedge dx_i \wedge \mu \otimes \left[\left(\frac{\partial \xi_V}{\partial x_i}\right)_H^\sim, \eta_V\right]$$

$$\text{(IV)} \qquad (-1)^r [\mathfrak{U}_V, [\chi, \mathfrak{V}_V]_H^\sim] = (-1)^{rs+s+1} \mu \wedge dx_i \wedge \left[\mathfrak{L}\left(\frac{\partial \eta_V}{\partial x_i}\right)_H^\sim \omega\right] \otimes \xi_V +$$

$$(-1)^{rs+s+1} \mu \wedge dx_i \wedge \omega \otimes \left[\left(\frac{\partial \eta_V}{\partial x_i}\right)_H^\sim, \xi_V\right]$$

$$\text{(V)} \quad (-1)^{rs} [\chi, \mathfrak{V}_V]_H^\sim \barwedge [\chi, \mathfrak{U}_V] = (-1)^{rs+r+s} \mu \wedge dx_i \wedge \left[i\left(\frac{\partial \eta_V}{\partial x_i}\right)_H^\sim \omega\right] \wedge dx_j \otimes \frac{\partial \xi_V}{\partial x_j} +$$

$$(-1)^{rs+s} \left\langle \left(\frac{\partial \eta_V}{\partial x_i}\right)_H^\sim, dx_j \right\rangle \mu \wedge dx_i \wedge \omega \otimes \frac{\partial \xi_V}{\partial x_j}$$

$$\text{(VI)} \qquad -[\chi, \mathfrak{U}_V]_H^\sim \barwedge [\chi, \mathfrak{V}_V] = (-1)^{r+s+1} \omega \wedge dx_i \wedge \left[i\left(\frac{\partial \xi_V}{\partial x_i}\right)_H^\sim \mu\right] \wedge dx_j \otimes \frac{\partial \eta_V}{\partial x_j} +$$

$$(-1)^{r+1} \left\langle \left(\frac{\partial \xi_V}{\partial x_i}\right)_H^\sim, dx_j \right\rangle \omega \wedge dx_i \wedge \mu \otimes \frac{\partial \eta_V}{\partial x_j}$$

where

$$i\left(\frac{\partial \xi_V}{\partial x_i}\right)_H^\sim = i\left[\left(\frac{\partial \xi_V}{\partial x_i}\right)_H\right]^\sim$$

and similarly for the other terms. Since $d\mu = 0$, we have

$$\mathfrak{L}\left(\frac{\partial \xi_V}{\partial x_i}\right)_H^\sim \mu = di\left(\frac{\partial \xi_V}{\partial x_i}\right)_H^\sim \mu$$

and similarly for

$$\mathfrak{L}\left(\frac{\partial \eta_V}{\partial x_i}\right)_H^\sim \omega .$$

Moreover, if we set

$$\left(\frac{\partial \xi_V}{\partial x_i}\right)_H^{\sim} = A_\ell \frac{\partial}{\partial x_\ell}$$

with $A_\ell \in i(\mathcal{O})$, then

$$\sum_j \left\langle \left(\frac{\partial \xi_V}{\partial x_i}\right)_H^{\sim}, dx_j \right\rangle \frac{\partial \eta_V}{\partial x_j} = \sum_j A_j \frac{\partial \eta_V}{\partial x_j}$$

and

$$\left[\left(\frac{\partial \xi_V}{\partial x_i}\right)_H^{\sim}, \eta_V\right] = \sum_\ell A_\ell \frac{\partial \eta_V}{\partial x_\ell}$$

since η_V is vertical. Hence the second term in (III) cancels with the second term in (VI) and similarly the second term in (IV) cancels with the second term in (V). The desired equality is now obvious.

Remark: Since $\tilde{D}u = [\![\chi_k, u]\!]$, $u \in \Lambda\underline{T}^* \otimes_{\mathcal{O}} \tilde{\mathcal{J}}_k T$, Lemma 6 is simply the Jacobi identity (in the general sense) for $\chi_k, u, v \in \Lambda\underline{T}^* \otimes_{\mathcal{O}} \check{\mathcal{J}}_k T$ with $u \in \Lambda^r\underline{T}^* \otimes_{\mathcal{O}} \tilde{\mathcal{J}}_k T$ and $v \in \Lambda^s\underline{T}^* \otimes_{\mathcal{O}} \tilde{\mathcal{J}}_k T$. In Section 14 we already remarked that the Jacobi identity does not hold in general for the bracket $[\![\ ,\ ]\!]$ defined on $\Lambda\underline{T}^* \otimes_{\mathcal{O}} \check{\mathcal{J}}_k T$.

A formula similar to that of Lemma 6 also holds for the operator D and the Nijenhuis bracket in $\Lambda\underline{T}^* \otimes_{\mathcal{O}} \mathcal{J}_k T$. This is a trivial consequence of the first part of Theorem 2 and the Jacobi identity (in the general sense) for the Nijenhuis bracket.

THEOREM 4. *The operator*

$$\tilde{D}: \Lambda\underline{T}^* \otimes_{\mathcal{O}} \tilde{\mathcal{J}}_k T \to \Lambda\underline{T}^* \otimes_{\mathcal{O}} \tilde{\mathcal{J}}_{k-1} T$$

is a derivation of degree one with respect to the Lie algebra morphism $\tilde{\rho}_{k-1}$, $k \geq 2$.

Lemma 5 implies that $[\chi_k, \chi_k] = [\![\chi_k, \chi_k]\!] = 0$. Moreover, $ad(\chi_k)\alpha = \mathcal{L}(\chi_k)\alpha = \mathcal{L}(\mathrm{Id})\alpha = d\alpha$ ($\chi_k = \pi_1^*\mathrm{Id}$) for any $\alpha \in \Lambda\underline{T}^*$ hence $ad(\chi_k) = d$.

LEMMA 7. *For any* $u \in \Lambda \underline{T}^* \otimes_{\mathcal{O}} \check{\mathcal{J}}_k T$ *we have* $[\mathit{ad}\chi_k, \mathit{ad}u] = \mathit{ad}([\![\chi_k, u]\!])$ *or, equivalently,* $[d, \mathit{ad}u] = \mathit{ad}(\tilde{D}u)$.

Proof: We assume that $u \in \Lambda^r \underline{T}^* \otimes_{\mathcal{O}} \check{\mathcal{J}}_k T$. Then

$$\mathit{ad}u = \mathcal{L}(u) + (-1)^{r+1} i(\tilde{D}u)$$

and

$$[d, \mathit{ad}u] = [d, \mathcal{L}(u)] + (-1)^{r+1}[d, i(\tilde{D}u)] = [i(\tilde{D}u), d] = \mathcal{L}(\tilde{D}u)$$

since $[d, \mathcal{L}(u)] = 0$. Furthermore,

$$\mathit{ad}(\tilde{D}u) = \mathcal{L}(\tilde{D}u) + (-1)^r i(\tilde{D}\tilde{D}u) = \mathcal{L}(\tilde{D}u)$$

since $\tilde{D}^2 = 0$.

Remark: The previous lemma completes Lemma 4 of Section 15. We infer that $\mathit{ad}[\![\ ,\ ]\!] = [\mathit{ad}\ ,\mathit{ad}\ ]$ for

a) any two elements $u, v \in \Lambda \underline{T}^* \otimes_{\mathcal{O}} \check{\mathcal{J}}_k T$,

b) any $u \in \Lambda \underline{T}^* \otimes_{\mathcal{O}} \check{\mathcal{J}}_k T$ and χ_k,

c) any two elements $u, v \in \Lambda \underline{T}^* \otimes_{\mathcal{O}} \underline{T}$ = horizontal component of $\Lambda \underline{T}^* \otimes_{\mathcal{O}} \check{\mathcal{J}}_k T$ since for these elements $\mathit{ad} = \mathcal{L}$ and $[\![\ ,\ ]\!] = [\ ,\ ]$.

The graded Lie algebra structures defined on $\Lambda \underline{T}^* \otimes_{\mathcal{O}} \tilde{\mathcal{J}}_k T$ by $[\ ,\ ]$ and $[\![\ ,\ ]\!]$ transport via ε_k to graded Lie algebra structures on $\Lambda \underline{T}^* \otimes_{\mathcal{O}} \mathcal{J}_k T$. The proof of Lemma 6 shows that if $u \in \Lambda^r \underline{T}^* \otimes_{\mathcal{O}} \mathcal{J}_k T$ and $v \in \Lambda^s \underline{T}^* \otimes_{\mathcal{O}} \mathcal{J}_k T$ are represented by $\mathfrak{U} \in \Phi_V^r$ and $\mathfrak{V} \in \Phi_V^s$ respectively, then the transported brackets are given by

$$[u, v] = [\mathfrak{U}, \mathfrak{V}] + [\mathfrak{U}, \mathfrak{V}'_H] + [\mathfrak{U}'_H, \mathfrak{V}] \ \mathit{mod}\ \mathcal{J}^{k+1}\Phi_V$$

and

$$[\![u, v]\!] = [u, v] + (-1)^{r+1}(Du)'_H \barwedge v + (-1)^{rs+s}(Dv)'_H \barwedge u$$

where $\mathfrak{U}'_H = (\varepsilon^{-1}\mathfrak{U})_H \in \Phi_{PH}^r$ and $(Du)'_H = (\tilde{D}u)_H \in \Lambda^{r+1}\underline{T}^* \otimes_{\mathcal{O}} \underline{T}$ = horizontal component of $\Lambda^{r+1}\underline{T}^* \otimes_{\mathcal{O}} \check{\mathcal{J}}_{k-1} T$. Since $\tilde{D}$ transports into D via ε_k we also have the formula

$$D[\![u, v]\!] = [\![Du, \rho_{k-1}v]\!] + (-1)^r[\![\rho_{k-1}u, Dv]\!].$$

CHAPTER IV

NON-LINEAR COMPLEXES

17. *Non-linear jet sheaves*

Let X be a manifold. Denote by $\mathcal{C}^\infty X$ the sheaf of germs of local C^∞-maps of X and by $\mathcal{A}ut\, X$ the sheaf of germs of local C^∞-diffeomorphisms of X. Let $J_k X$ be the manifold of k-jets of local C^∞-maps $f: U \to X$, U open in X. The *source* map

$$\alpha_k: J_k X \to X, \quad j_k f(x) \mapsto x ,$$

and the *target* map

$$\beta_k: J_k X \to X, \quad j_k f(x) \mapsto f(x) ,$$

are submersions onto X. $J_k X$ has a natural structure of small differentiable category, the composition being defined by

$$j_k g(y) \cdot j_k f(x) = j_k (g \circ f)(x)$$

with $y = f(x)$. The units are the elements $j_k \mathrm{Id}(x)$ which can be identified with the points of X. Let $\Pi_k X$ be the groupoid of invertible elements in the category $J_k X$. The elements of $\Pi_k X$ are the k-jets of local diffeomorphisms of X. Denote by $\underline{J_k X}$ (resp. $\underline{\Pi_k X}$) the sheaf of germs of local sections of $\alpha_k: J_k X \to X$ (resp. $\alpha_k: \Pi_k X \to X$). The injective sheaf map

$$j_k: \mathcal{C}^\infty X \to \underline{J_k X}, \quad [f]_a \mapsto [x \mapsto j_k f(x)]_a$$

restricts to a sheaf map

$$j_k: \mathcal{A}ut\, X \to \underline{\Pi_k X} .$$

An element $A \in \underline{\Pi_k X}$ will be called *admissible* if $\beta_k \circ A \in \mathcal{A}ut\ X$. The admissible elements are the germs of sections σ of $\alpha_k\colon \Pi_k X \to X$ such that $\beta_k \circ \sigma$ is a local diffeomorphism of X. Denote by $\Gamma_k X$ the subsheaf of admissible elements of $\underline{\Pi_k X}$. The map

$$j_k\colon \mathcal{A}ut\ X \to \underline{\Pi_k X}$$

actually takes values in $\Gamma_k X$, hence defines a sheaf map

$$j_k\colon \mathcal{A}ut\ X \to \Gamma_k X\ .$$

The category structure of $J_k X$ extends to $\underline{J_k X}$ in the following way: Two elements $[\tau]_b, [\sigma]_a \in \underline{J_k X}$ are composable if and only if $b = \beta_k \circ \sigma(a)$ and their composition is then given by

$$[\tau]_b \bullet [\sigma]_a = [x \mapsto \tau(\beta_k \circ \sigma(x)) \bullet \sigma(x)]_a$$

where $\tau(\ ) \bullet \sigma(\ )$ denotes the composition in the category $J_k X$. The units are the elements $[j_k \mathrm{Id}]_x$ which can be identified with the points of X. The *source* (right unit) and the *target* (left unit) maps in the category $\underline{J_k X}$ are given respectively by

$$a_k\colon \underline{J_k X} \to X,\ [\sigma]_x \mapsto x\ ,$$

and

$$b_k\colon \underline{J_k X} \to X,\ [\sigma]_x \mapsto \beta_k \circ \sigma(x)\ .$$

The map a_k is simply the *source* map for germs and b_k is the composite of β_k followed by the *target* map for germs. With this category structure on $\underline{J_k X}$ and the obvious category structure on $\mathcal{C}^{\infty} X$, the map $j_k\colon \mathcal{C}^{\infty} X \to \underline{J_k X}$ becomes a functor (morphism of categories). Furthermore, $\underline{\Pi_k X}$ is a subcategory of $\underline{J_k X}$ whose elements are not all invertible. In fact, the groupoid of invertible elements of $\underline{\Pi_k X}$ or $\underline{J_k X}$ is precisely $\Gamma_k X$. The map $j_k\colon \mathcal{A}ut\ X \to \Gamma_k X$ becomes an injective sheaf morphism of groupoids.

Having made this brief review of some classical notions introduced by Ehresmann, we shall proceed to an alternate description of the above sheaves. Following Malgrange, we shall make the description in terms of maps in X^2 in such a way that the infinitesimal counterpart will be precisely the description of the linear jet sheaves in terms of vertical or diagonal vector fields. Though not strictly indispensable, we shall also discuss the *vertical* approach in order to stress the advantages of the *diagonal* one. Operations requiring $k+1$ derivatives will be compensated by the diagonal procedure so as to need only k derivatives (e.g., the definitions of $[\ ,\ ]$ and $[\![\ ,\ ]\!]$).

a) *Vertical approach*

A local map $F\colon \mathcal{U} \to X^2$, $\mathcal{U}$ open in X^2, will be called *vertical* if $\pi_1 \circ F = \pi_1$. In terms of components, such a map reads $F\colon (x,y) \mapsto (x, \varphi(x,y))$. A local map $f\colon U \to X$, U open in X, lifts to a vertical map

$$jf\colon X \times U \to X^2, \quad (x,y) \mapsto (x, f(y)) .$$

Let $\mathcal{C}_V^\infty X^2$ be the sheaf of germs of vertical maps and denote by $[F]_{(x,y)}$ the germ of F at the point (x,y). Define the equivalence relation $\sim_k$ on $\mathcal{C}_V^\infty X^2$ by

i) $x = y$:

$[F]_{(x,x)} \sim_k [G]_{(x,x)}$ if and only if $j_k F|_\Delta$ and $j_k G|_\Delta$ define the same germ at (x,x) ($j_k F$ is a local map $\mathcal{U} \to J_k X^2$). In other words, the partial derivatives (with respect to some coordinate systems) of F and G agree to order $\leq k$ in a neighborhood of (x,x) in Δ.

ii) $x \neq y$:

$[F]_{(x,y)} \sim_k [G]_{(x,y)}$, $\forall F, G$.

The quotient sheaf $\mathcal{C}_V^\infty X^2/\sim_k$ is trivial outside the diagonal (each stalk has only one element) hence will be considered as a sheaf on Δ. Let $[F]_{(a,a)} \in \mathcal{C}_V^\infty X^2$ and define $F_x\colon y \mapsto F(x,y)$ for any x near a. Then F_x is a local map of X and the sheaf map

$$\mathcal{C}_V^\infty X^2|\Delta \to \underline{J_k X}, \quad [F]_{(a,a)} \mapsto [x \mapsto j_k F_x(x)]_a$$

factors to a sheaf isomorphism

$$\mathcal{C}^\infty_V X^2/\sim_k \to \underline{J_k X} .$$

The map $f \to jf$ induces the injective sheaf map

$$j_k : \mathcal{C}^\infty X \to \mathcal{C}^\infty_V X/\sim_k$$

which identifies with the j_k defined at the outset. If we restrict $\mathcal{C}^\infty_V X^2$ to $\mathcal{A}ut_V X^2$ = sheaf of germs of local vertical diffeomorphisms, then the above sheaf isomorphism restricts to the isomorphism

$$\mathcal{A}ut_V X^2/\sim_k \to \underline{\Pi_k X} .$$

We remark that local one parameter groups of vertical transformations are generated by vertical vector fields (sections of T_V).

b) *Diagonal approach*

We shall now consider local maps of X^2 which preserve the diagonal. As in Section 10, in order to obtain $\underline{J_k X}$ we shall have to consider a proper subsheaf of the sheaf of (germs of) diagonal transformations.

A local map $F: U \to X^2$, U open in X^2, will be called *admissible* if

i) F preserves the diagonal, i.e., $F(\Delta \cap U) \subset \Delta$

ii) F is π_1-projectable, i.e., there exists $f: U \to X$ such that $\pi_1 \circ F = f \circ \pi_1$.

In terms of components, F is admissible if and only if

$$F: (x, y) \mapsto (f(x), \varphi(x, y)) \quad \text{with} \quad f(x) = \varphi(x, x)$$

whenever $\varphi(x, x)$ is defined. Germs of admissible maps at the points of Δ are in one to one correspondence with germs of vertical maps. The admissible map $F = (f, \varphi)$ is transformed into a vertical map $(x, y) \mapsto (x, \varphi(x, y))$ and the vertical map $G = (\mathrm{Id}, \varphi)$ is transformed into the admissible map $(x, y) \to (\varphi(x, x), \varphi(x, y))$. However, this correspondence does not preserve (germs of) diffeomorphisms. If the admissible map F

is a local diffeomorphism then the corresponding vertical map G is also a local diffeomorphism. The converse is not always true as shown by the following example: In R^2, the map $(x, y) \mapsto (x, a+y-x)$ is a vertical diffeomorphism whereas the admissible image $(x, y) \mapsto (a, a+y-x)$ is constant on Δ. One parameter local groups of admissible transformations are generated by admissible vector fields (sections of $\tilde{\Theta}$). Let $F = (f, \varphi)$ and $G = (g, \psi)$ be two local admissible maps which are composable. Then $G \circ F = (g \circ f, \psi(f, \varphi))$ is again admissible but its vertical correspondent $(x, y) \mapsto (x, \psi(f(x), \varphi(x, y))$ is different from the composite of the vertical correspondents of F and G. This shows in particular that one parameter local groups of admissible (resp. vertical) transformations are not transformed into one parameter local groups of vertical (resp. admissible) transformations. Nevertheless, if (F_t) is a one parameter local family of admissible maps with $F_0 = \mathrm{Id}$ and

$$\xi = (dF_t/dt)_{t=0}$$

is the corresponding admissible vector field then

$$\varepsilon\xi = (dG_t/dt)_{t=0}$$

where (G_t) is the one parameter local family of vertical maps corresponding to (F_t). Hence the map $F \mapsto G$, with G the vertical correspondent of F, is a finite form of the map $\varepsilon: \tilde{\Theta} \to \Theta_V$. Its inverse is a finite form of ε^{-1}.

Let $\tilde{\mathcal{C}}^\infty X^2$ be the sheaf of germs of admissible maps. The quotient sheaf $\tilde{\mathcal{C}}^\infty X^2/\sim_k$ will be considered as a sheaf on Δ (the equivalence relation $\sim_k$ being defined in the same way as for vertical maps). Using the maps F_x one establishes a sheaf isomorphism

$$\tilde{\mathcal{C}}^\infty X^2/\sim_k \to \underline{J_k X}$$

and, transforming admissible maps into vertical maps, one establishes a sheaf isomorphism

$$\tilde{\mathcal{C}}^\infty X^2/\sim_k \to \mathcal{C}^\infty_V X^2/\sim_k .$$

The following diagram is commutative:

$$\begin{array}{ccc} \tilde{\mathcal{C}}^\infty X^2/\sim_k & \xrightarrow{\ \Lambda\ } & \mathcal{C}^\infty_V X^2/\sim_k \\ & {}_{\Lambda_1}\searrow \quad \swarrow_{\Lambda_2} & \\ & \underline{J_k X} & \end{array}$$

Let $\widetilde{\mathcal{A}ut}\, X^2$ be the sheaf of germs of admissible local diffeomorphisms, $\widetilde{\mathcal{A}ut}_V X^2$ the sheaf of germs of admissible local maps $F\colon U \to X^2$ such that, for each $(x, x) \in U \cap \Delta$, the germ $[y \mapsto F(x, y)]_x$ is invertible (as a local map of X) and finally $\overline{\mathcal{A}ut}_V X^2$ the sheaf of germs of local vertical diffeomorphisms F such that $x \mapsto \pi_2 \circ F(x, x)$ is a local diffeomorphism. All the above subsheaves are saturated for the relation $\sim_k$. Moreover, Λ restricts to the sheaf isomorphisms

$$\widetilde{\mathcal{A}ut}\ {}_V X^2/\sim_k \to \mathcal{A}ut\ {}_V X^2/\sim_k \quad \text{and} \quad \widetilde{\mathcal{A}ut}\ X_2/\sim_k \to \overline{\mathcal{A}ut}\ {}_V X^2/\sim_k\,,$$

Λ_1 restricts to

$$\widetilde{\mathcal{A}ut}\ {}_V X^2/\sim_k \to \underline{\Pi_k X} \quad \text{and} \quad \widetilde{\mathcal{A}ut}\ X^2/\sim_k \to \Gamma_k X$$

and Λ_2 restricts to

$$\mathcal{A}ut\ {}_V X^2/\sim_k \to \underline{\Pi_k X} \quad \text{and} \quad \overline{\mathcal{A}ut}\ {}_V X^2/\sim_k \to \Gamma_k X\,.$$

$\tilde{\mathcal{C}}^\infty X^2$ carries a natural structure of category, namely the composition of (germs of) admissible maps. This structure is compatible with $\sim_k$ hence induces a category structure on the quotient sheaf $\tilde{\mathcal{C}}^\infty X^2/\sim_k$ for which Λ_1 becomes a category isomorphism. The sheaf $\widetilde{\mathcal{A}ut}\ {}_V X^2/\sim_k$ is a subcategory which by Λ_1 is isomorphic to $\underline{\Pi_k X}$. Moreover, $\widetilde{\mathcal{A}ut}\ X^2/\sim_k$ is the groupoid of invertible elements of the previous quotient categories and Λ_1 transforms it isomorphically onto $\Gamma_k X$.

Corresponding to

$$j_k\colon \mathcal{C}^\infty X \to \underline{J_k X}$$

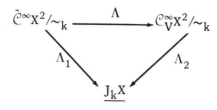

we have the injective sheaf maps

$$j_k: \mathcal{C}^\infty X \to \mathcal{C}^\infty_V X^2/\sim_k \quad \text{induced by} \quad j: f \in \mathcal{C}^\infty X \mapsto \mathrm{Id} \times f \in \mathcal{C}^\infty_V X^2$$

and

$$\tilde{j}_k: \mathcal{C}^\infty X \to \tilde{\mathcal{C}}^\infty X^2/\sim_k \quad \text{induced by} \quad \tilde{j}: f \in \mathcal{C}^\infty X \mapsto f \times f \in \tilde{\mathcal{C}}^\infty X^2 .$$

The map $\tilde{j}_k$ is a category morphism which restricts to the groupoid morphism

$$\tilde{j}_k: \mathcal{A}ut\, X \to \widetilde{\mathcal{A}ut}\, X^2/\sim_k .$$

The previous notations being rather awkward, we shall only retain $\underline{J_k X}$, $\underline{\Pi_k X}$ and $\Gamma_k X$. It will be clear from the context whether they represent germs of sections of α_k or quotients of vertical or admissible maps.

18. *Coordinates*

Let $F: U \to X^2$ be an admissible mapping defined in an open neighborhood U of the point $(a, a) \in X^2$ and let $G: U \to X^2$ be the corresponding vertical map. We can write

$$F: (x, y) \mapsto (f(x), \varphi(x, y))$$

and

$$G: (x, y) \mapsto (x, \varphi(x, y))$$

with $f(x) = \varphi(x, x)$. Let $(x_i, y_i) = (x_1, \dots, x_n, y_1, \dots, y_n)$ be a system of local coordinates defined in the neighborhood of (a, a) as in Section 5. Let (x'_i, y'_i) be a similar system of local coordinates defined in the neighborhood of $(b, b) = F(a, a)$ and (x_i, y'_i) a third system defined in the neighborhood of $G(a, a) = (a, b)$. F has the local expression

$$(x_i, y_i) \mapsto (f_i(x), \varphi_i(x, y))$$

and G has the local expression

$$(x_i, y_i) \mapsto (x_i, \varphi_i(x, y)) .$$

Taylor's expansion to order k with respect to y and along the diagonal gives for the components of F

$$(18.1)\qquad \varphi_i(x, y) \equiv \sum_{|\alpha|\leq k} a_{i,\alpha}(x) \frac{(y-x)^\alpha}{\alpha!} \; mod \; \mathcal{J}^{k+1}$$

$$(18.2)\qquad f_i(x) \equiv a_{i,0}(x) \; mod \; \mathcal{J}^{k+1}$$

with $a_{i,0} = f_i$ and $a_{i,\alpha} = D_y^\alpha \varphi_i(x, x)$. The relation (18.1) also gives the Taylor expansion for the components of the vertical map G. It follows that any element in $\tilde{\mathcal{C}}^\infty X^2/\sim_k$ or $\mathcal{C}_V^\infty X^2/\sim_k$ (i.e. in $\underline{J_k X}$) with source a is uniquely determined, with respect to given coordinate systems, by the germs at the point a of polynomial functions

$$\sum_{|\alpha|\leq k} a_{i,\alpha}(x) \frac{(y-x)^\alpha}{\alpha!} .$$

The element belongs to $\underline{\Pi_k X}$ if and only if the matrix $(a_{i,j}(a))$ is invertible where $j = (0, \dots, 1, \dots, 0)$ with 1 at the j-th position. The element belongs to $\Gamma_k X$ if further, the matrix

$$\left(\frac{\partial f_i}{\partial x_j}(a)\right)$$

is also invertible.

Assume now that F is an admissible transformation (local diffeomorphism) and let $F^{-1} = (f^{-1}, \psi)$ be its inverse. The relation

$$\psi(f(x), \varphi(x, y)) = y$$

yields by differentiation

$$(18.3)\qquad \sum_k \left(\frac{\partial \psi_i}{\partial x'_k} \circ F\right) \frac{\partial f_k}{\partial x_j} + \left(\frac{\partial \psi_i}{\partial y'_k} \circ F\right) \frac{\partial \varphi_k}{\partial x_j} = 0$$

$$(18.4)\qquad \sum_k \left(\frac{\partial \psi_i}{\partial y'_k} \circ F\right) \frac{\partial \varphi_k}{\partial y_j} = \delta^i_j \quad \text{or} \quad \left(\frac{\partial \psi}{\partial y'} \circ F\right) = \left(\frac{\partial \varphi}{\partial y}\right)^{-1} .$$

19. *The Lie algebra of the groupoid* $\Gamma_k X$

Let $\Gamma(X, \tilde{\mathcal{J}}_k T)$ be the set of global sections of $\tilde{\mathcal{J}}_k T$ and $\Gamma_c(X, \tilde{\mathcal{J}}_k T)$ the subset of those sections Ξ such that the support of $\tilde{\beta}_k \circ \Xi \in \mathfrak{X}(X)$ is compact. The bracket $[\![\ ,\ ]\!]$ on $\tilde{\mathcal{J}}_k T$ extends to $\Gamma(X, \tilde{\mathcal{J}}_k T)$ and defines a structure of Lie algebra. Let $\Gamma_a(X, \Gamma_k X)$ be the set of global sections F of $\Gamma_k X$ which are admissible i.e., $b_k \circ F$ is an automorphism of X with $b_k: \Gamma_k X \to X$ the target map. Then $\Gamma_a(X, \Gamma_k X)$ identifies with the set of global sections σ of $a_k: \Pi_k X \to X$ for which $\beta_k \circ \sigma$ is an automorphism of X (i.e., σ is admissible). The groupoid structure of $\Gamma_k X$ extends to a group structure on $\Gamma_a(X, \Gamma_k X)$ by setting

$$(F \bullet G)(x) = F(b_k \circ G(x)) \bullet G(x)$$

the right hand product being the composition in $\Gamma_k X$ and $F, G \in \Gamma_a(X, \Gamma_k X)$. If we identify F and G with admissible sections σ and τ of a_k then $F \bullet G$ identifies with the section

$$(\sigma \bullet \tau)(x) = \sigma(\beta_k \circ \tau(x)) \bullet \tau(x)$$

the right hand product being the composition in $\Pi_k X$. Denote by $\exp t\theta$ the 1-parameter local (or global) group of transformations generated by the vector field θ and let $\mathfrak{X}_c(X)$ be the set of vector fields with compact support.

We will show that the sheaf $\tilde{\mathcal{J}}_k T$ can be regarded as the *Lie algebra* of $\Gamma_k X$ in the same sense as $\underline{T}$ is considered the Lie algebra of $\mathcal{A}ut\ X$. That is, the following properties hold:

a) Let (F_t), $|t| < \varepsilon$ be a 1-parameter family of sections of $\Gamma_k X$ depending differentiably on the real parameter t and reducing to the identity section $j_k \mathrm{Id}$ at $t = 0$; then

$$\frac{d}{dt} F_t \Big|_{t=0} \in \Gamma(X, \tilde{\mathcal{J}}_k T)$$

b) There exists a differentiable map

$$\mathrm{Exp}: \Gamma_c(X, \tilde{\mathcal{J}}_k T) \to \Gamma_a(X, \Gamma_k X)$$

which satisfies the following properties:

1) For each $\Xi \in \Gamma_c(X, \tilde{\mathfrak{J}}_k T)$, the map

$$t \in \mathbf{R} \mapsto \operatorname{Exp} t\Xi \in \Gamma_a(X, \Gamma_k X)$$

is a 1-parameter subgroup of $\Gamma_a(X, \Gamma_k X)$ generated by Ξ, that is, we have the relations

$$\begin{aligned}
&\operatorname{Exp}(t+t')\Xi = (\operatorname{Exp} t\Xi)\bullet(\operatorname{Exp} t'\Xi)\\
&\operatorname{Exp} 0 = \tilde{j}_k \mathrm{Id}\\
&\operatorname{Exp}\text{-}t\Xi = (\operatorname{Exp} t\Xi)^{-1}\\
&\frac{d}{dt}\operatorname{Exp} t\Xi|_{t=0} = \Xi
\end{aligned}$$

2) $$\rho_h \circ \operatorname{Exp} t\Xi = \operatorname{Exp} t(\tilde{\rho}_h \circ \Xi),\ h \le k,\ \Xi \in \Gamma_c(X, \tilde{\mathfrak{J}}_k T)\ ,$$

$$\operatorname{Exp} t\, \tilde{j}_k\theta = \tilde{j}_k \exp t\theta,\ \theta \in \mathfrak{X}_c(X)\ .$$

In particular

$$b_k \circ \operatorname{Exp} t\Xi = \exp t(\beta_k \circ \Xi)\ .$$

3) The map Exp is uniquely determined by the property (1).

Let us start by proving (a). Since

$$\Gamma_k X \subset \underline{\Pi_k} X\ ,$$

then a section $F \in \Gamma(X, \Gamma_k X)$ identifies with a section σ of $\alpha_k\colon \Pi_k X \to X$ for which $\beta_k \circ \sigma\colon X \to X$ is locally a diffeomorphism (a diffeomorphism in the neighborhood of each point). The 1-parameter family (F_t), $|t| < \varepsilon$, is said to depend differentiably on t if the map $(t, x) \in]-\varepsilon, \varepsilon[\times X \mapsto \sigma_t(x) \in \Pi_k X$ is differentiable. If M is a manifold and (F_m) is a family of sections in $\Gamma(X, \Gamma_k X)$ parametrized by $m \in M$ then the differentiable dependence on m is defined similarly.

Let (x_i, y_i) be a coordinate system around the point $(a, a) \in \Delta$. Each F_t is locally represented by the functions $a_{i,\alpha,t}$, $|\alpha| \le k$, given by the formulas (18.1) and the differentiability condition for the family F_t is equivalent to the differentiability of

$$(t, x) \in]-\nu, \nu[\times U \longmapsto a_{i,\alpha,t}(x) \in \mathbf{R}$$

where the constant $\nu \leq \varepsilon$ and the neighborhood U of a in X are chosen sufficiently small.

We shall now lift F_t, locally, onto X^2. More precisely, there exists an open neighborhood $\mathcal{U}$ of (a, a) and a differentiable 1-parameter family (F_t), $|t| < \nu$, of admissible transformations defined on $\mathcal{U}$ such that $\mathrm{F}_0 = \mathrm{Id}$ and each F_t represents F_t in the neighborhood $\mathcal{U} \cap \Delta$ of (a, a). Such a family can be constructed using for example the polynomials in (18.3). Then

$$\xi = \frac{d}{dt} \mathrm{F}_t|_{t=0}$$

is an admissible local vector field defined on $\mathcal{U}$ and its class $\mathit{mod}\ \mathfrak{J}^{k+1}\Delta\tilde{\Theta}$ at each point $(x, x) \in \Delta$ depends only on

$$\frac{d}{dt}(D_y^\alpha \varphi_{i,t}(x, x))_{t=0} = \frac{d}{dt} a_{i,\alpha,t}(x)|_{t=0}, \ |\alpha| \leq k ,$$

(partial derivatives with respect to t and y_i commute) hence is uniquely determined by the family (F_t). The local sections

$$\xi \ \mathit{mod}\ \mathfrak{J}^{k+1}\Delta\tilde{\Theta} \in \Gamma(U, \tilde{\mathfrak{J}}_k T)$$

thus constructed agree in their common intersection hence define a global section which by definition is

$$\frac{d}{dt} F_t|_{t=0} .$$

The above argument shows also that any differentiable 1-parameter family $F_t \in \Gamma(U, \Gamma_k X)$ with $F_0 = \tilde{j}_k \mathrm{Id}|_U$, defines the section

$$\frac{d}{dt} F_t|_{t=0} \in \Gamma(U, \tilde{\mathfrak{J}}_k T) .$$

The proof of (b) follows essentially the same steps though the local uniqueness will require a somewhat more subtle argument. Recall first that $\tilde{\mathcal{J}}_k T$ is a locally free sheaf of finite rank hence can be identified with the sheaf of germs of sections of a vector bundle E on X (for example $J_k T$). A family (Ξ_m), $m \in M$, of sections in $\Gamma(X, \tilde{\mathcal{J}}_k T)$ is said to depend differentiably on m if the map

$$(m, x) \in M \times X \longmapsto \sigma_m(x) \in E$$

is differentiable where σ_m is the section of E which identifies with Ξ_m. The map Exp is differentiable in the sense that it transforms differentiable families of sections in $\Gamma_c(X, \tilde{\mathcal{J}}_k T)$ into differentiable families of sections in $\Gamma_a(X, \Gamma_k X)$.

Let $\Xi \in \Gamma(X, \tilde{\mathcal{J}}_k T)$. For each point $(a, a) \in \Delta$ there is an open neighborhood U of (a, a) in X^2 and an admissible vector field ξ defined on U such that

$$\Xi|_{U \cap \Delta} = \xi \mod \mathcal{J}^{k+1}\Delta\tilde{\Theta} .$$

Let (F_t) be the local 1-parameter group of admissible transformations generated by ξ. There is an open neighborhood V of (a, a) and $\varepsilon > 0$ such that F_t is defined on V for $|t| < \varepsilon$. We want, essentially, to set

$$\operatorname{Exp} t\Xi|_{V \cap \Delta} = F_t \mod \sim_k .$$

With that in mind, we must first prove that $F_t \mod \sim_k$ depends only on $\Xi|_{U \cap \Delta}$.

Let ξ and η be two admissible vector fields defined on U and assume that

$$\xi \equiv \eta \mod \mathcal{J}^{k+1}\Delta\tilde{\Theta} .$$

Let (F_t) and (G_t) be the local 1-parameter groups of admissible transformations generated by ξ and η respectively. Assume further that, on the open set V, F_t is defined for $|t| < \varepsilon$ and G_t is defined for $|t| < \nu$.

Let $\mu = min\{\varepsilon, \nu\}$. We shall prove that $F_t \sim_k G_t$ for $|t| < \mu$. Write $F_t = (f_t, \varphi_t)$ and $G_t = (g_t, \psi_t)$. Since $\xi \equiv \eta \ mod \ \mathcal{J}^{k+1}\Delta\tilde{\Theta}$, then

$$\xi|_\Delta = \eta|_\Delta ,$$

hence integration of this vector field on Δ yields

$$F_t(x, x) = G_t(x, x)$$

for any t and (x, x) or equivalently $f_t = g_t$. Let (x_i, y_i) be a coordinate system defined in a neighborhood of a point $(a, a) \in \Delta$. Restricting, if necessary, the open set V and the constant μ in such a way that V, $F_t(V)$ and $G_t(V)$ ($|t| < \mu$) are contained in the coordinate system, we shall first prove that

$$D_y^\alpha \varphi_t^i(x, x) = D_y^\alpha \psi_t^i(x, x)$$

for any $(x, x) \in V \cap \Delta$, $|t| < \mu$ and $0 < |\alpha| \leq k$ where the φ_t^i and ψ_t^i are the components of φ_t and ψ_t in the given coordinate system. Let $(x, x) \in V \cap \Delta$ be fixed and write

$$\varphi^{i,\alpha}(t) = D_y^\alpha \varphi_t^i(x, x), \ \psi^{i,\alpha}(t) = D_y^\alpha \psi_t^i(x, x) .$$

Differentiation with respect to t yields

$$\frac{d}{dt}\varphi^{i,\alpha}(t) = D_y^\alpha\left(\frac{d\varphi_t^i}{dt}\right)(t, x, x) .$$

Remark now that

$$\frac{dF_t}{dt} = \xi \circ F_t ,$$

hence

$$\frac{d\varphi_t^i}{dt} = \xi^i \circ F_t$$

where ξ^i is the component of ξ in the direction $\frac{\partial}{\partial y_i}$. It follows that

$$D_y^\alpha\left(\frac{d\varphi_t^i}{dt}\right)(t, x, x)$$

is equal to a polynomial $P^{i,\alpha}$ in the variables $\varphi^{j,\beta}(t)$, $1 \le j \le n$, $0 < |\beta| \le |\alpha|$, whose coefficients are polynomials in

$$D_y^\gamma \xi^i(f_t(x), f_t(x))$$

with integer coefficients. Since these coefficients only depend on the class of ξ *mod* $\mathcal{J}^{k+1}\Delta\tilde{\Theta}$ it follows that the functions $(\varphi^{i,\alpha})$ and $(\psi^{i,\alpha})$ are, for $|t| < \mu$, solutions of the ordinary system

$$\text{(19.1)} \qquad \frac{d}{dt}\varphi^{i,\alpha}(t) = P^{i,\alpha}(\varphi^{j,\beta}(t)) \quad 1 \le i, j \le n, \quad 0 < |\beta| \le |\alpha| \le k$$

with the same initial data

$$\varphi^{i,\alpha}(0) = \psi^{i,\alpha}(0) = D_y^\alpha y^i(x, x), \; (F_0 = G_0 = \mathrm{Id})$$

hence

$$\varphi^{i,\alpha} = \psi^{i,\alpha} \text{ for } |t| < \mu .$$

Let us now prove that $F_t \sim_k G_t$ for $|t| < \mu = \min\{\varepsilon, \nu\}$. The previous discussion shows that for every point $(a, a) \in V$ there is a neighborhood V_a in Δ and a constant $\mu_a > 0$ such that $j_k F_t(x, x) = j_k G_t(x, x)$ for any $(x, x) \in V_a$ and $|t| < \mu_a$. Let $(a, a) \in V$ be given and let $\gamma(t)$ be the integral curve of ξ with initial data $\gamma(0) = (a, a)$. Since $\xi|_\Delta = \eta|_\Delta$, then $\gamma(t)$ is also an integral curve of η with initial data (a, a). If $|\mu_1| < \mu$ then $C = \gamma([0, \mu_1])$ (or $\gamma([\mu_1, 0])$) is a compact subset of Δ and can be covered by a finite number of pairs (U_i, ε_i) such that each U_i is open in Δ and $j_k F_t(x, x) = j_k G_t(x, x)$ for $x \in U_i$ and $|t| < \varepsilon_i$. Let $\varepsilon = \min \varepsilon_i$ and $n \in N$ such that $|\mu_1| < n\varepsilon$. Then

$$j_k F_{\mu_1}(a, a) = j_k F_u(a_{n-1}, a_{n-1}) \bullet \dots \bullet j_k F_u(a_1, a_1) \bullet j_k F_u(a_0, a_0)$$

where $u = \frac{1}{n}\mu_1$, $(a_0, a_0) = (a, a)$ and $(a_i, a_i) = \gamma(iu)$. A similar formula holds for $j_k G_{\mu_1}(a, a)$, hence

$$j_k F_{\mu_1}(a,a) = j_k G_{\mu_1}(a,a)$$

for any $(a,a) \in V$ and $|\mu_1| < \mu$. It follows that $F_t \sim_k G_t$ for $|t| < \mu$.

Returning to the section Ξ, let us write $U = \Delta^{-1}(V)$ and $\mathbf{F}_t = F_t \ \mathit{mod} \sim_k$. We have shown that for every $a \in X$ there is an open neighborhood U of a, an $\varepsilon > 0$ (depending on U) and a canonically defined differentiable 1-parameter family $\mathbf{F}_t \in \Gamma_a(U, \Gamma_k X)$, $|t| < \varepsilon$, such that:

i) Two families coincide on the overlap of their neighborhoods U and U′ and for common values of t.

ii) $\frac{d}{dt}\mathbf{F}_t|_{t=0} = \Xi$.

iii) $\mathbf{F}_t$ is a local 1-parameter subgroup of the groupoid $\Gamma_{a,\,loc}(X, \Gamma_k X)$ of local admissible sections of $\Gamma_k X$. That is, we have the properties

1) $\mathbf{F}_0 = \tilde{j}_k \mathrm{Id}|_U$,

2) $\mathbf{F}_{t+t'}(x) = \mathbf{F}_t(y) \bullet \mathbf{F}_{t'}(x)$, $y = b_k \circ \mathbf{F}_{t'}(x)$, whenever both members are defined,

3) $\rho_h \circ \mathbf{F}_t = \mathrm{Exp}\, t(\tilde{\rho}_h \circ \Xi)$, $h \le k$,

4) $b_k \circ \mathbf{F}_t = \exp t(\tilde{\beta}_k \circ \Xi)$.

It is clear that each $\mathbf{F}_t$ is admissible since $b_k \circ \mathbf{F}_t = f_t$ where $F_t = (f_t, \varphi_t)$. Property (ii) follows directly from the construction since

$$\frac{d}{dt}\mathbf{F}_t|_{t=0} = \xi \ \mathit{mod}\ \mathcal{J}^{k+1}\Delta\tilde{\Theta} = \Xi|_U .$$

Properties (iii) (1) and (2) are implied by corresponding properties of the local group (F_t) and (3) is obvious by definition. Finally (4) follows from the fact that

$$b_k \circ \mathbf{F}_t = f_t, \quad \tilde{\beta}_k \circ \Xi = \pi_{1*}\xi \quad \text{and} \quad f_t = \exp t\, \pi_{1*}\xi .$$

Since the 1-parameter family $(\mathbf{F}_t)$ is given by the solutions of the ordinary system (19.1) and this system is entirely determined by the conditions (ii) and (iii) (1)-(2) it follows that $(\mathbf{F}_t)$ is locally unique in the sense that two such families, satisfying the above conditions, agree in their common domain. Observe also that (19.1) could be used to prove the existence of (F_t). Let us write

$$F_t = \operatorname{Exp} t\Xi .$$

Assume now that Ξ is *uniform* in the sense that there exists an $\varepsilon > 0$ such that for any $x \in X$ there is an open neighborhood U of x on which $\operatorname{Exp} t\Xi$ is defined for $|t| < \varepsilon$. In other words we have a one parameter family of sections

$$\operatorname{Exp} t\Xi \in \Gamma(X, \Gamma_k X)$$

defined for $|t| < \varepsilon$. The standard argument, used for globalizing local uniform 1-parameter groups generated by vector fields, can be applied in the present context. Given any $t \in \mathbf{R}$, we define the global section $\operatorname{Exp} t\Xi \in \Gamma(X, \Gamma_k X)$ by

$$\operatorname{Exp} t\Xi = (\operatorname{Exp} t_1\Xi) \bullet (\operatorname{Exp} t_2\Xi) \bullet \dots \bullet (\operatorname{Exp} t_n\Xi)$$

where $\Sigma t_i = t$, $|t_i| < \varepsilon$ and the product $\bullet$ is the composition in $\Gamma(X, \Gamma_k X)$ obtained by extending the composition of $\Gamma_k X$. The property (iii) (2) assures that $\operatorname{Exp} t\Xi$ is independent of the choice of t_i's. The map

$$t \in R \longrightarrow \operatorname{Exp} t\Xi \in \Gamma(X, \Gamma_k X)$$

clearly satisfies

$$\frac{d}{dt} \operatorname{Exp} t\Xi|_{t=0} = \Xi$$

since, for small t and in the neighborhood of each point, $\operatorname{Exp} t\Xi$ reduces to F_t in (ii). The 1-parameter group properties of this map follow immediately from the definition. In particular, for a fixed t, $\operatorname{Exp} t\Xi$ is an invertible element in $\Gamma(X, \Gamma_k X)$, hence the map

$$f = b_k \circ \operatorname{Exp} t\Xi : X \to X$$

is a diffeomorphism and

$$\operatorname{Exp} t\Xi \in \Gamma_a(X, \Gamma_k X) .$$

The relation

$$b_k \circ \operatorname{Exp} t\Xi = \exp t(\tilde{\beta}_k \circ \Xi)$$

follows by iteration of (iii) (3). If $\theta \in \mathfrak{X}(X)$ is uniform then

$$\xi = \tilde{j}\theta = i\theta + j\theta \in \tilde{\mathfrak{X}}(X^2)$$

is also uniform and $\exp t\xi = \exp t\theta \times \exp t\theta$. Since $\tilde{j}_k\theta$ is represented by ξ it follows that

$$\operatorname{Exp} t\tilde{j}_k\theta = \tilde{j}_k \exp t\theta .$$

LEMMA. *Let* $\Xi \in \Gamma(X, \tilde{\mathcal{J}}_k T)$ *and* $\theta = \tilde{\beta}_k \circ \Xi \in \mathfrak{X}(X)$. *Then* Ξ *is uniform if and only if* θ *is uniform.*

Proof. If Ξ is uniform then the relation $b_k \circ \operatorname{Exp} t\Xi = \exp t\theta$ shows that θ is uniform. Assume, conversely, that θ is uniform and let $\Delta: X \to X^2$ be the diagonal inclusion. If ξ is a local admissible vector field defined on $U \subset X^2$ that represents Ξ on $U \cap \Delta$ then $\xi|_{U\cap\Delta} = \Delta_*\theta|_{U\cap\Delta}$. By assumption θ is uniform hence $\Delta_*\theta$ is also uniform as a vector field on Δ. Let $\varepsilon > 0$ be given, take a point $(a, a) \in \Delta$ and let $\gamma(t)$ be the integral curve of $\Delta_*\theta$ with initial value $\gamma(0) = (a, a)$. Consider the compact subset $C = \{\gamma(t) \mid |t| \leq \varepsilon\}$ of Δ and let (U_i, ξ_i) be a finite family of open subsets of X^2 covering C and admissible vector field ξ_i defined on U_i such that each ξ_i represents Ξ on $U_i \cap \Delta$. We can assume further that each U_i is homeomorphic to an open set in R^n. Let (λ_i) be a partition of unity of $U = \cup U_i$ subordinated to the open cover (U_i). Then $\xi = \Sigma\lambda_i\Delta\xi_i$ is an admissible vector field defined on U and

$$\Xi|_{U\cap\Delta} = \xi \ \mathit{mod}\ \mathcal{J}^{k+1}\Delta\tilde{\Theta}$$

(recall that $\tilde{\mathcal{J}}_k T$ is a diagonal $\mathcal{O}_{X^2}$-module). Since $\gamma(t)$ is also an integral curve of ξ, it follows by the continuation lemma for integral curves that there is an open neighborhood V of (a, a) in X^2 such that the integral curves of ξ with initial value in V are all defined for $|t| \leq \varepsilon$ hence the 1-parameter local group $\exp t\xi$ is defined on V for $|t| \leq \varepsilon$. This implies that $\mathrm{Exp}\, t\Xi$ is also defined on $V \cap \Delta$ for $|t| \leq \varepsilon$.

The proof of the lemma gives in fact the following more general result. Let $\Xi \in \Gamma(X, \tilde{\mathcal{J}}_k T)$, $\theta = \tilde{\beta}_k \circ \Xi$ and $a \in X$. If $]r, s[$ is the domain of the maximal integral curve γ of θ with initial value $\gamma(0) = a$ then, for any $[r_o, s_o] \subset]r, s[$, there is an admissible local vector field ξ defined on an open set $U \subset X^2$ which represents Ξ on $U \cap \Delta$ and an open neighborhood V of (a, a) in X^2 such that $\exp t\xi$ is defined on V for any $t \in [r_o, s_o]$ and $\mathrm{Exp}\, t\Xi|_{\mathrm{V}} = \exp t\xi|_V \bmod \sim_k$ where $\mathrm{V} = \Delta^{-1} V$.

If X is paracompact then, using a partition of unity along the diagonal, we can represent any $\Xi \in \Gamma(X, \tilde{\mathcal{J}}_k T)$ by an admissible vector field η defined on an open neighborhood U of Δ. Since Δ is closed in X^2, there exists a function $\lambda\colon X^2 \to R$ whose support is contained in U and which is equal to 1 in a neighborhood of Δ. The global vector field $\xi = \lambda\Delta\eta$ is admissible and

$$\Xi = \xi \bmod \mathcal{J}^{k+1}\Delta\tilde{\Theta} .$$

Any global section Ξ of $\tilde{\mathcal{J}}_k T$ on a paracompact manifold X can be represented by an admissible global vector field ξ on X^2. Moreover, Ξ is uniform if and only if the restriction $\xi|_\Delta$ is uniform as a vector field on Δ. This being the case, then for any $t_o \in R_+$ there is an open neighborhood V of Δ in X^2 such that $\exp t\xi$ is defined on V for $|t| \leq t_o$ and

$$\mathrm{Exp}\, t\Xi = \exp t\xi|_V \bmod \sim_k .$$

We remark however that ξ need not be uniform on X^2.

Let $\Xi \in \Gamma_c(X, \tilde{\mathcal{J}}_k T)$. A compactness argument shows that θ and consequently Ξ is uniform, hence defines a 1-parameter group

$$t \in R \longmapsto \mathrm{Exp}\, t\Xi \in \Gamma_a(X, \Gamma_k X) .$$

We define the map

$$\mathrm{Exp}\colon \Gamma_c(X, \tilde{\mathfrak{J}}_k T) \to \Gamma_a(X, \Gamma_k X)$$

by $\mathrm{Exp}\,\Xi = \mathrm{Exp}\,1\Xi$. If we set $t\Xi = \Xi'$ then clearly $\mathrm{Exp}\,\Xi' = \mathrm{Exp}\,t\Xi$ (the left hand term being defined using 1-parameter local groups associated with Ξ'). The properties (b) (1) and (2) of Exp have all been already proved. Property (3) is a consequence of local uniqueness and the first formula in (1). It only remains to prove differentiability. Let (Ξ_m), $m \in M$, be a differentiable family of sections in $\Gamma_c(X, \tilde{\mathfrak{J}}_k T)$. We shall prove the differentiability of

$$(t, m, x) \in R \times M \times X \longmapsto \mathrm{Exp}\, t\Xi_m(x) \in \Pi_k X$$

which, for $t = 1$, will give the differentiability of

$$(m, x) \in M \times X \mapsto \mathrm{Exp}\,\Xi_m(x) \in \Pi_k X\ .$$

Let $a \in X$ and take a local coordinate system $(x_i, y_i, U \times U)$ in the neighborhood of (a, a). Each Ξ_m can be uniquely represented by a polynomial admissible vector field (cf. Section 11)

$$\xi_m = \sum_{|\alpha| \le k} a_{\alpha,i,m}(x) \frac{(y-x)^\alpha}{\alpha!} \frac{\partial}{\partial y_i} + a_{0,i,m}(x) \frac{\partial}{\partial x_i}$$

and the functions

$$(m, x) \in M \times U \mapsto a_{\alpha,i,m}(x) \in R$$

are differentiable. By a standard theorem on ordinary differential equations with parameters, there is an $\varepsilon > 0$, an open neighborhood $\mathcal{U}$ of (a, a), an open neighborhood $\mathcal{M}$ of m_o and a differentiable mapping

$$F\colon\ ]-\varepsilon, \varepsilon[\, \times \mathcal{M} \times \mathcal{U} \to X^2$$

such that for each fixed $m \in \mathcal{M}$ the family

$$(F_{m,t}),\ |t| < \varepsilon,\ F_{m,t}\colon w \in \mathcal{U} \mapsto F(t, m, w) \in X^2$$

is equal to $(\exp t\xi_m|_U)$. Now,

$$\operatorname{Exp} t\Xi_m|_U = \exp t\xi_m|_U \bmod \sim_k$$

and $D_y^\alpha \varphi_{m,t}^i$ depends differentiably on (t, m, w) where $F_{m,t} = (f_{m,t}, \varphi_{m,t})$. Hence the map

$$(t, m, x) \in]-\varepsilon, \varepsilon[\times M \times U \mapsto \operatorname{Exp} t\Xi_m(x) \in \Pi_k X$$

is differentiable which shows that $\operatorname{Exp} t\Xi_m(x)$ is differentiable for small t or more precisely in a neighborhood of $\{0\} \times M \times X$ in $R \times M \times X$. Let $\gamma(t)$ be the integral curve of

$$\tilde{\beta}_k \circ \Xi_{m_o}$$

with initial data $\gamma(0) = a$, take $t_o \in R$ and consider $C = \gamma([0, t_o])$ (or $\gamma([t_o, 0])$). Composition of a finite number of these local differentiable families $\operatorname{Exp} t\Xi_m$ along the compact subset C yields the differentiable map

$$]t_o-\varepsilon, t_o+\varepsilon[\times M \times U \to \Pi_k X,\ (t, m, x) \to \operatorname{Exp} t\Xi_m(x)$$

with U and M suitable neighborhoods of $a \in X$ and $m_o \in M$. It follows that $\operatorname{Exp} t\Xi_m(x)$ is differentiable everywhere.

We close this section with some useful remarks. Let $\Xi \in \Gamma(X, \tilde{\mathfrak{J}}_k T)$ be any section of $\tilde{\mathfrak{J}}_k T$. From the local construction of the sections $F_t = \operatorname{Exp} t\Xi$ we infer that for any relatively compact open subset U of X there exists an $\varepsilon > 0$ (depending on U) such that $\operatorname{Exp} t\Xi$ is defined on U for any $|t| < \varepsilon$. The map

$$t \in]-\varepsilon, \varepsilon[\longmapsto \operatorname{Exp} t\Xi \in \Gamma_a(U, \Gamma_k X)$$

satisfies the local 1-parameter group properties. Moreover there exist two open sets $\mathcal{U} \supset \mathcal{V}$ of X^2 and an admissible vector field ξ defined on $\mathcal{U}$ such that $\exp t\xi$ is defined on $\mathcal{V}$ for $|t| < \varepsilon$ and

$$\operatorname{Exp} t\Xi = \exp t\xi \bmod \sim_k .$$

Let (F_t), $|t-t_o| < \varepsilon$, be a differentiable 1-parameter family of sections in $\Gamma(U, \Gamma_k X)$ and assume that F_{t_o} is admissible i.e.,

$$f = b_k \circ F_{t_o} : U \to V$$

is a diffeomorphism of U onto the open set $V = f(U)$. Local lifting of (F_t) to a differentiable 1-parameter family of local admissible maps (F_t), $|t-t_o| < \nu$, on X^2 shows that (F_t) defines canonically a section

$$\frac{d}{dt} F_t|_{t=t_o} \in \Gamma(V, \tilde{\mathcal{J}}_k T) .$$

Moreover, since F_{t_o} is invertible, if we set

$$G_u = F_{t_o+u} \bullet F_{t_o}^{-1} ,$$

then

$$\frac{d}{dt} F_t|_{t=t_o} = \frac{d}{dt} G_u|_{u=0} .$$

As an example, we have

$$\frac{d}{dt} \operatorname{Exp} t\Xi|_{t=t_o} = \frac{d}{du} \operatorname{Exp} u\Xi|_{u=0} = \Xi .$$

20. *Actions of* $\Gamma_k X$

The groupoid $\Pi_k X$ acts differentiably on the bundle $J_{k-1}T$ in the obvious way:

$$j_k f(x) \bullet j_{k-1}\theta(x) = j_{k-1}(f_*\theta)(f(x))$$

where $f_*: TX \to TX$ is the tangent map to f and $f_*\theta = f_* \circ \theta \circ f^{-1}$ is a vector field defined in the neighborhood of f(x). This action extends to germs of sections and defines a covariant (left) action of the groupoid $\Gamma_k X$ on $\underline{J_{k-1}T}$. We are bound to loose one order of jets (i.e., the action is not defined on $\underline{J_k T}$) since f_* involves first order differentiation. On the X^2 level, the sheaf $\widetilde{\mathcal{A}ut}\, X^2$ operates on $\underline{T_V} = \Theta_V$ by

$$[F]_{(x,y)} \bullet [\xi]_{(x,y)} = [F_*\xi]_{F(x,y)} .$$

The vector field $F_*\xi$ is vertical since F is π_1-projectable and the action preserves $\Theta_V|_\Delta$ since F is diagonal. This action factors to a covariant action of $\Gamma_k X$ on the sheaf

$$\Theta_V / \mathcal{J}^k \Theta_V = \mathcal{J}_{k-1} T \equiv \underline{J_{k-1} T}$$

which identifies with the previous one. Similarly, $\widetilde{\mathcal{A}ut}\, X^2$ operates on $\tilde{\Theta}$ by

$$[F]_{(x,y)} \bullet [\tilde{\xi}]_{(x,y)} = [F_*\tilde{\xi}]_{F(x,y)} .$$

The vector field $F_*\tilde{\xi}$ is clearly admissible and the action preserves $\tilde{\Theta}|_\Delta$. Furthermore, this action factors to a covariant action of $\Gamma_k X$ on

$$\tilde{\Theta} / \mathcal{J}^{k+1} \Delta \tilde{\Theta} = \tilde{\mathcal{J}}_k T \equiv \underline{J_k T} .$$

Before proving this assertion let us examine how this latter action differs from the previous one. Let

$$F: (x, y) \longmapsto (f(x), \phi(x, y))$$

be an admissible transformation, ξ a vertical vector field and $\tilde{\xi} = \varepsilon^{-1}\xi$ the corresponding admissible vector field. Then

$$\tilde{\xi} = \xi + \xi'_H = \xi + i\theta$$

with θ a vector field on X and

$$F_*\tilde{\xi} = F_*\xi + F_*\xi'_H = F_*\xi + V \circ F_*\xi'_H + i(f_*\theta)$$

where $V: TX^2 \to T_V$ is the vertical projection. The action on $\tilde{\mathcal{J}}_k T$ transports, via ε_k, to an equivalent action on $\mathcal{J}_k T$ which is induced by

$$[F]_{(x,y)} \bullet [\xi]_{(x,y)} = [F_*\xi + V \circ F_*\xi'_H]_{F(x,y)} .$$

Hence the former action is corrected by the term $V \circ F_*\xi'_H$.

To prove the above assertion, let F and G be admissible transformations such that

$$[F]_{(a,a)} \sim_k [G]_{(a,a)}$$

and let $\tilde{\xi}$ and $\tilde{\eta}$ be admissible vector fields defined in the neighborhood of (a, a) such that

$$[\tilde{\xi}]_{(a,a)} \equiv [\tilde{\eta}]_{(a,a)} \ \mathrm{mod}\ \mathcal{J}^{k+1}\Delta\tilde{\Theta} .$$

Then

$$F_*\tilde{\xi} = F_* \circ \tilde{\xi} \circ F^{-1}, \ G_*\tilde{\eta} = G_* \circ \tilde{\eta} \circ G^{-1} ,$$

hence it will suffice to prove that $j_k(F_* \circ \tilde{\xi})$ and $j_k(G_* \circ \tilde{\eta})$ agree in a neighborhood of (a, a) in Δ. Since $j_k F$ and $j_k G$ agree in a neighborhood U of (a, a) in Δ it follows (taking coordinate systems) that the (k+1)-Fréchet derivatives of F and G satisfy the relation

$$F^{(k+1)}(x, x)(\delta, v_1, \ldots, v_k) = G^{(k+1)}(x, x)(\delta, v_1, \ldots, v_k)$$

for any $(x, x) \in U$ and any vectors $\delta, v_1, \ldots, v_k \in T_{(x,x)}X^2$ with δ diagonal. Moreover, $F_* \circ \tilde{\xi} = \langle \tilde{\xi}, F_* \rangle$ hence

$$(F_* \circ \tilde{\xi})^{(k)} = i(\tilde{\xi})\, F^{(k+1)} + \text{ terms involving only } F^{(\ell)} (\ell \leq k)$$

and similarly for G. Since $\tilde{\xi}|_V = \tilde{\eta}|_V$ in some neighborhood V of (a, a) in Δ, the previous remark implies that

$$i(\tilde{\xi})\, F^{(k+1)}|_{U \cap V} = i(\tilde{\eta})\, G^{(k+1)}|_{U \cap V}$$

and clearly the remaining terms all agree on $U \cap V$ since F and G as well as $\tilde{\xi}$ and $\tilde{\eta}$ agree to order k on $U \cap V$.

We define similarly an action of $\widetilde{\mathcal{A}ut}\, X^2$ on the sheaf $\check{\Theta}$ which factors to a covariant action of $\Gamma_k X$ on

$$\check{\Theta}/\mathcal{J}^k\Delta\check{\Theta} = \check{\mathcal{J}}_{k-1}T .$$

This action restricts to the previous actions on $\mathcal{J}_{k-1}T$ and $\tilde{\mathcal{J}}_{k-1}T$ but it is not the direct sum of the latter with the obvious action on $\underline{T}$ (recalling that $\check{\mathcal{J}}_{k-1}T = \mathcal{J}_{k-1}T \oplus \underline{T} = \tilde{\mathcal{J}}_{k-1}T \oplus \underline{T}$).

21. *The first non-linear complex*

In this section we construct the *first* non-linear complex using the naive operations F^* (transposition of vector forms) and $[\ ,\ ]$ (Nijenhuis bracket of vector forms). In the next section we shall define the *second* non-linear complex where the naive operations will be replaced by twisted operations $\mathcal{A}dF$ and $[\![\ ,\ ]\!]$. Since the main interest lies in the second complex, we shall only outline some of the properties of the first one. The proofs of these properties are essentially the same as those written in Section 22 for the second complex.

Let $F\colon \mathcal{U} \to X^2$ be an admissible transformation and set

$$\mathfrak{D}F = \chi - F^*\chi \tag{21.1}$$

where $F^*\chi$ is the transpose of the vector form χ, namely

$$F^*\chi(\xi) = F_*^{-1}(\langle F_*\xi, \chi\rangle)\,.$$

$\mathfrak{D}F$ is a vector form defined on $\mathcal{U}$ and we claim that it is a horizontal form with values in T_V. Since F is π_1-projectable, F_* preserves vertical vector fields hence F^* preserves horizontal forms. Moreover, if $F = (f, \varphi)$, $F^{-1} = (f^{-1}, \psi)$ and v is any horizontal vector, then

$$\mathfrak{D}F(v) = v - F_*^{-1}[(F_*v)_H] \quad \text{and} \quad F_*v = \varphi_*v + f_*v$$

(with some evident abuse of notation) where φ_*v is the vertical component and f_*v is the horizontal component of F_*v. It follows that

$$F_*^{-1}[(F_*v)_H] = \psi_*(f_*v) + v$$

hence

$$\mathfrak{D}F(v) = -\psi_*(f_*v)$$

is vertical. We infer that

$$\mathfrak{D}F(\xi) = -\psi_* \circ (f \times f)_*(\xi) \tag{21.2}$$

for any projectable horizontal vector field ξ. If $G: U \to X^2$ is another admissible transformation and

$$[F]_{(x,x)} \sim_{k+1} [G]_{(x,x)}$$

then (21.2) shows that

$$[\mathfrak{D}F]_{(x,x)} \equiv [\mathfrak{D}G]_{(x,x)} \ \mathrm{mod}\ \mathcal{J}^{k+1}\Phi^1_V$$

hence $\mathfrak{D}$ induces a sheaf map

$$\mathfrak{D}: \Gamma_{k+1}X \longrightarrow \underline{T}^* \otimes_{\mathcal{O}} \mathcal{J}_k T \tag{21.3}$$

by setting

$$\mathfrak{D}(F) = [\mathfrak{D}F]_{(x,x)} \ \mathrm{mod}\ \mathcal{J}^{k+1}\Phi^1_V$$

where $F = [F]_{(x,x)}\ \mathrm{mod} \sim_{k+1}$.

Recall now that $\Gamma_{k+1}X$ operates on $\check{\mathcal{J}}_k T$ and it obviously operates on $\underline{T}$ by

$$F \bullet [\theta]_x = f_*[\theta]_x$$

with $f = \beta_{k+1} \circ F$ hence on $\Lambda \underline{T}^*$ by transposition. The tensor product of these actions yields a contravariant (right) action of $\Gamma_{k+1}X$ on $\Lambda \underline{T}^* \otimes_{\mathcal{O}} \check{\mathcal{J}}_k T$, namely

$$F^* \bullet (\omega \otimes A) = (F^* \bullet \omega) \otimes (F^{-1} \bullet A)\ .$$

This action is simply the quotient of the standard contravariant action of $\widetilde{\mathcal{A}ut}\, X^2$ on $\check{\Phi}$. Let χ_k be the section of $\underline{T}^* \otimes_{\mathcal{O}} \check{\mathcal{J}}_k T$ induced by χ. The formula (21.1) implies that

$$\mathfrak{D}F = \chi_k(x) - F^* \bullet \chi_k(y) \tag{21.4}$$

with $x = a_{k+1}F$ = source of F and $y = b_{k+1}F$ = target of F. If we consider

$$\mathfrak{D}F \in \mathcal{H}om_{\mathcal{O}}(\underline{T}, \mathcal{J}_k T) \simeq \underline{T}^* \otimes_{\mathcal{O}} \mathcal{J}_k T$$

and take $\xi \in \underline{T}$, then the last formula can be rewritten by

$$<\xi, \mathfrak{D}F> = \xi - F^{-1} \bullet (f_* \xi) \tag{21.5}$$

where $f = \beta_{k+1} \circ F$ and $f_* \xi$ and ξ are considered as elements of the horizontal component in the direct sum $\check{\mathcal{J}}_k T = \mathcal{J}_k T \oplus \underline{T}$.

Take coordinates (x_i, y_i) in the neighborhood of the point $(a, a) \in U$ and (x'_i, y'_i) in the neighborhood of $F(a, a)$. Then

$$\mathfrak{D}F = dx_i \otimes \frac{\partial}{\partial x_i} - F^*\left(dx'_\mu \otimes \frac{\partial}{\partial x'_\mu}\right),$$

$$F^*\left(dx'_\mu \otimes \frac{\partial}{\partial x'_\mu}\right) = df_\mu \otimes F_*^{-1} \frac{\partial}{\partial x'_\mu},$$

$$F_*^{-1} \frac{\partial}{\partial x'_\mu} = \left(\frac{\partial f_i^{-1}}{\partial x'_\mu} \circ f\right) \frac{\partial}{\partial x_i} + \left(\frac{\partial \psi_i}{\partial x'_\mu} \circ F\right) \frac{\partial}{\partial y_i},$$

$$F^*\chi = dx_i \otimes \frac{\partial}{\partial x_i} + \left(\frac{\partial \psi_i}{\partial x'_\mu} \circ F\right) \frac{\partial f_\mu}{\partial x_j} dx_j \otimes \frac{\partial}{\partial y_i}$$

hence

$$\mathfrak{D}F = -\left(\frac{\partial \psi_i}{\partial x'_\mu} \circ F\right) \frac{\partial f_\mu}{\partial x_j} dx_j \otimes \frac{\partial}{\partial y_i} = \left(\frac{\partial \psi_i}{\partial y'_\mu} \circ F\right) \frac{\partial \varphi_\mu}{\partial x_j} dx_j \otimes \frac{\partial}{\partial y_i} \tag{21.6}$$

where the second equality follows from (18.3). If $F = f \times f$ with f a local diffeomorphism of X then $F^*\chi = \chi$ and $\mathfrak{D}F = 0$. Conversely, if $\mathfrak{D}F = 0$ then (21.6) shows that

$$\left(\frac{\partial \varphi_\mu}{\partial x_j}\right)$$

is the zero matrix since

$$\left(\frac{\partial \psi_i}{\partial y'_\mu} \circ F\right)$$

is invertible. Hence $\varphi_\mu(x, y) = f_\mu(y)$ in a neighborhood of $U \cap \Delta$, i.e., $F = f \times f$ (locally). It follows that $\mathfrak{D}F = 0$ is equivalent (locally) to $F = \tilde{j}f$.

Since $[\mathcal{X},\mathcal{X}] = 0$ and $[F^*\mathcal{X}, F^*\mathcal{X}] = F^*[\mathcal{X},\mathcal{X}]$ then, by Lemma 4 of Section 16, we obtain the formula

$$d_H\mathfrak{D}F - \tfrac{1}{2}[\mathfrak{D}F, \mathfrak{D}F] = 0$$

which factors to the *structure equation*

$$\text{(21.7)} \qquad D\mathfrak{D}F - \tfrac{1}{2}[\mathfrak{D}F, \mathfrak{D}F] = 0$$

where $F \in \Gamma_{k+1}X$ and $k \geq 1$. The structure equation is vacuous for $k = 0$. We thus obtain the *first non-linear complex*

$$(21.8)_{k+1} \quad 1 \longrightarrow \mathcal{A}ut\, X \xrightarrow{\tilde{j}_{k+1}} \Gamma_{k+1}X \xrightarrow{\mathfrak{D}} \underline{T}^* \otimes_{\mathcal{O}} \mathcal{J}_k T \xrightarrow{\mathfrak{D}_1} \Lambda^2 \underline{T}^* \otimes_{\mathcal{O}} \mathcal{J}_{k-1} T$$

where 1 is the trivial subsheaf of $\mathcal{A}ut\, X$ composed of the germs of Id and

$$\mathfrak{D}_1 u = Du - \tfrac{1}{2}[u,u], \quad u \in \underline{T}^* \otimes_{\mathcal{O}} \mathcal{J}_k T\,.$$

If $\mathcal{U} \in \Phi_V^1$ is a representative of u *mod* $\mathcal{J}^{k+1}\Phi_V^1$ then $\mathfrak{D}_1 u$ is represented *mod* $\mathcal{J}^k\Phi_V^2$ by

$$\mathfrak{D}_1\mathcal{U} = d_H\mathcal{U} - \tfrac{1}{2}[\mathcal{U},\mathcal{U}]\,.$$

It is clear that $\tilde{j}_{k+1}[f]_x = [\tilde{j}_{k+1}\mathrm{Id}]_x =$ the unit of $\Gamma_{k+1}X$ of source x if and only if $[f]_x = [\mathrm{Id}]_x$, hence $\tilde{j}_{k+1}$ is injective since it is a groupoid morphism. Moreover, since $\mathfrak{D}\tilde{j}f = 0$ for any local diffeomorphism f then $\mathfrak{D} \circ \tilde{j}_{k+1} = 0$. The relation $\mathfrak{D}_1 \circ \mathfrak{D} = 0$ is the structure equation. We shall now prove the exactness of $(21.8)_{k+1}$ at $\Gamma_{k+1}X$. The above complex is not exact at $\underline{T}^* \otimes_{\mathcal{O}} \mathcal{J}_k T$. In fact, using the coordinate expression (21.6), one proves that

$$\mathfrak{D}(\Gamma_{k+1}X) = \{u \in \underline{T}^* \otimes_{\mathcal{O}} \mathcal{J}_k T \mid \mathrm{Id} + \rho_0 u \in \mathcal{A}ut_{\mathcal{O}}(\underline{T}),\ \mathfrak{D}_1 u = 0\}$$

where $\rho_0\colon \underline{T}^* \otimes_{\mathcal{O}} \mathcal{J}_k T \to \underline{T}^* \otimes_{\mathcal{O}} \mathcal{J}_0 T \equiv \underline{T}^* \otimes_{\mathcal{O}} \underline{T} = \mathcal{H}om_{\mathcal{O}}(\underline{T})$ (see also Section 24).

Assume that $\mathfrak{D}F = 0$, $F \in \Gamma_{k+1}X$, and let F be an admissible transformation whose germ at the point (a,a) represents F. Then $\mathfrak{D}F = 0$ is equivalent to

$$[\mathfrak{D}F]_{(a,a)} \in \mathcal{J}^{k+1}\Phi^1_V$$

hence, by (21.6), is equivalent to

$$\left[\left(\frac{\partial\psi_i}{\partial y'_\mu} \circ F\right) \frac{\partial\varphi_\mu}{\partial x_j}\right]_{(a,a)} \in \mathcal{J}^{k+1} .$$

Since the matrix

$$\left(\frac{\partial\psi_i}{\partial y'_\mu} \circ F\right)$$

is invertible in the neighborhood of (a,a), it follows that

$$\left.\frac{\partial\varphi_\mu}{\partial x_j}\right|_\Delta$$

is zero around (a, a). Assume by induction that

$$\left.\left(D_x^\alpha \frac{\partial\varphi_\mu}{\partial x_j}\right)\right|_\Delta$$

is zero in a neighborhood U of (a, a) for $|\alpha| \leq h$, $h < k$ and all μ and j. Then, shrinking if necessary the neighborhood U, we have

$$0 = D_x^{\alpha+\varepsilon_i}\left[\left(\frac{\partial\psi_i}{\partial y'_\mu} \circ F\right) \frac{\partial\varphi_\mu}{\partial x_j}\right]\Bigg|_U = \left(\frac{\partial\psi_i}{\partial y'_\mu} \circ F\right)\left.\left(D_x^{\alpha+\varepsilon_i} \frac{\partial\varphi_\mu}{\partial x_j}\right)\right|_U$$

which implies that

$$\left.\left(D_x^{\alpha+\varepsilon_i} \frac{\partial\varphi_\mu}{\partial x_j}\right)\right|_U = 0 .$$

We have proved that $\left.\left(D_x^\alpha\varphi_\mu\right)\right|_\Delta$ is zero in the neighborhood of (a,a) for $1 \leq |\alpha| \leq k+1$. If $F = (f, \varphi)$ then

$$[F]_{(a,a)} \sim_{k+1} [f\times f]_{(a,a)}$$

since φ and jf agree in a neighborhood of (a,a) in Δ and $D_x^\alpha jf_\mu = 0$ for all α and μ. It follows that

$$F = \tilde{j}_{k+1}[f]_a \, .$$

The previous proof could also be accomplished using the first equality in (21.6). Since the matrix $\left(\frac{\partial f_\mu}{\partial x_j}\right)$ is invertible in a neighborhood of a, the same argument as above will show that $\left(D_x^\alpha \psi_i\right)|_\Delta$ is zero in a neighborhood of $F(a,a)$, hence

$$F^{-1} = \tilde{j}_{k+1}\,[f^{-1}]_{f(a)}$$

which implies that

$$F = \tilde{j}_{k+1}\,[f]_a \, .$$

We now state some fundamental identities for the operators $\mathfrak{D}$ and $\mathfrak{D}_1$. Let F and G be two admissible transformations which are composable and let $\mathfrak{U}$ be a local horizontal form on X^2 with values in T_V whose domain contains the image of F. Then

$$\mathfrak{D}(F \circ G) = G^*\mathfrak{D}F + \mathfrak{D}G$$

and

$$d_H F^*\mathfrak{U} = F^*(d_H\mathfrak{U} - [\mathfrak{D}F^{-1}, \mathfrak{U}]) \, .$$

These relations factor to

$$\mathfrak{D}(F \bullet G) = G^* \bullet \mathfrak{D}F + \mathfrak{D}G \tag{21.9}$$

$$D(F^* \bullet u) = F^* \bullet (Du - [\mathfrak{D}F^{-1}, u]) \tag{21.10}$$

where $F, G \in \Gamma_{k+1}X$ and $u \in \Lambda\underline{T}^* \otimes_{\mathcal{O}} \mathcal{J}_k T$. Assume now that $\mathfrak{U}$ is of degree one and define

$$\mathfrak{U}^F = F^*\mathfrak{U} + \mathfrak{D}F \, .$$

Then

$$\mathfrak{U}^{F \circ G} = (\mathfrak{U}^F)^G \quad \text{and} \quad \mathfrak{D}_1(\mathfrak{U}^F) = F^*(\mathfrak{D}_1 \mathfrak{U}) .$$

These relations factor to

$$u^F = F^* \bullet u + \mathfrak{D}F \tag{21.11}$$

$$u^{F \bullet G} = (u^F)^G \tag{21.12}$$

$$\mathfrak{D}_1(u^F) = F^* \bullet \mathfrak{D}_1 u \tag{21.13}$$

where $u \in \underline{T}^* \otimes_{\mathcal{O}} \mathcal{J}_k T$. The formula (21.12) shows that $\Gamma_{k+1} X$ operates contravariantly on $\underline{T}^* \otimes_{\mathcal{O}} \mathcal{J}_k T$ by

$$(u, F) \longmapsto u^F$$

and (21.13) shows that $\mathcal{ker}\, \mathfrak{D}_1$ is invariant by that action. The proofs are left to the reader.

The complex $(21.8)_{k+1}$ is a finite (integrated) form of

$$(21.14)_{k+1} \quad 0 \longrightarrow \underline{T} \xrightarrow{\tilde{j}_{k+1}} \tilde{\mathcal{J}}_{k+1} T \xrightarrow{D \circ \varepsilon_{k+1}} \underline{T}^* \otimes_{\mathcal{O}} \mathcal{J}_k T \xrightarrow{D} \Lambda^2 \underline{T} \otimes_{\mathcal{O}} \mathcal{J}_{k-1} T$$

in the sense that each operator in $(21.14)_{k+1}$ is the linearization of the corresponding operator in $(21.8)_{k+1}$. Let us start by proving that $\tilde{j}_{k+1}$ is the linearization of $\tilde{j}_{k+1}$. Observe first that a section $\sigma \in \Gamma(U, \mathcal{A}ut\ X)$ identifies with the map $f = b \circ \sigma \colon U \to X$ which is locally a diffeomorphism ($b \colon \mathcal{A}ut\ X \to X$ is the *target* map). In fact $\sigma(x) = [f]_x$. Furthermore, $\tilde{j}f = f \times f$ is an admissible map of X^2 which is locally a diffeomorphism hence $[\tilde{j}f]_{(x,x)} \in \widetilde{\mathcal{A}ut}\ X^2$ and

$$\tilde{j}_{k+1}\sigma = \tilde{j}f \ \mathcal{mod} \sim_{k+1} \ \in \Gamma(U, \Gamma_{k+1} X) .$$

Let $\theta \in \Gamma(U, \underline{T}) \equiv \Gamma(U, T)$ and (reducing U if necessary) let (σ_t), $|t| < \varepsilon$, be a differentiable 1-parameter family of sections in $\Gamma(U, \mathcal{A}ut\ X)$ generated by θ, i.e.,

$$\sigma_0 = \mathrm{Id}|_U \quad \text{and} \quad \frac{d}{dt}\sigma_t|_{t=0} = \theta$$

(an example is given by $\sigma_t = \exp t\theta$). Then $(\tilde{j}_{k+1}\sigma_t)$ is a differentiable

1-parameter family of sections in $\Gamma(U, \Gamma_{k+1}X)$ with $\tilde{j}_{k+1}\sigma_0$ the identity section on U and we have to prove that

(21.15) $$\frac{d}{dt}\tilde{j}_{k+1}\sigma_t|_{t=0} = \tilde{j}_{k+1}\theta .$$

Let $f_t\colon U \to X$ identify with σ_t. The family (f_t) is, by definition, a differentiable 1-parameter family of maps in X with $f_0 = \mathrm{Id}$, $(\tilde{j}f_t)$ is a differentiable 1-parameter family of admissible maps in X^2 with $\tilde{j}f_0 = \mathrm{Id}$, each $\tilde{j}f_t$ represents $\tilde{j}_{k+1}\sigma_t$ and

$$\frac{d}{dt}\tilde{j}f_t|_{t=0} = \tilde{j}\theta ,$$

hence we obtain the formula (21.15) (cf. the proof of (a) Section 19). More generally, let (σ_t), $|t-t_\circ| < \varepsilon$, be a differentiable 1-parameter family of sections in $\Gamma(U, \mathcal{A}ut\, X)$ and assume that $\sigma_{t_\circ}$ is admissible, i.e., $f_{t_\circ}: U \to V$ is a diffeomorphism with $V = f_{t_\circ}(U)$ open. Then

$$\theta = \frac{d}{dt}\sigma_t|_{t=t_\circ} \in \Gamma(V, \underline{T}) \quad \text{and} \quad \tilde{j}_{k+1}\theta = \frac{d}{dt}\tilde{j}_{k+1}\sigma_t|_{t=t_\circ} .$$

An example of this situation is given by the formula

$$\operatorname{Exp} t\tilde{j}_{k+1}\theta = \tilde{j}_{k+1}\exp t\theta$$

which by differentiation yields

$$\frac{d}{dt}\tilde{j}_{k+1}\exp t\theta|_{t=t_\circ} = \tilde{j}_{k+1}\theta .$$

Next, we prove that $D \circ \varepsilon_{k+1}$ linearizes $\mathfrak{D}$. Take $\Xi \in \Gamma(U, \tilde{\mathfrak{J}}_{k+1}T)$ and let (F_t), $|t| < \varepsilon$, be a differentiable 1-parameter family of sections in $\Gamma(U, \Gamma_{k+1}X)$ such that F_0 is the identity section and

$$\frac{d}{dt}F_t|_{t=0} = \Xi .$$

Let (F_t), $|t| < \nu$, be a local lifting of (F_t) to a differentiable 1-parameter family of admissible local transformations of X^2 with $F_0 = \mathrm{Id}$. Since

F_t represents locally F_t then

$$\mathcal{D}F_t = \chi - F_t^*\chi \quad \text{represents} \quad \mathcal{D}F_t \in \Gamma(U, \underline{T}^* \otimes_{\mathcal{O}} \mathcal{J}_k T),$$

$$\xi = \frac{d}{dt} F_t|_{t=0} \quad \text{represents} \quad \Xi$$

and

$$\frac{d}{dt}\mathcal{D}F_t|_{t=0} \quad \text{represents} \quad \frac{d}{dt}\mathcal{D}F_t|_{t=0}\,.$$

But

$$\frac{d}{dt}\mathcal{D}F_t|_{t=0} = -\mathcal{L}(\xi)\chi = [\chi, \xi] = d_H(\varepsilon\xi)$$

where the last equality is a special case of Lemma 4 of Section 16, hence

$$\frac{d}{dt}\mathcal{D}F_t|_{t=0} = D \circ \varepsilon_{k+1}\Xi\,. \tag{21.16}$$

More generally, let (F_t), $|t-t_\circ| < \varepsilon$, be a differentiable 1-parameter family of sections in $\Gamma(U, \Gamma_{k+1}X)$ with $F_{t_\circ}$ admissible and let

$$\Xi = \frac{d}{dt}F_t|_{t=t_\circ} \in \Gamma(V, \tilde{\mathcal{J}}_{k+1}T)$$

with $V = f_{t_\circ}(U)$. Write

$$G_u = F_{t_\circ+u} \bullet F_{t_\circ}^{-1}\,.$$

Since

$$\mathcal{D}(G_u \bullet F_{t_\circ}) = F_{t_\circ}^* \bullet \mathcal{D}G_u + \mathcal{D}F_{t_\circ}$$

it follows that

$$\frac{d}{dt}\mathcal{D}F_t|_{t=t_\circ} = F_{t_\circ}^* \bullet \left(\frac{d}{du}\mathcal{D}G_u\right)_{u=0}$$

hence

$$\frac{d}{dt}\mathcal{D}F_t|_{t=t_\circ} = F_{t_\circ}^* \bullet (D \circ \varepsilon_{k+1}\Xi)\,. \tag{21.17}$$

An example of such a situation is given by

$$\frac{d}{dt}\mathfrak{D}\,\mathrm{Exp}\,t\Xi|_{t=t_o} = (\mathrm{Exp}\,t_o\Xi)^* \bullet (D \circ \varepsilon_{k+1}\Xi)\,.$$

Finally let us show that D linearizes $\mathfrak{D}_1$. Let $u \in \Gamma(U, \underline{T}^* \otimes_{\mathcal{O}} \mathcal{J}_k T)$ and let (v_t), $|t| < \varepsilon$, be a differentiable 1-parameter family of sections in $\Gamma(U, \underline{T}^* \otimes_{\mathcal{O}} \mathcal{J}_k T)$ such that

$$v_0 = 0 \quad \text{and} \quad \frac{d}{dt} v_t|_{t=0} = u$$

(for example $t \mapsto tu$). Since D is linear and [,] is quadratic, it follows that

$$\frac{d}{dt} Dv_t|_{t=0} = Du \quad \text{and} \quad \frac{d}{dt}[v_t, v_t]|_{t=0} = 0$$

hence

(21.18)
$$\frac{d}{dt}\mathfrak{D}_1 v_t|_{t=0} = Du\,.$$

More generally, let (v_t), $|t-t_o| < \varepsilon$, be a differentiable 1-parameter family of sections in $\Gamma(U, \underline{T}^* \otimes_{\mathcal{O}} \mathcal{J}_k T)$ and set

$$u = \frac{d}{dt} v_t|_{t=t_o}\,.$$

Then

$$\frac{d}{dt} Dv_t|_{t=t_o} = Du \quad \text{and} \quad \frac{d}{dt}[v_t, v_t]|_{t=t_o} = 2[u, v_{t_o}]$$

hence

(21.19)
$$\frac{d}{dt}\mathfrak{D}_1 v_t|_{t=t_o} = Du - [u, v_{t_o}]\,.$$

The previous formulas can also be obtained by lifting the families and the operators to X^2 and computing with representatives.

The complex (21.14) has the unpleasant feature of involving simultaneously *diagonal* and *vertical* jet sheaves. This inconvenience is trivially remedied. In fact, we can transport, via ε_k, the Nijenhuis bracket of $\underline{T}^* \otimes_{\mathcal{O}} \mathcal{J}_k T$ to a bracket [,] on $\underline{T}^* \otimes_{\mathcal{O}} \tilde{\mathcal{J}}_k T$ which however is different from the Nijenhuis bracket on $\underline{T}^* \otimes_{\mathcal{O}} \tilde{\mathcal{J}}_k T$ (cf. end of Section 16). We can redefine (21.8) by

$$(21.20)_{k+1} \quad 1 \longrightarrow \mathcal{A}ut\, X \xrightarrow{\tilde{j}_{k+1}} \Gamma_{k+1} X \xrightarrow{\tilde{\mathcal{D}}} \underline{T}^* \otimes_{\mathcal{O}} \tilde{\mathcal{J}}_k T \xrightarrow{\tilde{\mathcal{D}}_1} \Lambda^2 \underline{T}^* \otimes_{\mathcal{O}} \tilde{\mathcal{J}}_{k-1} T$$

where $\tilde{\mathcal{D}} = \varepsilon_k^{-1} \circ \mathcal{D}$, $\tilde{\mathcal{D}}_1 u = \tilde{D}u - \tfrac{1}{2}[u,u]$, $u \in \underline{T}^* \otimes_{\mathcal{O}} \tilde{\mathcal{J}}_k T$ and [,] is the transported Nijenhuis bracket. Then, clearly, the complex (21.20) is a finite form of

$$0 \longrightarrow \underline{T} \xrightarrow{\tilde{j}_{k+1}} \tilde{\mathcal{J}}_{k+1} T \xrightarrow{\tilde{D}} \underline{T}^* \otimes_{\mathcal{O}} \tilde{\mathcal{J}}_k T \xrightarrow{\tilde{D}} \Lambda^2 \underline{T}^* \otimes_{\mathcal{O}} \tilde{\mathcal{J}}_{k-1} T \,.$$

A second remedy is the following. Consider $\Gamma_{k+1} X$ as the quotient $\overline{\mathcal{A}ut_V X^2}/\sim_{k+1}$. Observe next that if $F = (f, \varphi)$ is a local admissible transformation and $G = (\mathrm{Id}, \varphi)$ is the corresponding vertical transformation, then $F^*\chi = G^*\chi$ and therefore

$$\mathcal{D}F = \chi - G^*\chi \overset{\text{def}}{=\!=} \mathcal{D}G \,.$$

All the previously discussed properties for admissible transformations remain valid for *admissible* vertical transformations $(x \mapsto \pi_2 \circ G(x,x)$ is a local diffeomorphism). However it is rather awkward to define the groupoid structure and the groupoid action of $\Gamma_{k+1} X$ by means of vertical maps. Nevertheless we obtain the non-linear complex

$$(21.21)_{k+1} \quad 1 \longrightarrow \mathcal{A}ut\, X \xrightarrow{j_{k+1}} \Gamma_{k+1} X \xrightarrow{\mathcal{D}} \underline{T}^* \otimes_{\mathcal{O}} \mathcal{J}_k T \xrightarrow{\mathcal{D}_1} \Lambda^2 \underline{T}^* \otimes_{\mathcal{O}} \mathcal{J}_{k-1} T$$

which is the transform of $(21.20)_{k+1}$ by the verticalizing maps. This complex is a finite form of

$$0 \longrightarrow \underline{T} \xrightarrow{j_{k+1}} \mathcal{J}_{k+1} T \xrightarrow{D} \underline{T}^* \otimes_{\mathcal{O}} \mathcal{J}_k T \xrightarrow{D} \Lambda^2 \underline{T}^* \otimes_{\mathcal{O}} \mathcal{J}_{k-1} T$$

which is the initial portion of $(6.2)_{k+1}$ with $\mathcal{E} = \underline{T}$.

We finally remark that $\mathrm{Id} = \Omega + \mathcal{X}$, hence $\mathfrak{D}F = F^*\Omega - \Omega$. This definition of $\mathfrak{D}$ is (up to a simple identification) the one given by Malgrange [9(c)]. It is also (up to a more elaborate identification) the one given by Guillemin and Sternberg [6].

22. *The* $\mathfrak{Ad}$ *representation and the operator* $\tilde{\mathfrak{D}}$

We begin this section by defining a finite form for the differential operator

$$ad\colon \tilde{\mathfrak{J}}_k T \longrightarrow \mathfrak{Der}\, \Lambda \underline{T}^*$$

(cf. the theorem of Section 15).

Let $F \in \widetilde{\mathfrak{Aut}}\, X^2$, $\nu \in \Lambda T^*_{\underline{V}}$ and define

$$\mathfrak{Ad}\, F(\nu) = V^*[F^*\nu]\,.$$

If $\nu \in \mathfrak{J}\, \Lambda T^*_{\underline{V}}$ then $F^*\nu$ vanishes on the diagonal, hence $\mathfrak{Ad}\, F(\nu) \in \mathfrak{J}\, \Lambda T^*_{\underline{V}}$. If $F \sim_k G$, $k \geq 1$, then

$$\mathfrak{Ad}\, F(\nu) \equiv \mathfrak{Ad}\, G(\nu) \bmod \mathfrak{J}^k \Lambda T^*_{\underline{V}}\,.$$

Moreover, if F and G are composable then

$$\mathfrak{Ad}(F \circ G) = (\mathfrak{Ad}\, G) \circ (\mathfrak{Ad}\, F) \quad \text{and} \quad \mathfrak{Ad}\, \mathrm{Id} = \mathrm{Id}$$

hence $\widetilde{\mathfrak{Aut}}\, X^2$ operates contravariantly on $\Lambda T^*_{\underline{V}}$. If $g \in \mathcal{O}$ and $F = (f, \varphi)$ then

$$\mathfrak{Ad}\, F(g\nu) = (g \circ f)\, \mathfrak{Ad}\, F(\nu)$$

where $g\nu$ is left multiplication and, more generally,

$$\mathfrak{Ad}\, F(\nu \wedge \nu') = \mathfrak{Ad}\, F(\nu) \wedge \mathfrak{Ad}\, F(\nu')\,.$$

Recalling that

$$\Lambda \underline{T}^* \simeq \Lambda T^*_{\underline{V}} / \mathfrak{J}\, \Lambda T^*_{\underline{V}}$$

via j_0, the previous discussion shows that, for $k \geq 1$, $\mathcal{A}d\,F$ factors to an $\mathbb{R}$-linear map

$$\mathcal{A}d\,F\colon (\Lambda\underline{T}^*)_y \longrightarrow (\Lambda\underline{T}^*)_x$$

where $F \in \Gamma_k X$, $x = a_k F$ and $y = b_k F$. Furthermore, $\mathcal{A}d$ defines a contravariant action of $\Gamma_k X$ on the sheaf $\Lambda\underline{T}^*$ and the formulas

$$\mathcal{A}d\,F(g\alpha) = (g \circ f)\,\mathcal{A}d\,F(\alpha)$$

with $\alpha \in \Lambda\underline{T}^*$ and $f = \beta_k \circ F$ and

$$\mathcal{A}d\,F(\alpha \wedge \alpha') = \mathcal{A}d\,F(\alpha') \wedge \mathcal{A}d\,F(\alpha')$$

hold. If F represents F, then

$$\mathcal{A}d\,F(\alpha) = \Delta^* \circ V^* \circ F^* \circ \pi_2^* \alpha\ .$$

Let now $F \in \Gamma_a(U, \Gamma_k X)$ be an admissible section, $f = b_k \circ F$ and $V = f(U)$. The sheaf map

$$\mathcal{A}d\,F\colon \Lambda\underline{T}^*|_V \longrightarrow \Lambda\underline{T}^*|_U$$

is semilinear with respect to $f^*\colon \mathcal{O}|_V \to \mathcal{O}|_U$ hence is the extension, to germs, of a vector bundle map

$$\begin{array}{ccc} \Lambda T^* & \xrightarrow{\ \mathcal{A}d\,F\ } & \Lambda T^* \\ \downarrow & & \downarrow \\ V & \xrightarrow{\ f^{-1}\ } & U \end{array}$$

The map

$$\mathcal{A}d\colon \Gamma_k X \longrightarrow \mathcal{H}om_{\mathbb{R}} \Lambda\underline{T}^*$$

is a finite form of ad. In fact, let $\Xi \in \Gamma(U, \tilde{\mathcal{J}}_k T)$ and let (F_t) be a 1-parameter differentiable family of sections in $\Gamma(U, \Gamma_k X)$ generated by Ξ. Then

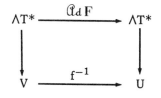

(22.1) $$\frac{d}{dt}\mathcal{A}d\, F_t|_{t=0} = ad\,\Xi ,$$

or equivalently,

$$\frac{d}{dt}\mathcal{A}d\, F_t|_{t=0}(\alpha) = \frac{d}{dt}\mathcal{A}d\, F_t(\alpha)|_{t=0} = ad\,\Xi(\alpha)$$

for any section $\alpha \in \Gamma(U, \Lambda\underline{T}^*)$. To see this, let (F_t) be a local lifting of (F_t) to a 1-parameter family of admissible maps on X^2. Then

$$\xi = \frac{d}{dt}F_t|_{t=0}$$

represents Ξ and

$$\frac{d}{dt}(\Delta^* \circ V^* \circ F_t^* \circ \pi_2^*\alpha)|_{t=0} = \Delta^* \circ V^* \circ \mathcal{L}(\xi) \circ \pi_2^*\alpha = L\Xi(\alpha) = ad\,\Xi(\alpha) .$$

More generally, let (F_t) be a 1-parameter differentiable family of sections in $\Gamma(U, \Gamma_k X)$ with F_{t_o} admissible. Then

(22.2) $$\frac{d}{dt}F_t|_{t=t_o} = \Xi \in \Gamma(V, \tilde{\mathcal{J}}_k T) \quad \text{and} \quad \frac{d}{dt}\mathcal{A}d\, F_t|_{t=t_o} = \mathcal{A}d\, F_{t_o} \circ ad\,\Xi .$$

We now extend the action $\mathcal{A}d$ to a contravariant action of $\Gamma_k X$, $k \geq 1$, on the sheaf $\Lambda\underline{T}^* \otimes_{\mathcal{O}} \check{\mathcal{J}}_{k-1}T$ by the formula

(22.3) $$\mathcal{A}d\, F(\alpha \otimes \xi) = \mathcal{A}d\, F(\alpha) \otimes F^{-1} \bullet \xi$$

where $\alpha \otimes \xi \in \Lambda\underline{T}^* \otimes_{\mathcal{O}} \check{\mathcal{J}}_{k-1}$ and $F^{-1} \bullet \xi$ is the quotient of the standard action of $\widetilde{\mathcal{A}ut}\, X^2$ on $\check{\Theta}$. Furthermore

$$\mathcal{A}d\, F(gu) = (g \circ f)\, \mathcal{A}d\, F(u) ,$$

where $g \in \mathcal{O}$, $u \in \Lambda\underline{T}^* \otimes_{\mathcal{O}} \check{\mathcal{J}}_{k-1}T$, $f = \beta_k \circ F$ and gu is the left multiplication. Hence, if $F \in \Gamma_a(U, \Gamma_k X)$ then $\mathcal{A}d\, F$ is the extension, to germs, of a vector bundle map. More generally,

$$\mathcal{A}d\, F(\alpha \dot{\wedge} u) = \mathcal{A}d\, F(\alpha) \wedge \mathcal{A}d\, F(u) .$$

Since $\Gamma_k X$ operates on $\tilde{\mathcal{J}}_k T$ (cf. Section 20), the above action, restricted to $\Lambda\underline{T}^* \otimes_{\mathcal{O}} \tilde{\mathcal{J}}_{k-1} T$, can be redefined, by the same formula, as a contravariant action of $\Gamma_k X$ on the sheaf $\Lambda\underline{T}^* \otimes_{\mathcal{O}} \mathcal{J}_k T$.

In Section 14 we defined the bracket $[\![\ ,\ ]\!]$ on $\Lambda\underline{T}^* \otimes_{\mathcal{O}} \check{\mathcal{J}}_k T$ with values in $\Lambda\underline{T}^* \otimes_{\mathcal{O}} \check{\mathcal{J}}_{k-1} T$. The same argument as in the proof of Lemma 3, Section 14, shows that if $u \in \Lambda\underline{T}^* \otimes_{\mathcal{O}} \tilde{\mathcal{J}}_k T$ and $v \in \Lambda\underline{T}^* \otimes_{\mathcal{O}} \check{\mathcal{J}}_{k-1}$ then $[\![u,v]\!]$ can be redefined to take values in $\Lambda\underline{T}^* \otimes_{\mathcal{O}} \check{\mathcal{J}}_{k-1}$. Setting $ad\, u(v) = [\![u, v]\!]$, we define the representation

$$ad\colon \Lambda\underline{T}^* \otimes_{\mathcal{O}} \tilde{\mathcal{J}}_k T \to \mathcal{H}om_{\mathbf{R}}(\Lambda\underline{T}^* \otimes_{\mathcal{O}} \check{\mathcal{J}}_{k-1} T)\,.$$

The restriction

$$ad\colon \tilde{\mathcal{J}}_k T \to \mathcal{H}om_{\mathbf{R}}(\Lambda\underline{T}^* \otimes_{\mathcal{O}} \check{\mathcal{J}}_{k-1} T) \tag{22.4$_k$}$$

is a Lie algebra representation ($\mathcal{H}om_{\mathbf{R}}(\ )$ with the commutator). If we replace $\check{\mathcal{J}}_{k-1} T$ by $\tilde{\mathcal{J}}_{k-1} T$ then ad can be redefined as the Lie algebra representation

$$ad\colon \Lambda\underline{T}^* \otimes_{\mathcal{O}} \tilde{\mathcal{J}}_k T \to \mathcal{D}er_{\mathbf{R}}(\Lambda\underline{T}^* \otimes_{\mathcal{O}} \tilde{\mathcal{J}}_k T)\,,$$

i.e., the adjoint representation to the bracket $[\![\ ,\ ]\!]$ on $\Lambda\underline{T}^* \otimes_{\mathcal{O}} \tilde{\mathcal{J}}_k T$. In particular

$$ad\colon \tilde{\mathcal{J}}_k T \to \mathcal{D}er_{\mathbf{R}}(\Lambda\underline{T}^* \otimes_{\mathcal{O}} \tilde{\mathcal{J}}_k T) \tag{22.4$'_k$}$$

is a Lie algebra representation into the derivations of degree zero.

In Section 15 we obtained an expression for $[\![\ ,\ ]\!]$ using the representation ad. If $\Xi \in \tilde{\mathcal{J}}_k T$ and $\alpha \otimes \eta \in \Lambda\underline{T}^* \otimes_{\mathcal{O}} \check{\mathcal{J}}_{k-1} T$, that expression reduces to

$$ad\,\Xi(\alpha \otimes \eta) = [\![\Xi, \alpha \otimes \eta]\!] = ad\,\Xi(\alpha) \otimes \eta + \alpha \otimes [\Xi, \eta]\,. \tag{22.5}$$

Let (F_t) be a differentiable 1-parameter family of sections in $\Gamma(U, \Gamma_k X)$ with F_{t_o} admissible, let

$$\Xi = \frac{d}{dt} F_t|_{t=t_o} \in \Gamma(V, \mathcal{J}_k T)$$

and take $\alpha \otimes \eta \in \Gamma(V, \Lambda \underline{T}^* \otimes_{\mathcal{O}} \check{\mathcal{J}}_{k-1} T)$. Then

$$\frac{d}{dt} \mathcal{A}d F_t(\alpha \otimes \eta)|_{t=t_o} = \frac{d}{dt} \mathcal{A}d F_t(\alpha)|_{t=t_o} \otimes F_{t_o}^{-1} \bullet \eta + \mathcal{A}d F_{t_o}(\alpha) \otimes \frac{d}{dt}(F_t^{-1} \bullet \eta)|_{t=t_o} =$$

$$\mathcal{A}d F_{t_o}(ad\Xi(\alpha) \otimes \eta + \alpha \otimes [\Xi, \eta]) ,$$

hence

$$\frac{d}{dt} \mathcal{A}d\, F_t|_{t=t_o} = \mathcal{A}d\, F_{t_o} \circ ad\,\Xi . \tag{22.6}$$

It follows that

$$\mathcal{A}d\colon \Gamma_k X \to \mathcal{H}om_R(\Lambda \underline{T}^* \otimes_{\mathcal{O}} \check{\mathcal{J}}_{k-1} T)$$

is a finite form of (22.4). If we restrict $\check{\mathcal{J}}_{k-1} T$ to $\tilde{\mathcal{J}}_{k-1} T$ then, in view of Section 20, $\mathcal{A}d$ can be redefined by (22.3) as an action of $\Gamma_k X$ on the sheaf $\Lambda \underline{T}^* \otimes_{\mathcal{O}} \tilde{\mathcal{J}}_k T$. The differential operator

$$\mathcal{A}d\colon \Gamma_k X \to \mathcal{H}om_R(\Lambda \underline{T}^* \otimes_{\mathcal{O}} \tilde{\mathcal{J}}_k T)$$

is a finite form of (22.4′).

We now define the second non-linear operator $\tilde{\mathfrak{D}}$ by the formula

$$\tilde{\mathfrak{D}} F = \chi_k(a) - \mathcal{A}d\, F(\chi_k(b)) \tag{22.7}$$

where $F \in \Gamma_{k+1} X$, $a = a_{k+1} F$, $b = b_{k+1} F$ and χ_k is the section of $\underline{T}^* \otimes_{\mathcal{O}} \check{\mathcal{J}}_k T$ induced by the vector form χ on X^2. This operator $\tilde{\mathfrak{D}}$ should not be confused with the operator $\tilde{\mathfrak{D}}$ in (21.20). We claim that $\tilde{\mathfrak{D}} F \in \underline{T}^* \otimes_{\mathcal{O}} \tilde{\mathcal{J}}_k T$. Let F be a local admissible transformation on X^2 defined in the neighborhood of (a, a) which represents F at that point. Let (x_i, y_i) be a system of coordinates around (a, a), (x'_i, y'_i) a system of coordinates around $(b, b) = F(a, a)$ and set $F = (f_i, \varphi_i)$, $F^{-1} = (f_i^{-1}, \psi_i)$. We shall denote by $\frac{\partial}{\partial x_i}$ the vector field on X^2 as well as the induced section in $\check{\mathcal{J}}_k T$. If $\alpha \in \underline{T}^*$ and $\alpha = a_\mu dx'_\mu$ then

(22.8) $$\mathfrak{A}d\, F(\alpha) = (a_\mu \circ f)\, \frac{\partial \varphi_\mu}{\partial y_j}(x, x) dx_j$$

$$F^{-1} \bullet \frac{\partial}{\partial x'_\mu} = \left(\frac{\partial f_i^{-1}}{\partial x'_\mu} \circ f\right) \frac{\partial}{\partial x_i} + \left(\frac{\partial \psi_i}{\partial x'_\mu} \circ F\right) \frac{\partial}{\partial y_i} \ \mathit{mod}\ \mathcal{J}^{k+1} \Delta \breve{\Phi}$$

hence

(22.9) $$\mathfrak{A}d\, F(\mathcal{X}_k) = \mathfrak{A}d\, F\left(dx'_\mu \otimes \frac{\partial}{\partial x'_\mu}\right) = \mathfrak{A}d\, F(dx'_\mu) \otimes F^{-1} \bullet \frac{\partial}{\partial x'_\mu} =$$

$$\frac{\partial \varphi_\mu}{\partial y_j}(x, x) \left(\frac{\partial f_i^{-1}}{\partial x'_\mu} \circ f\right) dx_j \otimes \frac{\partial}{\partial x_i} + \frac{\partial \varphi_\mu}{\partial y_j}(x, x) \left(\frac{\partial \psi_i}{\partial x'_\mu} \circ F\right) dx_j \otimes \frac{\partial}{\partial y_i} \ \mathit{mod}\ \mathcal{J}^{k+1} \Delta \breve{\Phi}\,.$$

Since $\varphi_\mu(x, x) = f_\mu(x)$ then

$$\frac{\partial \varphi_\mu}{\partial x_j}(x, x) + \frac{\partial \varphi_\mu}{\partial y_j}(x, x) = \frac{\partial f_\mu}{\partial x_j}(x)$$

and

(22.10) $$\mathfrak{A}dF(\mathcal{X}_k) = \frac{\partial f_\mu}{\partial x_j}\left(\frac{\partial f_i^{-1}}{\partial x'_\mu} \circ f\right) dx_j \otimes \frac{\partial}{\partial x_i} - \frac{\partial \varphi_\mu}{\partial x_j}(x, x) \left(\frac{\partial f_i^{-1}}{\partial x'_\mu} \circ f\right) dx_j \otimes \frac{\partial}{\partial x_i} +$$

$$\frac{\partial \varphi_\mu}{\partial y_j}(x, x) \left(\frac{\partial \psi_i}{\partial x'_\mu} \circ F\right) dx_j \otimes \frac{\partial}{\partial y_i} \ \mathit{mod}\ \mathcal{J}^{k+1} \Delta \breve{\Phi}$$

where the first term is equal to $\mathcal{X}$. It follows that

(22.11) $$\tilde{\mathfrak{D}}F = \mathcal{X}_k - \mathfrak{A}dF\mathcal{X}_k = -\frac{\partial \varphi_\mu}{\partial y_j}(x, x) \left(\frac{\partial \psi_i}{\partial x'_\mu} \circ F\right) dx_j \otimes \frac{\partial}{\partial y_i} +$$

$$\frac{\partial \varphi_\mu}{\partial x_j}(x, x) \left(\frac{\partial f_i^{-1}}{\partial x'_\mu} \circ f\right) dx_j \otimes \frac{\partial}{\partial x_i} \ \mathit{mod}\ \mathcal{J}^{k+1} \Delta \breve{\Phi}\,.$$

But

$$- \frac{\partial\varphi_\mu}{\partial y_j}(x,x)\left(\frac{\partial\psi_i}{\partial x'_\mu}\circ F\right)(x,x) = - \frac{\partial\varphi_\mu}{\partial y_j}(x,x)\,\frac{\partial\psi_i}{\partial x'_\mu}(f(x),f(x)) =$$

$$\frac{\partial\psi_i}{\partial x'_\mu}(f(x),f(x))\,\frac{\partial\varphi_\mu}{\partial x_j}(x,x) - \frac{\partial\psi_i}{\partial x'_\mu}(f(x),f(x))\,\frac{\partial f_\mu}{\partial x_j}(x) \underline{\underline{(18.3)}}$$

$$\frac{\partial\psi_i}{\partial x'_\mu}(f(x),f(x))\,\frac{\partial\varphi_\mu}{\partial x_j}(x,x) + \frac{\partial\psi_i}{\partial y'_\mu}(f(x),f(x))\,\frac{\partial\varphi_\mu}{\partial x_j}(x,x) = \frac{\partial f_i^{-1}}{\partial x'_\mu}(f(x))\,\frac{\partial\varphi_\mu}{\partial x_j}(x,x)$$

since $\psi_i(x',x') = f^{-1}(x')$. This computation shows that

$$\xi_j = -\frac{\partial\varphi_\mu}{\partial y_j}(x,x)\left(\frac{\partial\psi_i}{\partial x'_\mu}\circ F\right)\frac{\partial}{\partial y_i} + \frac{\partial\varphi_\mu}{\partial x_j}(x,x)\left(\frac{\partial f_i^{-1}}{\partial x'_\mu}\circ f\right)\frac{\partial}{\partial x_i} \in \tilde{\Theta}$$

hence $dx_j \otimes \xi_j \in \tilde{\Phi}$ and

$$\tilde{\mathfrak{D}}\colon \Gamma_{k+1}X \to \underline{T}^* \otimes_{\mathcal{O}} \mathcal{J}_k T\,.$$

Combining (22.11) with (18.3) we find the *Buttin formula* ([9(c)], formula (4.10))

$$\tilde{\mathfrak{D}}F = dx_j \otimes \left[M^i_j \Delta\left(\frac{\partial}{\partial y_i} + \frac{\partial}{\partial x_i}\right)\right]_{mod\ \mathcal{J}^{k+1}\Delta\tilde{\Phi}} \qquad (22.12)$$

where

$$M^i_j = \left(\frac{\partial\psi_i}{\partial y'_\lambda}\circ F\right)\frac{\partial\varphi_\lambda}{\partial x_k}\,\frac{\partial f_k^{-1}}{\partial x'_\mu}\left(\frac{\partial\varphi_\mu}{\partial y_j}(x,x)\right) = \left[\left(\frac{\partial\varphi}{\partial y}\right)^{-1}\bullet\frac{\partial\varphi}{\partial x}\bullet\left(\frac{\partial f}{\partial x}\right)^{-1}\bullet\frac{\partial\varphi}{\partial y}(x,x)\right]^i_j$$

Let $f \in \mathcal{A}ut\, X$, $F = \tilde{j}_{k+1} f \in \Gamma_{k+1} X$, and $\alpha \in \Lambda\underline{T}^*$; then $\mathcal{A}d\, F(\alpha) = F^* \bullet \alpha = f^*\alpha$ hence $\mathcal{A}d\, F(u) = F^* \bullet u$ for any $u \in \Lambda\underline{T}^* \otimes_{\mathcal{O}} \mathcal{J}_k T$. It follows that

$$\tilde{\mathfrak{D}} \circ \tilde{j}_{k+1} f = 0\,.$$

Assume, conversely, that $\tilde{\mathfrak{D}}F = 0$ and let $F \in \widetilde{\mathcal{A}ut}\, X^2$ represent F. Returning to formula (22.11), we see that this condition is equivalent to

$$\frac{\partial \varphi_\mu}{\partial y_j}(x,x)\left(\frac{\partial \psi_i}{\partial x'_\mu}\circ F\right) \in \mathcal{J}^{k+1} .$$

Since the matrix

$$\left(\frac{\partial \varphi_\mu}{\partial y_j}(x,x)\right)$$

is invertible, it follows by an argument similar to that in Section 21 that all the partial derivatives (with respect to x and y),

$$D^\alpha\left(\frac{\partial \psi_i}{\partial x'_\mu}\circ F\right)\Big|_\Delta , \ 0 \le |\alpha| \le k ,$$

vanish in a neighborhood of (a,a), hence all the partial derivatives

$$D^\alpha\left[\left(\frac{\partial \psi_i}{\partial x'_\mu}\circ F\right)\circ F^{-1}\right]\Big|_\Delta = D^\alpha \frac{\partial \psi_i}{\partial x'_\mu}\Big|_\Delta , \ 0 \le |\alpha| \le k ,$$

vanish in a neighborhood of (b,b). In particular we infer that

$$D^\alpha_{x'}\psi_i|_\Delta , \ 0 < |\alpha| \le k+1 ,$$

vanishes in a neighborhood of (b,b) hence $F^{-1} = \tilde{j}_{k+1} f^{-1}$ so that $F = \tilde{j}_{k+1} f$.

The sequence

$$1 \longrightarrow \mathfrak{Aut}\, X \xrightarrow{\tilde{j}_{k+1}} \Gamma_{k+1} X \xrightarrow{\tilde{\mathfrak{D}}} \underline{T}^* \otimes_{\mathcal{O}} \tilde{\mathcal{J}}_k T$$

is exact. We want to complete this sequence by an operator $\tilde{\mathfrak{D}}_1$ in such a way that $\tilde{\mathfrak{D}}_1 \circ \tilde{\mathfrak{D}} = 0$. This condition represents a *structure equation* for $\tilde{\mathfrak{D}}$. It will be obtained in the next section by studying the invariance of the bracket $[\![\ ,\ ]\!]$ under the action $\mathfrak{Ad}$.

23. *Invariance of* $[\![\ ,\ ]\!]$ *and the second non-linear complex*

It is not true in general that the bracket $[\![\ ,\ ]\!]$ defined on $\Lambda \underline{T}^* \otimes_{\mathcal{O}} \check{\mathcal{J}}_k T$ is invariant under the action of $\mathfrak{Ad}$. Nevertheless, we shall prove that the invariance holds on certain subsheaves and this is sufficient.

Let $u, v \in \Lambda \underline{T}^* \otimes_{\mathcal{O}} \check{\mathcal{J}}_k T$ and $F \in \Gamma_{k+1} X$. The bracket $[\![\ ,\ ]\!]$ is invariant under $\mathcal{A}d$ if

$$(23.1) \qquad \mathcal{A}d\, F([\![u, v]\!]) = [\![\mathcal{A}d\, F(u), \mathcal{A}d\, F(v)]\!]$$

Assume that u and v are decomposable, namely $u = \alpha \otimes \xi \in \Lambda^r \underline{T}^* \otimes_{\mathcal{O}} \check{\mathcal{J}}_k T$ and $v = \beta \otimes \eta \in \Lambda^s \underline{T}^* \otimes_{\mathcal{O}} \check{\mathcal{J}}_k T$. Then

$$\mathcal{A}dF([\![u,v]\!]) = \mathcal{A}dF[adu(\beta)] \otimes \check{\rho}_{k-1} F^{-1} \bullet \eta + (-1)^{rs+1} \mathcal{A}dF[adv(\alpha)] \otimes \check{\rho}_{k-1} F^{-1} \bullet \xi +$$
$$\mathcal{A}dF(\alpha \wedge \beta) \otimes F^{-1} \bullet [\xi, \eta]\,,$$

$$[\![\mathcal{A}dF(u), \mathcal{A}dF(v)]\!] = ad[\mathcal{A}dF(u)]\mathcal{A}dF(\beta) \otimes \check{\rho}_{k-1} F^{-1} \bullet \eta +$$
$$(-1)^{rs+1}\, ad[\mathcal{A}dF(v)]\, \mathcal{A}dF(\alpha) \otimes \check{\rho}_{k-1} F^{-1} \bullet \xi +$$
$$\mathcal{A}dF(\alpha) \wedge \mathcal{A}dF(\beta) \otimes [F^{-1} \bullet \xi, F^{-1} \bullet \eta]\,.$$

Since $F^{-1} \bullet [\xi, \eta] = [F^{-1} \bullet \xi, F^{-1} \bullet \eta]$ and $\mathcal{A}dF(\alpha \wedge \beta) = \mathcal{A}dF(\alpha) \wedge \mathcal{A}dF(\beta)$, the invariance relation is equivalent to the relations

$$(23.2) \qquad ad[\mathcal{A}dF(u)]\, \mathcal{A}dF(\beta) = \mathcal{A}dF[adu(\beta)]$$

$$(23.3) \qquad ad[\mathcal{A}dF(v)]\, \mathcal{A}dF(\alpha) = \mathcal{A}dF[adv(\alpha)]\,.$$

LEMMA 1. *The relation* (23.2) *holds for any* $u \in \Lambda \underline{T}^* \otimes_{\mathcal{O}} \tilde{\mathcal{J}}_k T$, $\beta \in \Lambda \underline{T}^*$, $F \in \Gamma_k X$ *and* $k \geq 1$.

Proof. We can assume that u is decomposable, $u = \alpha \otimes \xi$. Since

$$\mathcal{A}dF(u) = \mathcal{A}dF(\alpha) \otimes F^{-1} \bullet \xi$$

and ad restricted to $\Lambda \underline{T}^* \otimes_{\mathcal{O}} \tilde{\mathcal{J}}_k T$ is left $\Lambda \underline{T}^*$-linear then (23.2) becomes

$$\mathcal{A}dF(\alpha) \wedge [ad(F^{-1} \bullet \xi)\, \mathcal{A}dF(\beta)] = \mathcal{A}dF(\alpha) \wedge \mathcal{A}dF[ad\xi(\beta)]$$

hence it is enough to prove that

$$(23.4) \qquad ad(F^{-1} \bullet \xi)\, \mathcal{A}dF(\beta) = \mathcal{A}dF[ad\xi(\beta)]\,.$$

Extend F to an admissible section $F \in \Gamma_a(U, \Gamma_k X)$, extend ξ to a section $\xi \in \Gamma(V, \mathcal{J}_k T)$, $V = f(U)$, and extend β to a section $\beta \in \Gamma(V, \Lambda \underline{T}^*)$. Let (F_t) be a differentiable 1-parameter family of sections in $\Gamma_a(V, \Gamma_k X)$ such that F_0 is the identity section and

$$\frac{d}{dt} F_t|_{t=0} = \xi .$$

Then

$$ad\xi(\beta) = \frac{d}{dt} \mathcal{A}d F_t(\beta)|_{t=0}$$

and

$$F^{-1} \bullet \xi = \frac{d}{dt} G_t|_{t=0} \quad \text{with} \quad G_t = F^{-1} \bullet F_t \bullet F ,$$

hence

$$ad(F^{-1} \bullet \xi)[\mathcal{A}d F(\beta)] = \frac{d}{dt} \mathcal{A}d G_t[\mathcal{A}d F(\beta)]|_{t=0} .$$

Since

$$\mathcal{A}d G_t[\mathcal{A}d F(\beta)] = \mathcal{A}d(F \bullet G_t)\beta = \mathcal{A}d(F_t \bullet F)\beta = \mathcal{A}d F[\mathcal{A}d F_t(\beta)] ,$$

then

$$\frac{d}{dt} \mathcal{A}d F[\mathcal{A}d F_t \beta]|_{t=0} = \mathcal{A}d F \left[\frac{d}{dt} \mathcal{A}d F_t(\beta)\right]_{t=0} = \mathcal{A}d F[ad\xi(\beta)] .$$

The above lemma, or rather formula (23.2), expresses that $\mathcal{A}d$ commutes with ad when restricted to $\Lambda \underline{T}^* \otimes_{\mathcal{O}} \mathcal{J}_k T$, i.e., the following diagram is commutative for any $F \in \Gamma_k X$,

$$\begin{array}{ccc} \Lambda \underline{T}^* \otimes_{\mathcal{O}} \mathcal{J}_k T & \xrightarrow{\ ad\ } & \mathcal{D}er\, \Lambda \underline{T}^* \\ \Big\downarrow {\scriptstyle \mathcal{A}d F} & & \Big\downarrow {\scriptstyle AdF} \\ \Lambda \underline{T}^* \otimes_{\mathcal{O}} \mathcal{J}_k T & \xrightarrow{\ ad\ } & \mathcal{D}er\, \Lambda \underline{T}^* \end{array}$$

where

$$AdF\colon D \longmapsto \mathcal{A}d F \circ D \circ \mathcal{A}d F^{-1} .$$

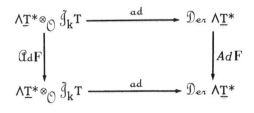

LEMMA 2. *For any* $F \in \Gamma_{k+1}X$, $\alpha \in \Lambda\underline{T}^*$ *and* $k \geq 1$ *the following relation holds*

$$ad[\mathfrak{a}dF(\chi_k)]\,\mathfrak{a}dF(\alpha) = \mathfrak{a}dF[ad\chi_k(\alpha)]\,.$$

Proof. Let $a = a_{k+1}F$, $b = b_{k+1}F$ and take coordinate systems (x_i, y_i) around (a, a) and (x'_i, y'_i) around (b, b). Consider the form

$$\overline{\chi}_k = \sum dx'_i \otimes \frac{\partial}{\partial y'_i} \bmod \mathcal{J}^{k+1}\Phi_V \in \underline{T}^* \otimes_{\mathcal{O}} \mathcal{J}_k T\,.$$

Since $\tilde{\chi}_k = \overline{\chi}_k + \chi_k \in \underline{T}^* \otimes_{\mathcal{O}} \tilde{\mathcal{J}}_k T$ then, by Lemma 1,

$$ad[\mathfrak{a}dF(\tilde{\chi}_k)]\,\mathfrak{a}dF(\alpha) = \mathfrak{a}dF[ad\tilde{\chi}_k(\alpha)]$$

hence the proof of Lemma 2 reduces to the proof of

$$(23.5)\qquad ad[\mathfrak{a}dF(\overline{\chi}_k)]\,\mathfrak{a}dF(\alpha) = \mathfrak{a}dF(ad\overline{\chi}_k(\alpha))\,.$$

Observe next that

$$ad\overline{\chi}_k = \mathfrak{L}[(\overline{\chi}_k)_H] + i[(\tilde{D}\overline{\chi}_k)_H] = 0$$

since $(\overline{\chi}_k)_H = 0$ and $D\overline{\chi}_k = 0$, hence (23.5) reduces to

$$(23.6)\qquad ad[\mathfrak{a}dF(\overline{\chi}_k)]\,\mathfrak{a}dF(\alpha) = 0\,.$$

Furthermore, since $\mathfrak{a}dF(\overline{\chi}_k) \in \underline{T}^* \otimes_{\mathcal{O}} \mathcal{J}_k T$, then $[\mathfrak{a}dF(\overline{\chi}_k)]_H = 0$ and

$$ad[\mathfrak{a}dF(\overline{\chi}_k)] = i([\tilde{D}\,\mathfrak{a}dF(\overline{\chi}_k)]_H)$$

hence (23.6) reduces to

$$(23.7)\qquad i([\tilde{D}\,\mathfrak{a}dF(\overline{\chi}_k)]_H)\,\mathfrak{a}dF(\alpha) = 0\,.$$

We shall now prove that

$$[\tilde{D}\,\mathfrak{a}dF(\overline{\chi}_k)]_H = 0\,.$$

Let F be an admissible transformation which represents F and write $F = (f, \varphi)$, $F^{-1} = (f^{-1}, \psi)$. Then

$$\mathfrak{A}d F(dx'_\mu) = \frac{\partial\varphi_\mu}{\partial y_\rho}(x,x)\,dx_\rho \quad\text{and}\quad F_*^{-1}\frac{\partial}{\partial y'_\mu} = \left(\frac{\partial\psi_\ell}{\partial y'_\mu}\circ F\right)\frac{\partial}{\partial y_\ell}$$

hence

$$\mathfrak{A}d F(\overline{X}_k) = \mathfrak{A}d F(dx'_\mu)\otimes F^{-1}\bullet\frac{\partial}{\partial y'_\mu} = dx_\rho\otimes\frac{\partial\varphi_\mu}{\partial y_\rho}(x,x)\left(\frac{\partial\psi_\ell}{\partial y'_\mu}\circ F\right)\frac{\partial}{\partial y_\ell}\ \mathrm{mod}\ \mathcal{J}^{k+1}\Phi_V .$$

Since $\mathfrak{A}d F(\overline{X}_k)\in \underline{T}^*\otimes_{\mathcal{O}}\mathcal{J}_k T$, then

$$\tilde{D}\,\mathfrak{A}d F(\overline{X}_k) = \varepsilon_{k-1}^{-1}\circ D\,\mathfrak{A}d F(\overline{X}_k) .$$

Now

$$d_H\left[dx_\rho\otimes\frac{\partial\varphi_\mu}{\partial y_\rho}(x,x)\left(\frac{\partial\psi_\ell}{\partial y'_\mu}\circ F\right)\frac{\partial}{\partial y_\ell}\right] = -\,dx_\rho\wedge d_H\left[\frac{\partial\varphi_\mu}{\partial y_\rho}(x,x)\left(\frac{\partial\psi_\ell}{\partial y'_\mu}\circ F\right)\frac{\partial}{\partial y_\ell}\right] =$$

$$-\,dx_\rho\wedge\frac{\partial}{\partial x_\lambda}\left[\frac{\partial\varphi_\mu}{\partial y_\rho}(x,x)\left(\frac{\partial\psi_\ell}{\partial y'_\mu}\circ F\right)\right]dx_\lambda\otimes\frac{\partial}{\partial y_\ell}$$

hence

$$[\tilde{D}\,\mathfrak{A}d F(\overline{X}_k)]_H = [D\,\mathfrak{A}d F(\overline{X}_k)]'_H = -\frac{\partial}{\partial x_\lambda}\left[\frac{\partial\varphi_\mu}{\partial y_\rho}(x,x)\left(\frac{\partial\psi_\ell}{\partial y'_\mu}\circ F\right)\right](x,x)\,dx_\rho\wedge dx_\lambda\otimes\frac{\partial}{\partial x_\ell}.$$

Furthermore,

$$-\frac{\partial}{\partial x_\lambda}\left[\frac{\partial\varphi_\mu}{\partial y_\rho}(x,x)\,\frac{\partial\psi_\ell}{\partial y'_\mu}(f(x),f(x))\right] = -\frac{\partial}{\partial x_\lambda}(\delta_\rho^\ell) = 0$$

implies that

$$-\frac{\partial}{\partial x_\lambda}\left[\frac{\partial\varphi_\mu}{\partial y_\rho}(x,x)\left(\frac{\partial\psi_\ell}{\partial y'_\mu}\circ F\right)\right](x,x) = \frac{\partial}{\partial y_\lambda}\left[\frac{\partial\varphi_\mu}{\partial y_\rho}(x,x)\left(\frac{\partial\psi_\ell}{\partial y'_\mu}\circ F\right)\right](x,x) =$$

$$\frac{\partial\varphi_\mu}{\partial y_\rho}(x,x)\,\frac{\partial}{\partial y_\lambda}\left(\frac{\partial\psi_\ell}{\partial y'_\mu}\circ F\right)(x,x) =$$

$$\frac{\partial\varphi_\mu}{\partial y_\rho}(x,x)\,\frac{\partial\psi_\ell}{\partial y'_\eta\partial y'_\mu}(f(x),f(x))\,\frac{\partial\varphi_\eta}{\partial y_\lambda}(x,x) .$$

It follows that

$$[D\,\mathfrak{A}d F(\overline{X}_k)]'_H = \frac{\partial\varphi_\mu}{\partial y_\rho}(x,x)\,\frac{\partial\psi_\ell}{\partial y'_\eta\partial y'_\mu}(f(x),f(x))\,\frac{\partial\varphi_\eta}{\partial y_\lambda}(x,x)\,dx_\rho\wedge dx_\lambda\otimes\frac{\partial}{\partial x_\ell} = 0$$

since the coefficients are symmetrical in η and μ hence in ρ and λ. This achieves the proof.

Lemma 2 can also be expressed by the commutative diagram

$$\begin{array}{ccc} \Lambda\underline{T}^* & \xrightarrow{\ ad\chi_k\ } & \Lambda\underline{T}^* \\ {\scriptstyle \mathfrak{A}d F}\downarrow & & \downarrow{\scriptstyle \mathfrak{A}d F} \\ \Lambda\underline{T}^* & \xrightarrow{\ ad[\mathfrak{A}d F(\chi_k)]\ } & \Lambda\underline{T}^* \end{array}$$

THEOREM 1. *For any* $F \in \Gamma_k X$ *and* $k \geq 1$, *the map* $\mathfrak{A}d F$ *is a germ of automorphism of the Lie algebra sheaf* $\Lambda\underline{T}^* \otimes_{\mathcal{O}} \tilde{\mathfrak{J}}_k T$ *with the bracket* $[\![\ ,\ ]\!]$. *The groupoid representation*

$$\mathfrak{A}d: \Gamma_k X \to \mathfrak{A}ut_R(\Lambda\underline{T}^* \otimes_{\mathcal{O}} \tilde{\mathfrak{J}}_k T)$$

is a finite form of the Lie algebra representation (22.4′).

THEOREM 2. *The subsheaf*

$$\Lambda_k = (\Lambda\underline{T}^* \otimes_{\mathcal{O}} \tilde{\mathfrak{J}}_k T) \oplus_R R\chi_k \subset \Lambda\underline{T}^* \otimes_{\mathcal{O}} \check{\mathfrak{J}}_k T,\ k \geq 1\ ,$$

is invariant under the groupoid action $\mathfrak{A}d$ *of* $\Gamma_{k+1}X$. *For any* $F \in \Gamma_{k+1}X$ *and* $u, v \in \Lambda_k$ *we have*

$$\mathfrak{A}d F([\![u, v]\!]) = [\![\mathfrak{A}d F(u), \mathfrak{A}d F(v)]\!]\ .$$

Proof. The sheaf $\Lambda\underline{T}^* \otimes_{\mathcal{O}} \tilde{\mathfrak{J}}_k T$ is invariant by $\mathfrak{A}d F$ and $\mathfrak{A}d F(\chi_k) = -\tilde{\mathfrak{D}}F + \chi_k$ with $\tilde{\mathfrak{D}}F \in \Lambda\underline{T}^* \otimes_{\mathcal{O}} \tilde{\mathfrak{J}}_k T$. The invariance relation for the bracket is a consequence of the lemmas.

Remark. One proves similarly that Λ_k is invariant under the Lie algebra representation $(22.4)_{k+1}$. If $u \in \Lambda\underline{T}^* \otimes_{\mathcal{O}} \tilde{\mathfrak{J}}_k T$, $F \in \Gamma_{k+1}X$ and $\Xi \in \mathfrak{J}_{k+1}T$ then the relation

$$\mathfrak{A}d F([\![u, \chi_k]\!]) = [\![\mathfrak{A}d F(u), \mathfrak{A}d F(\chi_k)]\!]$$

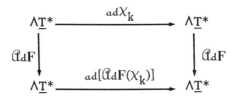

is a finite form of

$$ad\Xi([\![u, \mathcal{X}_k]\!]) = [\![ad\Xi(u), \mathcal{X}_k]\!] + [\![u, ad\Xi(\mathcal{X}_k)]\!]$$

the latter being simply the Jacobi identity (in the general sense) for Ξ, u and $\mathcal{X}_k$ (see the remark that follows the proof of Lemma 6, Section 16). Taking a differentiable 1-parameter family (F_t) of admissible local sections of $\Gamma_{k+1}X$, one can derive the second relation from the first one by differentiation with respect to t.

We can now derive the *structure equation* for $\tilde{\mathfrak{D}}F$. Since $[\![\mathcal{X}_k, \mathcal{X}_k]\!] = 0$, then

$$[\![\mathcal{A}d F(\mathcal{X}_k), \mathcal{A}d F(\mathcal{X}_k)]\!] = \mathcal{A}d F([\![\mathcal{X}_k, \mathcal{X}_k]\!]) = 0$$

for any $F \in \Gamma_{k+1}X$. It follows by Theorem 2 of Section 16 that

$$\tilde{D}\tilde{\mathfrak{D}}F - \tfrac{1}{2}\tilde{\rho}_{k-1}\,[\![\tilde{\mathfrak{D}}F, \tilde{\mathfrak{D}}F]\!] = 0 \tag{23.8}$$

for any $F \in \Gamma_{k+1}X$ and $k \geq 1$. The structure equation is vacuous for $k = 0$.

Define the *second non-linear complex* to be

$$(23.9)_{k+1} \quad 1 \longrightarrow \mathcal{A}ut\,X \xrightarrow{\tilde{j}_{k+1}} \Gamma_{k+1}X \xrightarrow{\tilde{\mathfrak{D}}} \underline{T}^* \otimes_{\mathcal{O}} \tilde{\mathfrak{J}}_k T \xrightarrow{\tilde{\mathfrak{D}}_1} \Lambda^2 \underline{T}^* \otimes_{\mathcal{O}} \tilde{\mathfrak{J}}_{k-1} T$$

where

$$\tilde{\mathfrak{D}}_1(u) = \tilde{D}u - \tfrac{1}{2}\tilde{\rho}_{k-1}[\![u, u]\!],\ u \in \underline{T}^* \otimes_{\mathcal{O}} \tilde{\mathfrak{J}}_k T\,.$$

This complex is exact at $\mathcal{A}ut\,X$ and $\Gamma_{k+1}X$. The non-exactness at $\underline{T}^* \otimes_{\mathcal{O}} \tilde{\mathfrak{J}}_k T$ will be discussed later. Presently, we prove some fundamental identities for the operators $\tilde{\mathfrak{D}}$ and $\tilde{\mathfrak{D}}_1$ (the latter should not be confused with the operator $\tilde{\mathfrak{D}}_1$ in $(21.20)_{k+1}$). The first identity reads

$$\mathcal{A}d F(\alpha) = F^* \bullet \alpha - i(\tilde{\mathfrak{D}}F)(F^* \bullet \alpha) = f^*\alpha - i([\tilde{\mathfrak{D}}F]_H) f^*\alpha \tag{23.10}$$

where $F \in \Gamma_{k+1}X$, $\alpha \in \underline{T}^*$ and $f = \beta_{k+1} \circ F$. In fact, if $\alpha = \alpha_i dx'_i$ then

$$\mathcal{A}d F(\alpha) = (\alpha_j \circ f)\, \mathcal{A}d F(dx'_j)\,,$$

$$(\tilde{\mathfrak{D}}F)_H = (\chi_k)_H - \left(\mathfrak{A}dF(dx'_i) \otimes F^{-1} \bullet \frac{\partial}{\partial x'_i}\right)_H = \mathrm{Id} - \mathfrak{A}dF(dx'_i) \otimes f_*^{-1} \frac{\partial}{\partial x'_i}$$

and

$$i([\tilde{\mathfrak{D}}F]_H) f^*\alpha = f^*\alpha - (\alpha_i \circ f)\, \mathfrak{A}dF(dx'_i)$$

hence the relation (23.10). Since $\mathfrak{A}dF$ is a morphism of exterior algebras, then

$$(23.10') \qquad \mathfrak{A}dF = (F^* - i(\tilde{\mathfrak{D}}F) \circ F^*)^{\wedge}$$

where $(\ )^{\wedge}$ is the standard extension, to $\wedge \underline{T}^*$, of the semilinear map

$$F^* - i(\tilde{\mathfrak{D}}F) \circ F^* : \underline{T}^* \longrightarrow \underline{T}^* .$$

Observe that (23.10′) is a finite form of $\mathfrak{ad}\Xi = \mathfrak{L}(\Xi) - i(\tilde{D}\Xi) \in \mathfrak{Der}\ \wedge\underline{T}^*$, $\Xi \in \tilde{\mathfrak{J}}_{k+1}T$. The next identity is

$$(23.11) \qquad \mathfrak{ad}(\tilde{\mathfrak{D}}F) = d - \mathfrak{A}dF \circ d \circ \mathfrak{A}dF^{-1} \in \mathfrak{Der}\ \wedge\underline{T}^*$$

where $F \in \Gamma_{k+1}X$ and $k \geq 1$. The right hand side F can actually be replaced by $G = \rho_k F$. This relation is a consequence of Lemma 2 and the relation $\mathfrak{ad}\chi_k = d$ (cf. end of Section 16). In fact

$$\mathfrak{ad}[\chi_k - \mathfrak{A}dF(\chi_k)] = \mathfrak{ad}\chi_k - \mathfrak{ad}[\mathfrak{A}dF(\chi_k)] = d - \mathfrak{A}dF \circ d \circ \mathfrak{A}dF^{-1} .$$

The formula (23.11) is a finite form of $\mathfrak{ad}(\tilde{D}\Xi) = [d, \mathfrak{ad}\Xi]$ (cf. Lemma 7, Section 16). Thirdly, we have

$$(23.12) \qquad \tilde{\mathfrak{D}}(F \bullet G) = \mathfrak{A}dG(\tilde{\mathfrak{D}}F) + \tilde{\mathfrak{D}}G$$

where $F, G \in \Gamma_{k+1}X$ are composable. In fact,

$$\tilde{\mathfrak{D}}(F \bullet G) = \chi_k - \mathfrak{A}d(F \bullet G)\chi_k = \chi_k - \mathfrak{A}dG \circ \mathfrak{A}dF(\chi_k) =$$

$$\chi_k + \mathfrak{A}dG[\chi_k - \mathfrak{A}dF(\chi_k)] - \mathfrak{A}dG(\chi_k) = \mathfrak{A}dG(\tilde{\mathfrak{D}}F) + \tilde{\mathfrak{D}}G .$$

If $k \geq 1$ then (23.12) can be rewritten as

$$(23.12') \qquad \tilde{\mathfrak{D}}(F \bullet G) = \mathfrak{A}d\rho_k G(\tilde{\mathfrak{D}}F) + \tilde{\mathfrak{D}}G .$$

Since $F \bullet F^{-1} = 1$ and $\tilde{\mathfrak{D}}(1) = 0$, the formula (23.12) yields

$$\tilde{\mathfrak{D}}F^{-1} = -\,\mathfrak{A}dF^{-1}(\tilde{\mathfrak{D}}F) \tag{23.13}$$

which in fact is obvious from the definition. If $k \geq 1$ then $\mathfrak{A}dF^{-1}$ can be replaced by $\mathfrak{A}d\rho_k F^{-1}$. The next identity is

$$\tilde{D}[\mathfrak{A}dF(u)] = \mathfrak{A}dF(\tilde{D}u - [\![\tilde{\mathfrak{D}}F^{-1}, \tilde{\rho}_k u]\!]) \in \Lambda\underline{T}^* \otimes_{\mathcal{O}} \tilde{\mathfrak{J}}_k T \tag{23.14}$$

where $F \in \Gamma_{k+1}X$, $u \in \Lambda\underline{T}^* \otimes_{\mathcal{O}} \check{\mathfrak{J}}_{k+1}T$ and $k \geq 1$. We have already remarked that if $u \in \Lambda\underline{T}^* \otimes_{\mathcal{O}} \tilde{\mathfrak{J}}_{k+1}T$ and $v \in \Lambda\underline{T}^* \otimes_{\mathcal{O}} \check{\mathfrak{J}}_k T$ then $[\![u, v]\!] \in \Lambda\underline{T}^* \otimes_{\mathcal{O}} \check{\mathfrak{J}}_{k-1}T$ can be redefined to take values in $\Lambda\underline{T}^* \otimes_{\mathcal{O}} \check{\mathfrak{J}}_k T$. Moreover, if $v' \in \Lambda\underline{T}^* \otimes_{\mathcal{O}} \check{\mathfrak{J}}_{k+1}T$, $v = \check{\rho}_k v'$ and $k \geq 1$ then $[\![u, v']\!] = [\![u, v]\!] \in \Lambda\underline{T}^* \otimes_{\mathcal{O}} \check{\mathfrak{J}}_k T$. With this in mind, we shall prove that

$$\tilde{D}[\mathfrak{A}dF(u)] - \mathfrak{A}dF(\tilde{D}u) = -\,\mathfrak{A}dF([\![\tilde{\mathfrak{D}}F^{-1}, \tilde{\rho}_k u]\!])\,.$$

In fact,

$$\tilde{D}[\mathfrak{A}dF(u)] = [\![\chi_{k+1}, \mathfrak{A}dF(u)]\!] = [\![\chi_k, \mathfrak{A}dF(u)]\!] \in \Lambda\underline{T}^* \otimes_{\mathcal{O}} \tilde{\mathfrak{J}}_k T \quad \text{and}$$

$$\mathfrak{A}dF(\tilde{D}u) = \mathfrak{A}dF([\![\chi_{k+1}, u]\!]) = \mathfrak{A}dF([\![\chi_k, u]\!]) = [\![\mathfrak{A}dF(\chi_k), \mathfrak{A}dF(u)]\!] \in \Lambda\underline{T}^* \otimes_{\mathcal{O}} \tilde{\mathfrak{J}}_k T\,,$$

hence

$$\tilde{D}[\mathfrak{A}dF(u)] - \mathfrak{A}dF(\tilde{D}u) = [\![\tilde{\mathfrak{D}}F, \mathfrak{A}dF(u)]\!] = [\![\tilde{\mathfrak{D}}F, \mathfrak{A}dF(\tilde{\rho}_k u)]\!]$$

since $\tilde{\mathfrak{D}}F \in \Lambda\underline{T}^* \otimes_{\mathcal{O}} \tilde{\mathfrak{J}}_k T$. Accordingly,

$$-\,\mathfrak{A}dF([\![\tilde{\mathfrak{D}}F^{-1}, \tilde{\rho}_k u]\!]) = -\,[\![\mathfrak{A}dF(\tilde{\mathfrak{D}}F^{-1}), \mathfrak{A}dF(\tilde{\rho}_k u)]\!] = [\![\tilde{\mathfrak{D}}F, \mathfrak{A}dF(\tilde{\rho}_k u)]\!]\,.$$

If $F \in \Gamma_{k+2}X$, $u \in \Lambda\underline{T}^* \otimes_{\mathcal{O}} \tilde{\mathfrak{J}}_{k+1}T$ and $k \geq 1$ then (23.14) can be rewritten as

$$\tilde{D}[\mathfrak{A}dF(u)] = \mathfrak{A}dF(\tilde{D}u - \tilde{\rho}_k[\![\tilde{\mathfrak{D}}F^{-1}, u]\!] \in \Lambda\underline{T}^* \otimes_{\mathcal{O}} \tilde{\mathfrak{J}}_k T\,. \tag{23.14'}$$

Let $u \in \underline{T}^* \otimes_{\mathcal{O}} \check{\mathfrak{J}}_k T$, $F \in \Gamma_{k+1}X$ and define

$$u^F = \mathfrak{A}dF(u) + \tilde{\mathfrak{D}}F \in \underline{T}^* \otimes_{\mathcal{O}} \check{\mathfrak{J}}_k T\,. \tag{23.15}$$

Then

(23.16) $$u^{F\bullet G} = (u^F)^G, \quad u^1 = u \quad \text{and} \quad u^{F^{-1}} = \mathcal{A}d F^{-1}(u - \tilde{\mathfrak{D}}F)$$

In fact

$$u^{F\bullet G} = \mathcal{A}d(F\bullet G)u + \tilde{\mathfrak{D}}(F\bullet G) = \mathcal{A}dG \circ \mathcal{A}dF(u) + \tilde{\mathfrak{D}}G + \mathcal{A}dG(\tilde{\mathfrak{D}}F) =$$
$$\mathcal{A}dG(\mathcal{A}dF(u) + \tilde{\mathfrak{D}}F) + \tilde{\mathfrak{D}}G = (u^F)^G .$$

The two other relations are obvious. Since $\mathcal{A}dF$ is semilinear, $u \mapsto u^F$ is affine. Moreover, if $F \in \Gamma_a(U, \Gamma_{k+1}X)$ then the sheaf map

$$u \in (\underline{T}^* \otimes_{\mathcal{O}} \check{\mathcal{J}}_k T)|_V \mapsto u^F \in (\underline{T}^* \otimes_{\mathcal{O}} \check{\mathcal{J}}_k T)|_U ,$$

where $V = b_{k+1} \circ F(U)$, is the extension, to germs, of an affine map of vector bundles. The first formula (23.16) shows that $\Gamma_{k+1}X$ operates contravariantly on the sheaf $\underline{T}^* \otimes_{\mathcal{O}} \check{\mathcal{J}}_k T$ by $(u, F) \mapsto u^F$ and this action is a representation of $\Gamma_{k+1}X$ into the groupoid of germs of affine motions of the sheaf $\underline{T}^* \otimes_{\mathcal{O}} \check{\mathcal{J}}_k T$. If $u \in \underline{T}^* \otimes_{\mathcal{O}} \mathcal{J}_k T$ then $u^F \in \underline{T}^* \otimes_{\mathcal{O}} \mathcal{J}_k T$. The final identity is

(23.17) $$\tilde{\mathfrak{D}}_1(u^F) = \mathcal{A}dF(\tilde{\mathfrak{D}}_1 u)$$

where $u \in \underline{T}^* \otimes_{\mathcal{O}} \mathcal{J}_k T$ and $F \in \Gamma_{k+1}X$. In fact,

$$\tilde{\mathfrak{D}}_1(u^F) = \tilde{\mathfrak{D}}_1(\mathcal{A}dF(u) + \tilde{\mathfrak{D}}F) =$$
$$\tilde{D}[\mathcal{A}dF(u)] + \tilde{D}(\tilde{\mathfrak{D}}F) - \tfrac{1}{2}\tilde{\rho}_{k-1}[\![\mathcal{A}dF(u) + \tilde{\mathfrak{D}}F, \mathcal{A}dF(u) + \tilde{\mathfrak{D}}F]\!] =$$
$$\tilde{D}[\mathcal{A}dF(u)] - \tfrac{1}{2}\tilde{\rho}_{k-1}[\![\mathcal{A}dF(u), \mathcal{A}dF(u)]\!] - \tilde{\rho}_{k-1}[\![\mathcal{A}dF(u), \tilde{\mathfrak{D}}F]\!] =$$
$$\mathcal{A}dF(\tilde{D}u - [\![\tilde{\rho}_{k-1}\tilde{\mathfrak{D}}F^{-1}, \tilde{\rho}_{k-1}u]\!]) - \tfrac{1}{2}\tilde{\rho}_{k-1}\mathcal{A}dF([\![u,u]\!]) - \tilde{\rho}_{k-1}[\![\mathcal{A}dF(u), \tilde{\mathfrak{D}}F$$

But

$$-\mathcal{A}dF([\![\tilde{\rho}_{k-1}\tilde{\mathfrak{D}}F^{-1}, \tilde{\rho}_{k-1}u]\!]) = \tilde{\rho}_{k-1}[\![-\mathcal{A}dF(\tilde{\mathfrak{D}}F^{-1}), \mathcal{A}dF(u)]\!] =$$
$$\tilde{\rho}_{k-1}[\![\tilde{\mathfrak{D}}F, \mathcal{A}dF(u)]\!] = \tilde{\rho}_{k-1}[\![\mathcal{A}dF(u), \tilde{\mathfrak{D}}F]\!]$$

hence

$$\tilde{\mathfrak{D}}_1(u^F) = \mathcal{A}dF(\tilde{D}u - \tfrac{1}{2}\tilde{\rho}_{k-1}[\![u, u]\!]) = \mathcal{A}dF(\tilde{\mathfrak{D}}_1 u) .$$

The relation (23.17) shows that $\mathcal{k}er\,\tilde{\mathfrak{D}}_1$ is invariant under the action $(u, F) \mapsto u^F$.

The complex $(23.9)_{k+1}$ is a finite form of

$$(23.18)_{k+1} \qquad 0 \longrightarrow \underline{T} \xrightarrow{\tilde{j}_{k+1}} \tilde{\mathfrak{J}}_{k+1}T \xrightarrow{\tilde{D}} \underline{T}^* \otimes_{\mathcal{O}} \tilde{\mathfrak{J}}_k T \xrightarrow{\tilde{D}} \Lambda^2 \underline{T}^* \otimes_{\mathcal{O}} \tilde{\mathfrak{J}}_{k-1} T .$$

We first prove that $\tilde{D}$ linearizes $\tilde{\mathfrak{D}}$. Let (F_t), $|t-t_o| < \varepsilon$, be a differentiable 1-parameter family of sections in $\Gamma(U, \Gamma_{k+1}X)$ with F_{t_o} admissible and let

$$\Xi = \frac{d}{dt} F_t|_{t=t_o} \in \Gamma(V, \tilde{\mathfrak{J}}_{k+1}T)$$

where $V = f_{t_o}(U)$. Theorem 2 of Section 16 and the relation (22.6) imply that

$$\frac{d}{dt}\tilde{\mathfrak{D}}F_t|_{t=t_o} = -\frac{d}{dt}\mathcal{A}d F_t(\chi_k)|_{t=t_o} = -\mathcal{A}d F_{t_o} \circ ad\Xi(\chi_k) = -\mathcal{A}d F_{t_o}([\![\Xi, \chi_k]\!]) =$$
$$-\mathcal{A}d F_{t_o}([\![\Xi, \chi_{k+1}]\!]) = \mathcal{A}d F_{t_o}([\![\chi_{k+1}, \Xi]\!]) = \mathcal{A}d F_{t_o}(\tilde{D}\Xi)$$

hence

$$(23.19) \qquad \frac{d}{dt}\tilde{\mathfrak{D}}F_t|_{t=t_o} = \mathcal{A}d F_{t_o}(\tilde{D}\Xi) .$$

In particular if $t_o = 0$ and $F_{t_o} = 1$ then

$$(23.20) \qquad \frac{d}{dt}\tilde{\mathfrak{D}}F_t|_{t=0} = \tilde{D}\Xi$$

If $F_t = \operatorname{Exp} t\Xi$ then

$$(23.21) \qquad \frac{d}{dt}\tilde{\mathfrak{D}} \operatorname{Exp} t\Xi|_{t=t_o} = \mathcal{A}d(\operatorname{Exp} t_o\Xi)(\tilde{D}\Xi) .$$

Next we prove that $\tilde{D}$ linearizes $\tilde{\mathfrak{D}}_1$. Let (v_t), $|t-t_o| < \varepsilon$, be a differentiable 1-parameter family of sections in $\Gamma(U, \underline{T}^* \otimes_{\mathcal{O}} \tilde{\mathfrak{J}}_k T)$ and set $u = \frac{d}{dt} v_t|_{t=t_o}$. Since $\tilde{D}$ is linear and $[\![\ ,\ ]\!]$ is quadratic it follows that

$$(23.22) \qquad \frac{d}{dt}\tilde{\mathfrak{D}}_1 v_t|_{t=t_o} = \tilde{D}u - \tilde{\rho}_{k-1}[\![u, v_{t_o}]\!] .$$

In particular if $t_o = 0$ and $v_{t_o} = 0$ then

$$(23.23) \qquad \frac{d}{dt}\tilde{\mathfrak{D}}_1 v_t|_{t=0} = \tilde{D}u .$$

24. *Partial exactness of the second non-linear complex*

In this section we examine the deviation from exactness of $(23.9)_{k+1}$ at $\underline{T}^* \otimes_{\mathcal{O}} \tilde{\mathcal{J}}_k T$. Define the sheaf $\gamma_{k+1}X$ by the exact sequence

$$1 \longrightarrow \gamma_{k+1}X \longrightarrow \Gamma_{k+1}X \xrightarrow{\rho_k} \Gamma_k X \longrightarrow 1$$

where ρ_k is the standard projection

$$[F]_{(x,x)} \; mod \sim_{k+1} \longmapsto [F]_{(x,x)} \; mod \sim_k$$

with F a local admissible transformation of X^2. If $F \in \Gamma_{k+1}X$ and $[F]_{(a,a)}$ represents F then $\rho_k F = 1 \in \Gamma_k X$ if and only if $[F]_{(a,a)} \sim_k [Id]_{(a,a)}$ i.e., F is the identity mapping up to terms of order $\geq k+1$ in a neighborhood of (a,a) in Δ. Since ρ_k is a sheaf and groupoid morphism which reduces to the identity on units it follows that $\gamma_{k+1}X$ is a sheaf of groups. Moreover, if $k \geq 1$, then $\gamma_{k+1}X$ is a subsheaf of abelian groups. In fact, if $F, G \in \gamma_{k+1}X$ are represented by $F = (f, \varphi)$, $G = (g, \psi)$ and $F \bullet G \in \gamma_{k+1}X$ is represented by $F \circ G = (h, \mu)$, we have in a neighborhood of (a, a) in Δ:

a) $F(x, x) = G(x, x) = F \circ G(x, x) = (x, x)$ or equivalently, $f = g = h = Id$ and, in terms of coordinates,

b) $D^\alpha \mu_j(x, x) = D^\alpha y_j(x, x)$, $|\alpha| = 1$
$D^\alpha \mu_j(x, x) = 0$, $1 < |\alpha| \leq k$
$D^\alpha \mu_j(x, x) = D^\alpha \varphi_j(x, x) + D^\alpha \psi_j(x, x)$, $|\alpha| = k+1$.

On the infinitesimal level, the exact sequence (cf. Section 6)

$$0 \longrightarrow \wedge \underline{T}^* \otimes_{\mathcal{O}} S^{k+1} \underline{T}^* \otimes_{\mathcal{O}} \underline{T} \xrightarrow{\zeta_{k+1}} \wedge \underline{T}^* \otimes_{\mathcal{O}} \mathcal{J}_{k+1} \otimes_{\mathcal{O}} \underline{T} \xrightarrow{\rho_k} \wedge \underline{T}^* \otimes_{\mathcal{O}} \mathcal{J}_k \otimes_{\mathcal{O}} \underline{T} \longrightarrow 0$$

yields, via the identification

$$\varepsilon_k \colon \tilde{\mathcal{J}}_k T \longrightarrow \mathcal{J}_k T$$

(or simply observing that $\mathcal{ker}(\tilde{\mathcal{J}}_k T \xrightarrow{\tilde{\beta}_k} \underline{T}) = \mathcal{ker}(\mathcal{J}_k T \xrightarrow{\beta_k} \underline{T}) = \tilde{\mathcal{J}}_k T \cap \mathcal{J}_k T$ as subsheaves of $\check{\mathcal{J}}_k T$), the exact sequence

$$0 \longrightarrow \wedge\underline{T}^* \otimes_{\mathcal{O}} S^{k+1}\underline{T}^* \otimes_{\mathcal{O}} \underline{T} \xrightarrow{\zeta_{k+1}} \wedge\underline{T}^* \otimes_{\mathcal{O}} \tilde{\mathcal{J}}_{k+1}T \xrightarrow{\tilde{\rho}_k} \wedge\underline{T}^* \otimes_{\mathcal{O}} \tilde{\mathcal{J}}_k T \longrightarrow 0$$

and

$$\tilde{D}|\mathcal{ker}\,\tilde{\rho}_k = \underline{\delta}\,.$$

Moreover, $\mathcal{ker}\,\tilde{\rho}_k$ is a subsheaf of Lie algebras of $\wedge\underline{T}^* \otimes_{\mathcal{O}} \tilde{\mathcal{J}}_{k+1}T$ with respect to the bracket $[\![\ ,\ ]\!]$ (and $[\ ,\ ]$).

It is easy to check that $\mathcal{ker}\,\tilde{\rho}_k$ is abelian for $k \geq 1$. In particular,

$$S^{k+1}\underline{T}^* \otimes_{\mathcal{O}} \underline{T} = \mathcal{ker}(\tilde{\mathcal{J}}_{k+1}T \xrightarrow{\tilde{\rho}_k} \tilde{\mathcal{J}}_k T)$$

is an abelian subalgebra of $\tilde{\mathcal{J}}_{k+1}T$ for $k \geq 1$ and $[\![\ ,\ ]\!]\,|\,\mathcal{ker}\,\tilde{\rho}_0$ is simply the commutator in $\underline{T}^* \otimes_{\mathcal{O}} \underline{T} \simeq \mathcal{H}om_{\mathcal{O}}(\underline{T})$.

We shall first prove that, for $k \geq 1$, the diagram

(24.1)
$$\begin{array}{ccc} S^{k+1}\underline{T}^* \otimes_{\mathcal{O}} \underline{T} & \xrightarrow{\quad\underline{\delta}\quad} & \underline{T}^* \otimes_{\mathcal{O}} S^k\underline{T}^* \otimes_{\mathcal{O}} \underline{T} \\ \Big\downarrow \mathrm{Exp} & \nearrow \tilde{\mathcal{D}} & \\ \gamma_{k+1}X & & \end{array}$$

is commutative. Let us start by explaining the vertical arrow Exp. Take $\Xi \in S^{k+1}\underline{T}^* \otimes_{\mathcal{O}} \underline{T}$, i.e., $\Xi \in \tilde{\mathcal{J}}_{k+1}T$ and $\tilde{\rho}_k\Xi = 0$, and let ξ be a local admissible vector field on X^2 such that

$$\Xi = [\xi]_{(a,a)}\ \mathcal{mod}\ \mathcal{J}^{k+2}\Delta\tilde{\Theta}\,.$$

The condition on Ξ implies in particular that $\xi|_\Delta = 0$ in a neighborhood V of (a,a) in Δ. The point (a,a) is a fixed point for the local 1-parameter group generated by ξ hence the integral curve γ_a of ξ with initial data $\gamma_a(0) = (a,a)$ is defined for all t (in fact $\gamma_a(t) = (a,a)$ for any t). The continuation lemma implies that for each constant $K > 0$ there exists an open neighborhood U_K of (a,a) in X^2 such that the local 1-parameter group (φ_t) generated by ξ is defined on U_k for

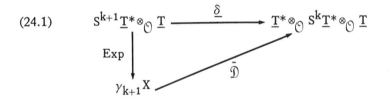

$|t| < K$ (U_k can even be chosen to contain V). It follows that the germ $[\varphi_t]_{(a,a)}$ is defined for arbitrary $t \in R$. We set

$$\operatorname{Exp} t\Xi = [\varphi_t]_{(a,a)} \bmod \sim_{k+1} \quad \text{and} \quad \operatorname{Exp}\Xi = \operatorname{Exp} 1\Xi .$$

If Σ is a local section of $\tilde{\mathfrak{J}}_{k+1}T$ such that $\Sigma(a) = \Xi$ then $\operatorname{Exp} t\Xi = (\operatorname{Exp} t\Sigma)(a)$ where $\operatorname{Exp} t\Sigma$ is the local 1-parameter group defined in Section 19. It is clear that $\operatorname{Exp} t\Xi \in \gamma_{k+1}X$ since

$$\rho_k \operatorname{Exp} t\Xi = \operatorname{Exp} t\tilde{\rho}_k\Xi = \operatorname{Exp} 0 = 1 .$$

Let now Ξ be a local section of $\tilde{\mathfrak{J}}_{k+1}T$ with $\tilde{\rho}_k\Xi = 0$, i.e., $\Xi \in \Gamma(U, S^{k+1}\underline{T}^* \otimes_{\mathcal{O}} \underline{T})$. The previous local argument shows that Ξ is uniform on U. In fact, for each $x \in U$ we can find a local admissible vector field ξ defined in a neighborhood V of (x, x) such that

$$\Xi|_{V \cap \Delta} = \xi \bmod \mathfrak{J}^{k+2}\Delta\tilde{\Theta}$$

and for each $K > 0$ we can find a neighborhood W_K of (x, x) such that the 1-parameter local group (φ_t) generated by ξ is defined on W_K for $|t| < K$ hence $\operatorname{Exp} t\Xi$ is defined on $W_k \cap \Delta$ for $|t| < K$. Since the neighborhoods $W_K \cap \Delta$ cover U, $\operatorname{Exp} t\Xi$ is defined on U for $|t| < K$, hence for all $t \in R$. Observe also that $b_{k+1} \circ \operatorname{Exp} t\Xi = \operatorname{Id}|_U$.

If $F \in \gamma_{k+1}X$ then

$$\tilde{\mathfrak{D}}F \in \operatorname{ker}(\underline{T}^* \otimes_{\mathcal{O}} \mathfrak{J}_k T \xrightarrow{\tilde{\rho}_{k-1}} \underline{T}^* \otimes_{\mathcal{O}} \mathfrak{J}_{k-1}T) = \underline{T}^* \otimes_{\mathcal{O}} S^k\underline{T}^* \otimes_{\mathcal{O}} \underline{T}$$

since

$$\rho_k(F) = 1,\ \tilde{\rho}_{k-1}(\tilde{\mathfrak{D}}F) = \tilde{\mathfrak{D}}(\rho_k F) \quad \text{and} \quad \tilde{\mathfrak{D}}1 = 0 .$$

We shall prove the commutativity of (24.1) for local sections. This will imply the commutativity for germs.

Let $\Xi \in \Gamma(U, S^{k+1}\underline{T}^* \otimes_{\mathcal{O}} \underline{T})$. The map

$$t \in R \longmapsto \operatorname{Exp} t\Xi \in \Gamma(U, \gamma_{k+1}X)$$

is a differentiable 1-parameter subgroup of $\Gamma(U, \gamma_{k+1}X)$, hence

$$\mathrm{Exp}(t'+t)\Xi = (\mathrm{Exp}\, t'\Xi)\bullet(\mathrm{Exp}\, t\Xi)\ .$$

The formula (23.12) and the relation $\rho_k(\mathrm{Exp}\, t\Xi) = 1$ imply that

$$\tilde{\mathfrak{D}}[\mathrm{Exp}(t'+t)\Xi] = \tilde{\mathfrak{D}}(\mathrm{Exp}\, t'\Xi) + \tilde{\mathfrak{D}}(\mathrm{Exp}\, t\Xi)\ .$$

We have shown that the map

$$\nu\colon \mathbf{R} \longrightarrow \Gamma(U, \underline{T}^* \otimes_{\mathcal{O}} S^k\underline{T}^* \otimes_{\mathcal{O}} \underline{T}),\ t \longmapsto \tilde{\mathfrak{D}}(\mathrm{Exp}\, t\Xi)$$

is additive and it is clearly a differentiable 1-parameter family of sections of $\underline{T}^* \otimes_{\mathcal{O}} S^k\underline{T}^* \otimes_{\mathcal{O}} \underline{T}$. We can prove further that ν is $\mathbf{R}$-linear. In fact, if $n \in \mathbf{N}$ then $\nu(n\frac{t}{n}) = n\nu(\frac{t}{n})$ hence $\nu(\frac{t}{n}) = \frac{1}{n}\nu(t)$ and this implies that $\nu(rt) = r\nu(t)$ for any $r \in \mathbf{Q}_+$. Since $\nu(-t) = -\nu(t)$ it follows that ν is linear over $\mathbf{Q}$ hence, by continuity, ν is linear over $\mathbf{R}$. If we set $\nu_1 = \nu(1)$ then $\nu\colon t \mapsto t\nu_1$. Applying formula (23.21) and observing that

$$\rho_k(\mathrm{Exp}\, t_o\Xi) = 1 \in \Gamma(U, \Gamma_k X)$$

we obtain

$$\tilde{\mathfrak{D}}(\mathrm{Exp}\Xi) = \nu_1 = \frac{d}{dt}\tilde{\mathfrak{D}}(\mathrm{Exp}\, t\Xi)|_{t=t_o} = \tilde{D}\Xi = \underline{\delta}\Xi$$

which is the desired relation.

Let us start by examining the complex $(23.9)_{k+1}$ for $k = 0$, namely

$$(23.9)_1 \qquad 1 \longrightarrow \mathcal{A}ut\, X \xrightarrow{\tilde{j}_1} \Gamma_1 X \xrightarrow{\tilde{\mathfrak{D}}} \underline{T}^* \otimes_{\mathcal{O}} \tilde{\mathcal{J}}_0 T \longrightarrow 0\ .$$

Recall that $\underline{T}^* \otimes_{\mathcal{O}} \tilde{\mathcal{J}}_0 T$ identifies canonically with $\underline{T}^* \otimes_{\mathcal{O}} \underline{T} \simeq \mathcal{H}om_{\mathcal{O}}(\underline{T})$. This identification can for example be obtained as follows. Let $u \in \underline{T}^* \otimes_{\mathcal{O}} \tilde{\mathcal{J}}_0 T$ and let $\mathfrak{U} \in \tilde{\Phi}^1$ be a representative of u; then

$$u \in \underline{T}^* \otimes_{\mathcal{O}} \tilde{\mathcal{J}}_0 T \longmapsto \Delta^*\mathfrak{U} \in \underline{T}^* \otimes_{\mathcal{O}} \underline{T}$$

is the desired $\mathcal{O}$-linear isomorphism. Recall also that

$$\Lambda \underline{T}^* \otimes_{\mathcal{O}} \check{\mathcal{J}}_k T = (\Lambda \underline{T}^* \otimes_{\mathcal{O}} \mathcal{J}_k T) \oplus (\Lambda \underline{T}^* \otimes_{\mathcal{O}} \underline{T})$$

and let $u \mapsto u_H$ be the projection on the horizontal component $\Lambda \underline{T}^* \otimes_{\mathcal{O}} \underline{T}$. The previous isomorphism is also given by

$$u \in \underline{T}^* \otimes_{\mathcal{O}} \tilde{\mathcal{J}}_0 T \longmapsto u_H \in \underline{T}^* \otimes_{\mathcal{O}} \underline{T} .$$

If $\mathfrak{U}$ represents u then $u_H = \mathfrak{U}_H \in \Phi^1_{PH} \simeq \underline{T}^* \otimes_{\mathcal{O}} \underline{T}$ (cf. Section 14).

Let $F \in \Gamma_1 X$ be represented by the local admissible transformation $F = (f, \varphi)$. The formula (22.11) together with the above identifications give

$$\tilde{\mathfrak{D}}F = \frac{\partial \varphi_\mu}{\partial x_j}(x, x)\left(\frac{\partial f_i^{-1}}{\partial x'_\mu} \circ f\right) dx_j \otimes \frac{\partial}{\partial x_i} \in \underline{T}^* \otimes_{\mathcal{O}} \underline{T} . \tag{24.2}$$

Since

$$\frac{\partial \varphi_\mu}{\partial x_j}(x, x) + \frac{\partial \varphi_\mu}{\partial y_j}(x, x) = \frac{\partial f_\mu}{\partial x_j}$$

(24.2) becomes

$$\tilde{\mathfrak{D}}F = \left[\delta^i_j - \left(\frac{\partial f_i^{-1}}{\partial x'_\mu} \circ f\right) \frac{\partial \varphi_\mu}{\partial y_j}(x, x)\right] dx_j \otimes \frac{\partial}{\partial x_i} .$$

Write

$$A^i_j(x) = \delta^i_j - \left(\frac{\partial f_i^{-1}}{\partial x'_\mu} \circ f\right) \frac{\partial \varphi_\mu}{\partial y_j}(x, x)$$

or, in matrix notation,

$$A(x) = \mathrm{Id} - \left(\frac{\partial f^{-1}}{\partial x'} \circ f\right) \frac{\partial \varphi}{\partial y}(x, x) .$$

Since the matrices $\frac{\partial f^{-1}}{\partial x'}$ and $\frac{\partial \varphi}{\partial y}(x, x)$ are invertible it follows that $\mathrm{Id} - A(x)$ is an invertible matrix hence $\mathrm{Id} - \tilde{\mathfrak{D}}F \in \mathcal{H}om_{\mathcal{O}}(\underline{T})$ is an invertible germ of homomorphism, i.e.,

$$\mathrm{Id} - \tilde{\mathfrak{D}}F \in \mathcal{A}ut_{\mathcal{O}}(\underline{T}) . \tag{24.3}$$

The intrinsic version of the above *coordinate* argument is the following. If $F \in \Gamma_1 X$ then

$$\tilde{\mathfrak{D}}F = \chi_0 - \mathfrak{A}dF(\chi_0)$$

which, by the identification $\underline{T}^* \otimes_{\mathcal{O}} \mathcal{J}_0 T \simeq \underline{T}^* \otimes_{\mathcal{O}} \underline{T}$, gives

$$\tilde{\mathfrak{D}}F \equiv (\tilde{\mathfrak{D}}F)_H = \mathrm{Id} - [\mathfrak{A}dF(\chi_0)]_H \in \underline{T}^* \otimes_{\mathcal{O}} \underline{T} . \tag{24.4}$$

The relation (24.3) reads in this context

$$[\mathfrak{A}dF(\chi_0)]_H \in \mathfrak{A}ut_{\mathcal{O}}(\underline{T}) .$$

For any $F \in \Gamma_{k+1} X$ we have by definition

$$\mathfrak{A}dF(\alpha \otimes \xi) = \mathfrak{A}dF(\alpha) \otimes F^{-1} \bullet \xi$$

hence

$$[\mathfrak{A}dF(\alpha \otimes \xi)]_H = \mathfrak{A}dF(\alpha) \otimes f_*^{-1}(\xi_H) .$$

Furthermore, the definition of

$$\mathfrak{A}dF\colon (\Lambda \underline{T}^*)_b \to (\Lambda \underline{T}^*)_a$$

shows, by duality, that this map is the transpose of a semilinear map $(\Lambda \underline{T})_a \to (\Lambda \underline{T})_b$ (with respect to $f_*\colon \mathcal{O}_a \to \mathcal{O}_b$) which is equal to the exterior product of the semilinear map

$$\theta \in (\underline{T})_a \longmapsto \pi_{2*}[(F_* j\theta)|_\Delta] \in (\underline{T})_b$$

where $a = a_1 F$, $b = b_1 F$ and F is a local admissible transformation which represents F. It follows that

$$[\mathfrak{A}dF(\chi_k)]_H = \mathfrak{A}dF(dx'_i) \otimes f_*^{-1} \frac{\partial}{\partial x'_i}$$

is equal to the map

$$\theta \in (\underline{T})_a \longmapsto f_*^{-1}[\pi_{2*}(F_* j\theta)|_\Delta] = \pi_{2*} \circ (\tilde{jf}^{-1})_*(F_* j\theta)|_\Delta \in (\underline{T})_a$$

which by inspection is seen to be invertible.

Conversely let

$$u = a^i_j(x)\, dx_j \otimes \frac{\partial}{\partial x_i} \in \underline{T}^* \otimes_{\mathcal{O}} \underline{T}$$

be a germ at the point $a \in X$ and assume that $\mathrm{Id} - u$ is invertible. Then there exists $F \in \Gamma_1 X$ such that $\tilde{\mathfrak{D}}F = u$. In fact, let $\varphi = (\varphi_1, \ldots, \varphi_n)$ be a function defined in the neighborhood of (a, a) in X^2 which satisfies the conditions

$$\varphi_i(x, x) = x_i \quad \text{and} \quad \frac{\partial \varphi_i}{\partial y_j}(x, x) = \delta^i_j - a^i_j(x).$$

For example

$$\varphi_i(x, y) = x_i + \sum_j [\delta^i_j - a^i_j(x)]\,(y_j - x_j), \quad \text{(cf. (18.1))} .$$

Then $F = (\mathrm{Id}, \varphi)$ is an admissible transformation in a neighborhood of (a, a) and $\tilde{\mathfrak{D}}F = u$ with $F = [F]_{(a,a)} \bmod \sim_1$, since $f = \mathrm{Id}$ and

$$\frac{\partial \varphi_i}{\partial x_j}(x, x) = a^i_j(x) .$$

This argument shows that

$$\text{(24.5)} \qquad \tilde{\mathfrak{D}}(\Gamma_1 X) = \tilde{\mathfrak{D}}(\gamma_1 X) = \{u \in \underline{T}^* \otimes_{\mathcal{O}} \tilde{\mathcal{J}}_0 T \mid \mathrm{Id} - u \in \mathcal{A}ut_{\mathcal{O}}(\underline{T})\} .$$

Remark. If $u, v \in \wedge \underline{T}^* \otimes_{\mathcal{O}} \tilde{\mathcal{J}}_k T$ then $\tilde{\rho}_h[u, v] = [\tilde{\rho}_h u, \tilde{\rho}_h v]$ for any $h \leq k$ but $\tilde{\rho}_h [\![u, v]\!] = [\![\tilde{\rho}_h u, \tilde{\rho}_h v]\!]$ is only true for $h \geq 1$, not for $h = 0$. In other words, $\tilde{\rho}_h$ is a Lie algebra morphism with respect to $[\ ,\ ]$ for any $h \leq k$ and with respect to $[\![\ ,\ ]\!]$ for any $1 \leq h \leq k$. Nevertheless, if

$$u, v \in \mathcal{K}er\,(\wedge \underline{T}^* \otimes_{\mathcal{O}} \tilde{\mathcal{J}}_k T \xrightarrow{\tilde{\rho}_0} \wedge \underline{T}^* \otimes_{\mathcal{O}} \tilde{\mathcal{J}}_0 T)$$

then

$$\tilde{\rho}_0 [\![u, v]\!] = [\![\tilde{\rho}_0 u, \tilde{\rho}_0 v]\!] = [\tilde{\rho}_0 u, \tilde{\rho}_0 v] = 0 .$$

Moreover, if $u, v \in \wedge \underline{T}^* \otimes_{\mathcal{O}} \tilde{\mathcal{J}}_k T$ and if we set $\{u, v\} = \tilde{\rho}_{k-1}[\![u, v]\!]$, then

$$\tilde{\rho}_{h-1}\{u, v\} = \{\tilde{\rho}_h u, \tilde{\rho}_h v\}$$

for any $h \le k$. We infer that

$$\tilde{\rho}_{h-1} \circ \tilde{\mathfrak{D}}_1(u) = \tilde{\mathfrak{D}}_1 \circ \tilde{\rho}_h(u)$$

for any $h \le k$ and $u \in \wedge \underline{T}^* \otimes_{\mathcal{O}} \tilde{\mathcal{J}}_k T$, i.e., $\tilde{\mathfrak{D}}_1$ commutes with $\tilde{\rho}_h$. Since $\mathcal{A}d F$ commutes with ρ_h and $\tilde{\rho}_h$ (for $h \ge 1$) it follows that $\tilde{\mathfrak{D}}$ commutes with ρ_h and $\tilde{\rho}_{h-1}$ for any $h \le k+1$ (we set $\tilde{\mathfrak{D}} = 0$ on $\Gamma_0 X \simeq \mathcal{A}ut\, X$ and $\tilde{\mathfrak{D}}_1 = 0$ on $\underline{T}^* \otimes_{\mathcal{O}} \tilde{\mathcal{J}}_0 T$). We remark finally that

$$\tilde{\mathfrak{D}}_1 \,|\, \underline{T}^* \otimes_{\mathcal{O}} S^k \underline{T}^* \otimes_{\mathcal{O}} \underline{T} = \tilde{\mathfrak{D}}_1 \,|\, \ker \tilde{\rho}_{k-1} = \underline{\delta}$$

for any $k \ge 0$ since

$$\tilde{D} \,|\, \ker \tilde{\rho}_{k-1} = \underline{\delta} \quad \text{and} \quad \tilde{\rho}_{k-1}[\![u, v]\!] = [\![\tilde{\rho}_{k-1} u, \tilde{\rho}_{k-1} v]\!] = 0$$

for any $u, v \in \ker \tilde{\rho}_{k-1}$.

THEOREM. *For any* $k \ge 0$

$$\tilde{\mathfrak{D}}(\Gamma_{k+1} X) = \{u \in \underline{T}^* \otimes_{\mathcal{O}} \tilde{\mathcal{J}}_k T \mid Id - \tilde{\rho}_0 u \in \mathcal{A}ut_{\mathcal{O}}(\underline{T}),\ \tilde{\mathfrak{D}}_1 u = 0\}\,.$$

Proof. If $F \in \Gamma_{k+1} X$, the structure equation implies that $\tilde{\mathfrak{D}}_1(\tilde{\mathfrak{D}} F) = 0$. Moreover, since $\tilde{\mathfrak{D}}$ commutes with ρ_h and $\tilde{\rho}_{h-1}$, $h \le k$, (24.5) implies that $Id - \tilde{\rho}_0 \tilde{\mathfrak{D}} F = Id - \tilde{\mathfrak{D}}(\rho_1 F) \in \mathcal{A}ut_{\mathcal{O}}(\underline{T})$. The converse will be proved by induction on k. If $k = 0$ the second condition is void ($\tilde{\mathfrak{D}}_1 = 0$) and the theorem reduces to (24.5). Assume that the assertion holds for an integer $k \ge 0$. The previous remark and the commutativity of (24.1) implies that the following diagram is commutative:

(24.6)

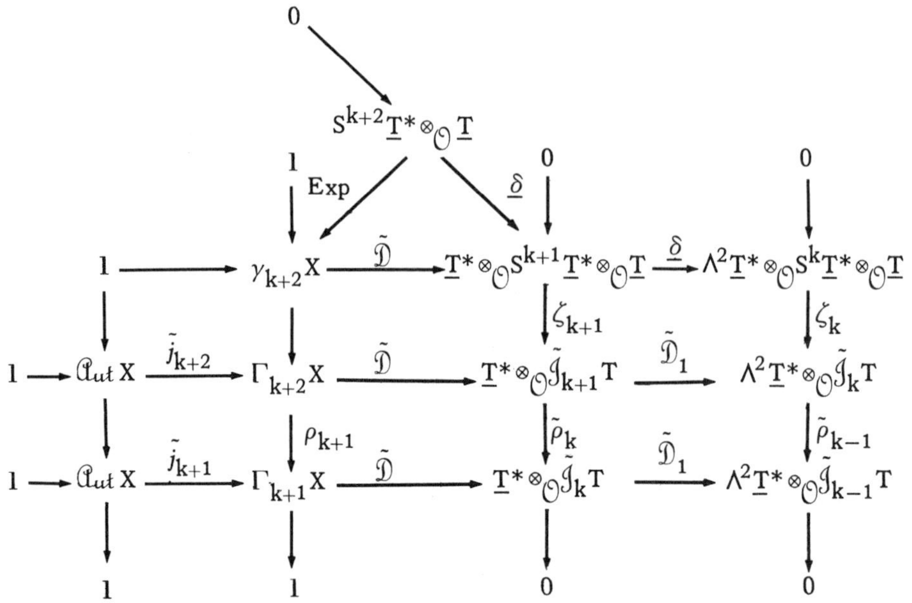

We distinguish two cases.

a) $k = 0$.

Let $u \in \underline{T}^* \otimes_{\mathcal{O}} \tilde{\mathfrak{J}}_1 T$ and assume that $\tilde{\mathfrak{D}}_1 u = 0$ and $\mathrm{Id} - \tilde{\rho}_0 u \in \mathfrak{Aut}_{\mathcal{O}}(\underline{T})$. By (24.5) there exists $G \in \Gamma_1 X$ such that $\tilde{\mathfrak{D}}G = \tilde{\rho}_0 u$. Let $F \in \Gamma_2 X$ be a lifting of G, i.e., $\rho_1 F = G$. Then

$$u - \tilde{\mathfrak{D}}F \in \underline{T}^* \otimes_{\mathcal{O}} S^1 \underline{T}^* \otimes_{\mathcal{O}} \underline{T}$$

and we claim that

$$\underline{\delta}(u - \tilde{\mathfrak{D}}F) = 0 .$$

In fact,

$$\begin{aligned}\underline{\delta}(u-\tilde{\mathfrak{D}}F) &= \tilde{\mathfrak{D}}_1(u-\tilde{\mathfrak{D}}F) = \tilde{D}(u-\tilde{\mathfrak{D}}F) - \tfrac{1}{2}\tilde{\rho}_0[\![u-\tilde{\mathfrak{D}}F, u-\tilde{\mathfrak{D}}F]\!] = \\ &\tilde{D}u - \tfrac{1}{2}\tilde{\rho}_0[\![u,u]\!] - \tilde{D}(\tilde{\mathfrak{D}}F) - \tfrac{1}{2}\tilde{\rho}_0[\![\tilde{\mathfrak{D}}F, \tilde{\mathfrak{D}}F]\!] + \tilde{\rho}_0[\![u,\tilde{\mathfrak{D}}F]\!] = \tilde{\rho}_0[\![u-\tilde{\mathfrak{D}}F, \tilde{\mathfrak{D}}F]\!] = \\ &\tilde{\rho}_0\{[u-\tilde{\mathfrak{D}}F, \tilde{\mathfrak{D}}F] + \tilde{D}(u-\tilde{\mathfrak{D}}F) \barwedge \tilde{\mathfrak{D}}F + \tilde{D}(\tilde{\mathfrak{D}}F) \barwedge (u-\tilde{\mathfrak{D}}F)\} = \underline{\delta}(u-\tilde{\mathfrak{D}}F) \barwedge \tilde{\rho}_0 u\end{aligned}$$

since

$$\tilde{\rho}_0[u-\tilde{\mathfrak{D}}F,\tilde{\mathfrak{D}}F] = [\tilde{\rho}_0(u-\tilde{\mathfrak{D}}F),\tilde{\rho}_0\tilde{\mathfrak{D}}F] = [0,\tilde{\rho}_0\tilde{\mathfrak{D}}F] = 0 ,$$

$$\tilde{D}(u-\tilde{\mathfrak{D}}F) = \underline{\delta}(u-\tilde{\mathfrak{D}}F) ,$$

$$\tilde{\rho}_0(\tilde{D}(u-\tilde{\mathfrak{D}}F)\barwedge\tilde{\mathfrak{D}}F) = \underline{\delta}(u-\tilde{\mathfrak{D}}F)\barwedge\tilde{\rho}_0\tilde{\mathfrak{D}}F = \underline{\delta}(u-\tilde{\mathfrak{D}}F)\barwedge\tilde{\rho}_0 u$$

and

$$\tilde{\rho}_0\{\tilde{D}(\tilde{\mathfrak{D}}F)\barwedge(u-\tilde{\mathfrak{D}}F)\} = \tilde{D}(\tilde{\mathfrak{D}}F)\barwedge\tilde{\rho}_0(u-\tilde{\mathfrak{D}}F) = \tilde{D}(\tilde{\mathfrak{D}}F)\barwedge 0 = 0 .$$

We infer that

$$0 = \underline{\delta}(u-\tilde{\mathfrak{D}}F) - \underline{\delta}(u-\tilde{\mathfrak{D}}F)\barwedge\tilde{\rho}_0 u = \underline{\delta}(u-\tilde{\mathfrak{D}}F) - (\tilde{\rho}_0 u)\circ\underline{\delta}(u-\tilde{\mathfrak{D}}F)$$

or, equivalently,

$$(\mathrm{Id} - \tilde{\rho}_0 u)(\underline{\delta}(u-\tilde{\mathfrak{D}}F)) = 0 .$$

Since $\mathrm{Id} - \tilde{\rho}_0 u$ is invertible it follows that $\underline{\delta}(u-\tilde{\mathfrak{D}}F) = 0$. The exactness of the $\underline{\delta}$-complex (cf. the first line of $(6.6)_k$) implies that there exists $w \in S^2\underline{T}^*\otimes_{\mathcal{O}}\underline{T}$ such that

$$\underline{\delta}(w) = u-\tilde{\mathfrak{D}}F .$$

Let $F_1 = \mathrm{Exp}\, w \in \gamma_2 X$. Then

$$\tilde{\mathfrak{D}}F_1 = \underline{\delta}(w) = u-\tilde{\mathfrak{D}}F$$

and

$$\tilde{\mathfrak{D}}(F\bullet F_1) = \tilde{\mathfrak{D}}F_1 + \mathfrak{Ad}F_1(\tilde{\mathfrak{D}}F) = \tilde{\mathfrak{D}}F_1 + \tilde{\mathfrak{D}}F = u$$

since $\rho_1 F_1 = 1$.

b) $k > 0$.

Let $u \in \underline{T}^*\otimes_{\mathcal{O}}\tilde{\mathfrak{J}}_{k+1}T$ and assume that $\tilde{\mathfrak{D}}_1 u = 0$ and $\mathrm{Id}-\tilde{\rho}_0 u \in \mathfrak{Aut}_{\mathcal{O}}(\underline{T})$. Let $v = \tilde{\rho}_k u \in \underline{T}^*\otimes_{\mathcal{O}}\tilde{\mathfrak{J}}_k T$. Then $\tilde{\mathfrak{D}}_1 v = 0$ and $\mathrm{Id}-\tilde{\rho}_0 v = \mathrm{Id}-\tilde{\rho}_0 u \in \mathfrak{Aut}_{\mathcal{O}}(\underline{T})$; hence, by induction, there exists $G \in \Gamma_{k+1}X$ such that $\tilde{\mathfrak{D}}G = v$. Let $F \in \Gamma_{k+2}X$ be a lifting of G. Then

$$u-\tilde{\mathfrak{D}}F \in \underline{T}^* \otimes_{\mathcal{O}} S^{k+1}\underline{T}^* \otimes_{\mathcal{O}} \underline{T}$$

and we claim that $\underline{\delta}(u-\tilde{\mathfrak{D}}F) = 0$. In fact, since $k > 0$ then

$$\underline{\delta}(u-\tilde{\mathfrak{D}}F) = \tilde{\mathfrak{D}}_1(u-\tilde{\mathfrak{D}}F) = \tilde{\rho}_k[\![u-\tilde{\mathfrak{D}}F, \tilde{\mathfrak{D}}F]\!] = [\![\tilde{\rho}_k(u-\tilde{\mathfrak{D}}F), \tilde{\rho}_k\tilde{\mathfrak{D}}F]\!] = [\![0, \tilde{\rho}_k\tilde{\mathfrak{D}}F]\!] = 0 \,.$$

Take $w \in S^{k+2}\underline{T}^* \otimes_{\mathcal{O}} \underline{T}$ such that $\underline{\delta}(w) = u-\tilde{\mathfrak{D}}F$. Then $u = \tilde{\mathfrak{D}}(F \bullet F_1)$ with $F_1 = \operatorname{Exp} w$ since $\rho_{k+1}F_1 = 1$.

The two conditions in the theorem are independent. In fact, if we take

$$u = \tilde{\chi}_k = \sum dx_i \otimes \left(\frac{\partial}{\partial x_i} + \frac{\partial}{\partial y_i}\right) \mathrm{mod}\ \mathfrak{g}^{k+1} \Delta \tilde{\Phi} \in \underline{T}^* \otimes_{\mathcal{O}} \mathfrak{J}_k T$$

then $\tilde{\mathfrak{D}}_1\tilde{\chi}_k = 0$ and $\tilde{\rho}_0\tilde{\chi}_k = \mathrm{Id}$ hence $\mathrm{Id} - \tilde{\rho}_0\tilde{\chi}_k = 0$. It follows that $\tilde{\chi}_k$ is not in the image of $\tilde{\mathfrak{D}}$.

Remark. If $u \in \underline{T}^* \otimes_{\mathcal{O}} \tilde{\mathfrak{J}}_k T$ satisfies the conditions of the theorem then we can actually choose $F \in \Gamma_{k+1}X$ such that $\rho_0 F = \mathrm{Id} \in \mathfrak{Aut}\ X$ and $\tilde{\mathfrak{D}}F = u$.

We shall now derive some consequences of the theorem. For $k \geq 0$, the sequence

$$1 \longrightarrow \gamma_{k+1}X \xrightarrow{\tilde{\mathfrak{D}}} \underline{T}^* \otimes_{\mathcal{O}} S^k\underline{T}^* \otimes_{\mathcal{O}} \underline{T}$$

is exact since

$$\gamma_{k+1}X \cap \tilde{j}_{k+1}(\mathfrak{Aut}\ X) = 1 \,.$$

If $k \geq 1$ and $u \in \underline{T}^* \otimes_{\mathcal{O}} S^k\underline{T}^* \otimes_{\mathcal{O}} \underline{T}$ then $\tilde{\rho}_0 u = 0$, hence $\mathrm{Id} - \tilde{\rho}_0 u = \mathrm{Id} \in \mathfrak{Aut}_{\mathcal{O}}(\underline{T})$. We infer that the sequence

$$(24.7) \quad 1 \longrightarrow \gamma_{k+2}X \xrightarrow{\tilde{\mathfrak{D}}} \underline{T}^* \otimes_{\mathcal{O}} S^{k+1}\underline{T}^* \otimes_{\mathcal{O}} \underline{T} \xrightarrow{\underline{\delta}} \Lambda^2\underline{T}^* \otimes_{\mathcal{O}} S^k\underline{T}^* \otimes_{\mathcal{O}} \underline{T} \xrightarrow{\underline{\delta}} \cdots$$
$$\xrightarrow{\underline{\delta}} \Lambda^n\underline{T}^* \otimes_{\mathcal{O}} S^{k+2-n}\underline{T}^* \otimes_{\mathcal{O}} \underline{T} \longrightarrow 0$$

is exact for $k \geq 0$. The property does not hold for $k = -1$ (cf. (24.5)).

The map

$$\mathrm{Exp}\colon S^{k+2}\underline{T}^* \otimes_{\mathcal{O}} \underline{T} \longrightarrow \gamma_{k+2}X$$

is bijective for $k \geq 0$. In fact, from the commutativity of (24.1) and the injectiveness of $\underline{\delta}$ we infer that Exp is injective. If $F \in \gamma_{k+2}X$ then

$$\underline{\delta}(\tilde{\mathfrak{D}}F) = \tilde{\mathfrak{D}}_1(\tilde{\mathfrak{D}}F) = 0$$

hence there exists $w \in S^{k+2}\underline{T}^* \otimes_{\mathcal{O}} \underline{T}$ such that

$$\underline{\delta}(w) = \tilde{\mathfrak{D}}F = \tilde{\mathfrak{D}}(\mathrm{Exp}\, w)\ .$$

Since $\tilde{\mathfrak{D}}$ is injective it follows that $F = \mathrm{Exp}\, w$. Let us prove, next, that Exp is an isomorphism from the additive structure of $S^{k+2}\underline{T}^* \otimes_{\mathcal{O}} \underline{T}$ onto the abelian group structure of $\gamma_{k+2}X$. For this, we simply observe that $\rho_{k+1}(\gamma_{k+2}X) = 1$, hence

$$\tilde{\mathfrak{D}}(F \bullet G) = \tilde{\mathfrak{D}}F + \tilde{\mathfrak{D}}G$$

for any $F, G \in \gamma_{k+2}X$. $\tilde{\mathfrak{D}}$ is an injective morphism with respect to the group structure of $\gamma_{k+2}X$ and the additive structure of $\underline{T}^* \otimes_{\mathcal{O}} S^{k+1}\underline{T}^* \otimes_{\mathcal{O}} \underline{T}$. The commutativity of (24.1) yields the equation

$$\tilde{\mathfrak{D}}[\mathrm{Exp}(u+v)] = \underline{\delta}(u) + \underline{\delta}(v) = \tilde{\mathfrak{D}}(\mathrm{Exp}\, u) + \tilde{\mathfrak{D}}(\mathrm{Exp}\, v) = \tilde{\mathfrak{D}}[(\mathrm{Exp}\, u) \bullet (\mathrm{Exp}\, v)]$$

for any $u, v \in S^{k+2}\underline{T}^* \otimes_{\mathcal{O}} \underline{T}$, hence

$$\mathrm{Exp}(u+v) = (\mathrm{Exp}\, u) \bullet (\mathrm{Exp}\, v)\ .$$

If we identify $S^{k+2}\underline{T}^* \otimes_{\mathcal{O}} \underline{T}$ with $\gamma_{k+2}X$ via the Exponential map then $\tilde{\mathfrak{D}}$ becomes $\underline{\delta}$ and (24.7) becomes the $\underline{\delta}$-complex. The previous results do not hold for $k = -1$.

It is instructive to give a proof of the above properties without using $\tilde{\mathfrak{D}}$ and $\tilde{\mathfrak{D}}_1$. Let us start by proving that Exp is a morphism. Let

$$\Xi, \Sigma \in \Gamma(U, S^{k+2}\underline{T}^* \otimes_{\mathcal{O}} \underline{T})$$

and take two local admissible vector fields ξ, σ defined in a neighborhood

V of (a, a) in X^2 ($a \in U$) which represent Ξ and Σ on $V \cap \Delta$. Let (φ_t) and (ψ_t) be the local 1-parameter groups generated by ξ and σ and set $\mu_t = \psi_t \circ \varphi_t$. We observe that

$$\sigma + \xi = \frac{d}{dt}\mu_t|_{t=0} . \tag{24.8}$$

The 1-parameter family (μ_t) is not a local 1-parameter group on X^2. However, since φ_t and ψ_t are equivalent to Id up to terms of order $\geq k+2$ along Δ, it follows that $\varphi_t \circ \psi_{t'}$ is equivalent to $\psi_{t'} \circ \varphi_t$ up to terms of order $\geq k+3$ hence (μ_t) is a local 1-parameter group up to terms of order $\geq k+3$ along Δ or, in view of (24.8), (μ_t) is equivalent to the local 1-parameter group (η_t) generated by $\sigma + \xi$ up to terms of order $\geq k+3$ along Δ. More precisely, if we set

$$F_t = \mu_t \ \mathit{mod} \sim_{k+2} = (\operatorname{Exp} t\Sigma) \bullet (\operatorname{Exp} t\Xi)$$

then, since $\gamma_{k+2}X$ is abelian, it follows that (F_t) is a differentiable local 1-parameter group of sections in $\Gamma_{loc}(X, \gamma_{k+2}X)$ which satisfies on $V = V \cap \Delta$ the condition

$$\frac{d}{dt}F_t|_{t=0} = \sigma+\xi \ \mathit{mod} \ \mathcal{J}^{k+3}{}_{\Delta}\tilde{\Theta} = (\Sigma+\Xi)|_V .$$

The previous discussion (or the uniqueness property of the local Exponential map) implies that

$$F_t = \eta_t \ \mathit{mod} \sim_{k+2} = \operatorname{Exp} t(\Sigma+\Xi) .$$

In the beginning of this section we noticed that $\operatorname{Exp} t\Xi$ and $\operatorname{Exp} t\Sigma$ are defined on U for all $t \in \mathbf{R}$. Since $\gamma_{k+2}X$ is abelian, the family $(\operatorname{Exp} t\Sigma) \bullet (\operatorname{Exp} t\Xi)$ is a differentiable 1-parameter group of sections in $\Gamma(U, \gamma_{k+2}X)$ and the previous local discussion shows that

$$\frac{d}{dt}(\operatorname{Exp} t\Xi) \bullet (\operatorname{Exp} t\Xi)|_{t=0} = \Sigma + \Xi .$$

We infer from the uniqueness property of the Exponential map that

$$(\mathrm{Exp}\, t\Sigma)\bullet(\mathrm{Exp}\, t\Xi) = \mathrm{Exp}\, t(\Sigma+\Xi)\ ,$$

hence the desired relation for $t = 1$.

Let us now examine the bijectivity of Exp. Recall that

$$J_{k+1}X \xrightarrow{\rho_k} J_kX$$

is an affine bundle with respect to the vector bundle

$$J_kX \times_X (S^{k+1}T^* \otimes T)\ ,$$

the fibre product being taken with respect to $\beta_k\colon J_kX \to X$ and the standard projection $S^{k+1}T^*\otimes T \to X$ (see [5(b)], [7]). Next observe that, for $k \geq 1$, the bundle

$$\Pi_{k+1}X \xrightarrow{\rho_k} \Pi_kX$$

is equal to the restriction

$$J_{k+1}X \xrightarrow{\rho_k} \Pi_kX\ ,$$

hence it also carries an affine bundle structure over the vector bundle

$$\Pi_kX \times_X (S^{k+1}T^* \otimes T)\ .$$

It follows that

$$\Gamma_{k+1}X \xrightarrow{\rho_k} \Gamma_kX$$

is an affine sheaf with respect to the $\mathcal{O}$-linear sheaf

$$\Gamma_kX \times_X (S^{k+1}\underline{T}^* \otimes_{\mathcal{O}} \underline{T})$$

for any $k \geq 1$. Correspondingly, the sheaf of groups $\gamma_{k+1}X$ operates to the left on $\Gamma_{k+1}X$ and this action is simply transitive on the stalks of

$$\Gamma_{k+1}X \xrightarrow{\rho_k} \Gamma_kX\ .$$

It can be proved (see [7]) that, for $k \geq 1$, this group action is equivalent to the previous affine action via the Exponential map, i.e.,

$$(\mathrm{Exp}\Xi)\bullet F = \Xi(F)$$

for any $\Xi \in S^{k+1}\underline{T}^* \otimes_{\mathcal{O}} \underline{T}$ and $F \in \Gamma_{k+1}X$. In particular,

$$\mathrm{Exp}\Xi = \Xi(1_{k+1})$$

where $1_{k+1} \in \Gamma_{k+1}X$ and the right hand operation is the affine action of $S^{k+1}\underline{T}^* \otimes_{\mathcal{O}} \underline{T}$ on the stalk of $\Gamma_{k+1}X$ over 1_k, i.e., on the stalk of $\gamma_{k+1}X$ over 1_k. This action, commencing at a fixed element, is bijective. The Exponential map

$$\mathrm{Exp}\colon S^{k+1}\underline{T}^* \otimes_{\mathcal{O}} \underline{T} \to \gamma_{k+1}X,\ k \geq 1$$

is simply the affine action of $S^{k+1}\underline{T}^* \otimes \underline{T}$ on $\Gamma_{k+1}X$ commenced at the identity section 1_{k+1}.

Remark. Let $\mathcal{R}_k$ be a linear Lie equation and $\Gamma(\mathcal{R}_k)$ the corresponding non-linear equation. Then, as in the linear case, the complexes $(23.9)_{\ell+1}$ restrict to $\Gamma(\mathcal{R}_k)$ and its prolongations. Since the proof of this fact is somewhat more elaborate than in the linear case, we postpone a detailed account to the second part of the present monograph. The same remark holds for the first non-linear complex.

25. *The third non-linear complex*

In this section we define a non-linear complex which is a finite form of the initial portion of the second linear complex (Section 7).

Define the quotient sheaves

$$\tilde{\mathcal{C}}^p_k T = \Lambda^p \underline{T}^* \otimes_{\mathcal{O}} \tilde{\mathcal{J}}_k T / \zeta_{p,k} \circ \underline{\delta}_{p-1,k+1}(\Lambda^{p-1}\underline{T}^* \otimes_{\mathcal{O}} S^{k+1}\underline{T} \otimes_{\mathcal{O}} \underline{T}) .$$

Then, $\tilde{\mathcal{C}}^0_k T = \tilde{\mathcal{J}}_k T$ for $k \geq 0$, $\tilde{\mathcal{C}}^p_k T = 0$ for $p > n$, $\tilde{\mathcal{C}}^p_k T = 0$ for $k < 0$ and $\tilde{\mathcal{C}}^p_0 T = 0$ for $p > 0$. The sheaves $\tilde{\mathcal{C}}^p_k T$ inherit a diagonal module structure over the left and right $\mathcal{O}$-algebra $\mathcal{J}_k$. Moreover, $\tilde{\mathcal{C}}^p_k T$ is canonically isomorphic to $\mathcal{C}^p_k T$ via the quotient of

$$\varepsilon_k : \Lambda^p \underline{T}^* \otimes_{\mathcal{O}} \tilde{\mathcal{J}}_k T \to \Lambda^p \underline{T}^* \otimes_{\mathcal{O}} \mathcal{J}_k T \simeq \Lambda^p \underline{T}^* \otimes_{\mathcal{O}} \mathcal{J}_k \otimes \underline{T} .$$

The complex $(7.1)_k$ with $\mathcal{E} = \underline{T}$ is transformed, via ε_k^{-1}, into the exact complex

$$(25.1)_k \qquad 0 \longrightarrow \underline{T} \xrightarrow{\tilde{j}_k} \tilde{\mathcal{C}}^0_k T \xrightarrow{\tilde{D}'} \tilde{\mathcal{C}}^1_k T \xrightarrow{\tilde{D}'} \cdots \xrightarrow{\tilde{D}'} \tilde{\mathcal{C}}^n_k T \longrightarrow 0 .$$

Consider the sheaf

$$\tilde{\mathcal{C}}^{\bullet}_k T = \bigoplus_p \tilde{\mathcal{C}}^p_k T = \Lambda \underline{T}^* \otimes_{\mathcal{O}} \tilde{\mathcal{J}}_k T / \zeta_k \circ \underline{\delta}_{k+1} (\Lambda \underline{T}^* \otimes_{\mathcal{O}} S^{k+1} \underline{T}^* \otimes_{\mathcal{O}} \underline{T})$$

the quotienting space being a homogeneous submodule of $\Lambda \underline{T}^* \otimes_{\mathcal{O}} \tilde{\mathcal{J}}_k T$. The bracket $[\![\ ,\ ]\!]$ defined on $\Lambda \underline{T}^* \otimes_{\mathcal{O}} \tilde{\mathcal{J}}_k T$ factors to $\tilde{\mathcal{C}}^{\bullet}_k T$ and defines on this sheaf a structure of graded Lie algebra. In fact, we shall prove that

$$\zeta_k \circ \underline{\delta}_{k+1} (\Lambda \underline{T}^* \otimes_{\mathcal{O}} S^{k+1} \underline{T}^* \otimes_{\mathcal{O}} \underline{T})$$

is an ideal in $\Lambda \underline{T}^* \otimes_{\mathcal{O}} \tilde{\mathcal{J}}_k T$. If $k = 0$ the result is obvious since the first sheaf is equal to

$$\bigoplus_{p>0} (\Lambda^p \underline{T}^* \otimes_{\mathcal{O}} \tilde{\mathcal{J}}_0 T)$$

and $[\![\ ,\ ]\!] = [\ ,\ ]$ on $\Lambda \underline{T}^* \otimes_{\mathcal{O}} \tilde{\mathcal{J}}_0 T \equiv \Lambda \underline{T}^* \otimes_{\mathcal{O}} \underline{T}$. We infer that $\tilde{\mathcal{C}}^{\bullet}_0 T = \tilde{\mathcal{J}}_0 T \equiv \underline{T}$ and the quotient bracket identifies with the ordinary bracket on $\underline{T}$. Assume that $k > 0$. Since

$$\tilde{\rho}_k : \Lambda \underline{T}^* \otimes_{\mathcal{O}} \tilde{\mathcal{J}}_{k+1} T \to \Lambda \underline{T}^* \otimes_{\mathcal{O}} \tilde{\mathcal{J}}_k T$$

is surjective, it is enough to prove the

LEMMA. *For any* $k \geq 1$, *the bracket* $[\![\ ,\ ]\!]$ *commutes with* $\underline{\delta}$ *and* $\tilde{\rho}_k$ *in the sense that*

$$\underline{\delta} [\![u, v]\!] = [\![\underline{\delta} u, \tilde{\rho}_k v]\!]$$

where $u \in \Lambda \underline{T}^* \otimes_{\mathcal{O}} S^{k+1}\underline{T}^* \otimes_{\mathcal{O}} \underline{T} = \ker \tilde{\rho}_k$ *and* $v \in \Lambda \underline{T}^* \otimes_{\mathcal{O}} \tilde{\mathcal{J}}_{k+1} T$ *(we omit* the ζ). *The property does not hold for* $k = 0$.

Proof. Assume that $u \in \Lambda^p \underline{T}^* \otimes_{\mathcal{O}} S^{k+1}\underline{T}^* \otimes_{\mathcal{O}} \underline{T}$. Applying Lemma 6 of Section 16 and observing that $\underline{\delta} u = \tilde{D} u$ we obtain

$$\tilde{D}[\![u, v]\!] = [\![\tilde{D}u, \tilde{\rho}_k v]\!] + (-1)^p [\![0, \tilde{D}v]\!] .$$

Since $k \geq 1$, $\tilde{\rho}_k [\![u, v]\!] = [\![\tilde{\rho}_k u, \tilde{\rho}_k v]\!] = 0$ and therefore

$$[\![u, v]\!] \in \Lambda \underline{T}^* \otimes_{\mathcal{O}} S^{k+1}\underline{T}^* \otimes_{\mathcal{O}} \underline{T}$$

and the previous relation reduces to $\underline{\delta}[\![u, v]\!] = [\![\underline{\delta} u, \tilde{\rho}_k v]\!]$.

Let us now define the third non-linear complex. In the diagram (24.6) consider the operator

$$\tilde{\mathfrak{D}}: \Gamma_{k+2} X \longrightarrow \underline{T}^* \otimes_{\mathcal{O}} \tilde{\mathcal{J}}_{k+1} T .$$

Let $F, G \in \Gamma_{k+2} X$ and assume that $\rho_{k+1} F = \rho_{k+1} G$ or, equivalently, $\rho_{k+1}(F^{-1} \bullet G) = 1_{k+1}$. If we write $F^{-1} \bullet G = H$ then $G = F \bullet H$ and, since $\rho_{k+1} H = 1_{k+1}$, we obtain

$$\tilde{\mathfrak{D}} G = \tilde{\mathfrak{D}} F + \tilde{\mathfrak{D}} H .$$

Furthermore, since $H \in \gamma_{k+2} X$, there exists $w \in S^{k+2}\underline{T}^* \otimes_{\mathcal{O}} \underline{T}$ such that $H = \operatorname{Exp} w$ hence $\tilde{\mathfrak{D}} H = \zeta_{k+1} \circ \underline{\delta}(w)$. We infer that $\tilde{\mathfrak{D}}$ factors to a first order operator

$$\tilde{\mathfrak{D}}': \Gamma_{k+1} X \longrightarrow \tilde{\mathcal{C}}^1_{k+1} T .$$

Since $\tilde{\mathfrak{D}} \circ \tilde{j}_{k+2} = 0$, it is clear that $\tilde{\mathfrak{D}}' \circ \tilde{j}_{k+1} = 0$. Conversely, assume that $\tilde{\mathfrak{D}}'(F) = 0$ for $F \in \Gamma_{k+1} X$. Let $G \in \Gamma_{k+2} X$ be a lifting of F. By definition,

$$\tilde{\mathfrak{D}}' F = \tilde{\mathfrak{D}} G \bmod \zeta_{k+1} \circ \underline{\delta}(S^{k+2}\underline{T}^* \otimes_{\mathcal{O}} \underline{T})$$

hence $\tilde{\mathfrak{D}}'(F) = 0$ is equivalent to $\tilde{\mathfrak{D}} G = \zeta_{k+1} \circ \underline{\delta}(w)$ with $w \in S^{k+2}\underline{T}^* \otimes_{\mathcal{O}} \underline{T}$.

The commutativity of (24.6) implies that $\tilde{\mathfrak{D}}G = \tilde{\mathfrak{D}} \circ \mathrm{Exp}(w)$, hence by (23.13) and the fact that $\rho_{k+1}\mathrm{Exp}(-w) = 1_{k+1}$ we obtain

$$\tilde{\mathfrak{D}}(G \bullet \mathrm{Exp}(-w)) = \tilde{\mathfrak{D}}G + \tilde{\mathfrak{D}}(\mathrm{Exp}(-w)) = \tilde{\mathfrak{D}}G - \tilde{\mathfrak{D}}(\mathrm{Exp}\, w) = 0 .$$

From the exactness of the second non-linear complex at $\Gamma_{k+2}X$ we infer that $G \bullet \mathrm{Exp}(-w) = \tilde{j}_{k+2}f$ with $f \in \mathcal{A}ut\, X$, hence

$$F = \rho_{k+1}G = \rho_{k+1}(G \bullet \mathrm{Exp}(-w)) = \rho_{k+1} \circ \tilde{j}_{k+2}f = \tilde{j}_{k+1}f .$$

Consider next the commutative diagram

(25.2)

$$\begin{array}{ccccc}
 & & 0 & & \\
 & & \downarrow & & \\
 & & S^{k+3}\underline{T}^*\otimes_{\mathcal{O}}\underline{T} & & \\
1 & \swarrow^{\mathrm{Exp}} & & \searrow^{\underline{\delta}} \quad 0 & \qquad 0 \\
\downarrow & & & \downarrow & \qquad \downarrow \\
\gamma_{k+3}X & \xrightarrow{\tilde{\mathfrak{D}}} & \underline{T}^*\otimes_{\mathcal{O}}S^{k+2}\underline{T}^*\otimes_{\mathcal{O}}\underline{T} & \xrightarrow{\underline{\delta}} & \Lambda^2\underline{T}^*\otimes_{\mathcal{O}}S^{k+1}\underline{T}^*\otimes_{\mathcal{O}}\underline{T} \\
\downarrow & & \downarrow \zeta_{k+2} & & \downarrow \zeta_{k+1} \\
\Gamma_{k+3}X & \xrightarrow{\tilde{\mathfrak{D}}} & \underline{T}^*\otimes_{\mathcal{O}}\tilde{\mathcal{J}}_{k+2}T & \xrightarrow{\tilde{\mathfrak{D}}_1} & \Lambda^2\underline{T}^*\otimes_{\mathcal{O}}\tilde{\mathcal{J}}_{k+1}T \\
\downarrow \rho_{k+2} & & \downarrow \tilde{\rho}_{k+1} & & \downarrow \tilde{\rho}_k \\
\Gamma_{k+2}X & \xrightarrow{\tilde{\mathfrak{D}}} & \underline{T}^*\otimes_{\mathcal{O}}\tilde{\mathcal{J}}_{k+1}T & \xrightarrow{\tilde{\mathfrak{D}}_1} & \Lambda^2\underline{T}^*\otimes_{\mathcal{O}}\tilde{\mathcal{J}}_kT \\
\downarrow & & \downarrow & & \downarrow \\
1 & & 0 & & 0
\end{array}$$

and take $u, v \in \underline{T}^*\otimes_{\mathcal{O}}\tilde{\mathcal{J}}_{k+2}T$ such that $\tilde{\rho}_{k+1}u = \tilde{\rho}_{k+1}v$, i.e., $v-u = \zeta_{k+2}(w)$ with $w \in \underline{T}^*\otimes_{\mathcal{O}} S^{k+2}\underline{T}^*\otimes_{\mathcal{O}}\underline{T}$. Then

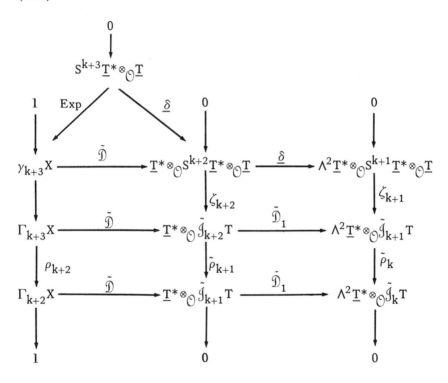

$$\begin{aligned}\tilde{\mathfrak{D}}_1(v) = \tilde{\mathfrak{D}}_1(u+\zeta_{k+2}(w)) &= \tilde{D}u + \tilde{D}\zeta_{k+2}w - \tfrac{1}{2}\tilde{\rho}_{k+1}[\![u+\zeta_{k+2}w, u+\zeta_{k+2}w]\!] = \\ &\tilde{\mathfrak{D}}_1(u) + \zeta_{k+1}\circ\underline{\delta}(w) - \tfrac{1}{2}[\![\tilde{\rho}_{k+1}\circ\zeta_{k+2}w, \tilde{\rho}_{k+1}\circ\zeta_{k+2}w]\!] - [\![\tilde{\rho}_{k+1}u, \tilde{\rho}_{k+1}\circ\zeta_{k+2}w \\ &\tilde{\mathfrak{D}}_1(u) + \zeta_{k+1}\circ\underline{\delta}(w)\ .\end{aligned}$$

It follows that $\tilde{\mathfrak{D}}_1$ factors to a first order operator

$$\tilde{\mathfrak{D}}_1^1 : \underline{T}^* \otimes_{\mathcal{O}} \tilde{\mathcal{J}}_{k+1}T \longrightarrow \tilde{\mathcal{C}}^2_{k+1}T\ .$$

Since $\tilde{\mathfrak{D}}_1 \circ \tilde{\mathfrak{D}} = 0$ (second line of (25.2)) then $\tilde{\mathfrak{D}}_1^1 \circ \tilde{\mathfrak{D}} = 0$ (third line $\tilde{\mathfrak{D}}$) hence $\tilde{\mathfrak{D}}_1^1$ vanishes on the image of $\tilde{\mathfrak{D}}$ and consequently it will also vanish on

$$\zeta_{k+1}\circ\underline{\delta}(S^{k+2}\underline{T}^* \otimes_{\mathcal{O}} \underline{T}) = \tilde{\mathfrak{D}} \circ \mathrm{Exp}(S^{k+2}\underline{T}^* \otimes_{\mathcal{O}} \underline{T})\ .$$

Let $u \in \underline{T}^* \otimes_{\mathcal{O}} \tilde{\mathcal{J}}_{k+1}T$ and $v = \zeta_{k+1}\circ\underline{\delta}(w) \in \underline{T}^* \otimes_{\mathcal{O}} \tilde{\mathcal{J}}_{k+1}T$ with $w \in S^{k+2}\underline{T}^* \otimes_{\mathcal{O}} \underline{T}$. We shall show that

$$\tilde{\mathfrak{D}}_1^1(u) = \tilde{\mathfrak{D}}_1^1(u+v)$$

hence $\tilde{\mathfrak{D}}_1^1$ factors to a (non-linear) operator of order 1

$$\tilde{\mathfrak{D}}'_1 : \tilde{\mathcal{C}}^1_{k+1}T \to \tilde{\mathcal{C}}^2_{k+1}T\ .$$

Let $u', v' \in \underline{T}^* \otimes_{\mathcal{O}} \tilde{\mathcal{J}}_{k+2}T$ be liftings of u, v. Then

$$\tilde{\mathfrak{D}}_1^1(u+v) = \tilde{\mathfrak{D}}_1(u'+v') \ \mathit{mod}\ \zeta_{k+1}\circ\underline{\delta}(\underline{T}^* \otimes_{\mathcal{O}} S^{k+2}\underline{T}^* \otimes_{\mathcal{O}} \underline{T})$$

and

$$\tilde{\mathfrak{D}}_1(u'+v') = \tilde{\mathfrak{D}}_1(u') + \tilde{\mathfrak{D}}_1(v') - [\![u, v]\!]$$

hence

$$\tilde{\mathfrak{D}}_1^1(u+v) = \tilde{\mathfrak{D}}_1^1(u) + \tilde{\mathfrak{D}}_1^1(v) + \{[\![u, v]\!] \ \mathit{mod}\ \zeta_{k+1}\circ\underline{\delta}(\ )\}\ .$$

We observed earlier that $\tilde{\mathfrak{D}}_1^1(v) = 0$. As for the last term, it also vanishes since $[\![u, v]\!] \in \zeta_{k+1}\circ\underline{\delta}(\ )$. This completes the proof.

The relation $\tilde{\mathfrak{D}}_1^1 \circ \tilde{\mathfrak{D}} = 0$ (third line $\tilde{\mathfrak{D}}$) also implies that $\tilde{\mathfrak{D}}'_1 \circ \tilde{\mathfrak{D}}' = 0$ since $\tilde{\mathfrak{D}}'_1$ is a quotient of $\tilde{\mathfrak{D}}_1^1$ and $\tilde{\mathfrak{D}}'$ is a quotient of $\tilde{\mathfrak{D}}$. For $k \geq 0$, we have thus constructed the *third non-linear complex*

$$(25.3)_{k+1} \qquad 1 \longrightarrow \mathcal{A}ut\ X \xrightarrow{\tilde{j}_{k+1}} \Gamma_{k+1}X \xrightarrow{\tilde{\mathfrak{D}}'} \tilde{\mathcal{C}}^1_{k+1}T \xrightarrow{\tilde{\mathfrak{D}}'_1} \tilde{\mathcal{C}}^2_{k+1}T.$$

Since $\tilde{D}'$ is obtained similarly as a quotient of $\tilde{D}$ (cf. Section 7) and since $\tilde{\rho}_{k+1}[\![u, v]\!] = [\![\tilde{\rho}_{k+1}u, \tilde{\rho}_{k+1}v]\!]$ for $u, v \in \Lambda\underline{T}^* \otimes_{\mathcal{O}} \tilde{\mathcal{J}}_{k+2}T$, we infer that

$$(25.4) \qquad \tilde{\mathfrak{D}}'_1 u = \tilde{D}'u - \tfrac{1}{2}[\![u, u]\!]$$

where $u \in \tilde{\mathcal{C}}^1_{k+1}T$ and $[\![\ ,\ ]\!]$ is the quotient bracket on $\tilde{\mathcal{C}}^\bullet_{k+1}T$. The *third structure equation* reads

$$(25.5) \qquad \tilde{D}'\tilde{\mathfrak{D}}'F - \tfrac{1}{2}[\![\tilde{\mathfrak{D}}'F, \tilde{\mathfrak{D}}'F]\!] = 0$$

where $F \in \Gamma_{k+1}X$ and $k \geq 0$. The previous construction does not hold for $k = -1$. The important feature in (25.4-5) is that $\tilde{\rho}_k$ has been eliminated. In other words, the operations involved preserve the order of jet sheaves. The same phenomenon happens in the complex $(25.3)_{k+1}$. Furthermore, Lemma 6 of Section 16 yields the relation

$$(25.6) \qquad \tilde{D}'[\![u, v]\!] = [\![\tilde{D}'u, v]\!] + (-1)^r[\![u, \tilde{D}'v]\!]$$

where $u \in \tilde{\mathcal{C}}^r_kT$, $v \in \tilde{\mathcal{C}}^\bullet_kT$, $k \geq 1$ and $[\![\ ,\ ]\!]$ is the bracket in $\tilde{\mathcal{C}}^\bullet_kT$. We obtain the

THEOREM. *The map*

$$\tilde{D}' : \tilde{\mathcal{C}}^\bullet_kT \to \tilde{\mathcal{C}}^\bullet_kT$$

is a derivation of degree one of the graded Lie algebra $\tilde{\mathcal{C}}^\bullet_kT$, $k \geq 1$.

We already proved that $(25.3)_{k+1}$ is exact at $\Gamma_{k+1}X$. However, it is not exact at $\tilde{\mathcal{C}}^1_{k+1}T$ and it satisfies only a partial exactness similar to the one verified by $(23.9)_{k+1}$. The map

$$\underline{T}^* \otimes_{\mathcal{O}} \tilde{\mathcal{J}}_{k+1}T \xrightarrow{\tilde{\rho}_0} \underline{T}^* \otimes_{\mathcal{O}} \tilde{\mathcal{J}}_0 T \simeq \underline{T}^* \otimes_{\mathcal{O}} \underline{T}$$

factors to a map

$$\tilde{\mathcal{C}}^1_{k+1}T \xrightarrow{\tilde{\rho}_0} \underline{T}^* \otimes_{\mathcal{O}} \tilde{\mathcal{J}}_0 T$$

since $\tilde{\rho}_0 \circ \zeta_{k+1} = 0$.

THEOREM. *For any* $k \geq 0$

$$\tilde{\mathfrak{D}}'(\Gamma_{k+1}X) = \{u \in \tilde{\mathcal{C}}^1_{k+1}T \mid \mathrm{Id} - \tilde{\rho}_o u \in \mathcal{A}ut_{\mathcal{O}}(\underline{T}), \tilde{\mathfrak{D}}'_1 u = 0\}.$$

Proof. If $F \in \Gamma_{k+1}X$ then

$$\tilde{\mathfrak{D}}'F = \tilde{\mathfrak{D}}G \ \mathit{mod}\ \zeta_{k+1} \circ \underline{\delta}(S^{k+2}\underline{T}^* \otimes_{\mathcal{O}} \underline{T})$$

where $G \in \Gamma_{k+2}X$ is any lifting of F, hence $\tilde{\rho}_0 \tilde{\mathfrak{D}}'F = \tilde{\rho}_0 \tilde{\mathfrak{D}}G$ and

$$\mathrm{Id} - \tilde{\rho}_0 \tilde{\mathfrak{D}}'F \in \mathcal{A}ut_{\mathcal{O}}(\underline{T}).$$

The condition $\tilde{\mathfrak{D}}'_1(\tilde{\mathfrak{D}}'F) = 0$ is the structure equation. Conversely, let $u \in \tilde{\mathcal{C}}^1_{k+1}T$ satisfy the two conditions and take a representative $v \in \underline{T}^* \otimes_{\mathcal{O}} \tilde{\mathcal{J}}_{k+1}T$ of u, i.e.,

$$u = v \ \mathit{mod}\ \zeta_{k+1} \circ \underline{\delta}(S^{k+2}\underline{T}^* \otimes_{\mathcal{O}} \underline{T}).$$

Let $v' \in \underline{T}^* \otimes_{\mathcal{O}} \tilde{\mathcal{J}}_{k+2}T$ be a lifting of v. By definition, the condition $\tilde{\mathfrak{D}}'_1 u = 0$ is equivalent to

$$\tilde{\mathfrak{D}}_1 v' \in \zeta_{k+1} \circ \underline{\delta}(\underline{T}^* \otimes_{\mathcal{O}} S^{k+2}\underline{T}^* \otimes_{\mathcal{O}} \underline{T})$$

hence $\tilde{\mathfrak{D}}_1 v = \tilde{\rho}_k \circ \tilde{\mathfrak{D}}_1 v' = 0$. Moreover, since $\tilde{\rho}_0 v = \tilde{\rho}_0 u$, $\mathrm{Id} - \tilde{\rho}_0 v \in \mathcal{A}ut_{\mathcal{O}}(\underline{T})$. By Theorem of Section 24, there exists $F \in \Gamma_{k+2}X$ such that $\tilde{\mathfrak{D}}F = v$; hence, by definition,

$$u = \tilde{\mathfrak{D}}F \ \mathit{mod}\ \zeta_{k+1} \circ \underline{\delta}(S^{k+2}\underline{T}^* \otimes_{\mathcal{O}} \underline{T}) = \tilde{\mathfrak{D}}'G$$

where $G = \rho_{k+1}F \in \Gamma_{k+1}X$.

Let $F \in \Gamma_{k+1}X$. Then $\mathcal{A}d F$ operates on $\Lambda^p \underline{T}^* \otimes_{\mathcal{O}} \tilde{\mathcal{J}}_{k+1} T$ and we claim that

$$\zeta_{p,k+1} \circ \underline{\delta}_{p-1,k+2}(\Lambda^{p-1}\underline{T}^* \otimes_{\mathcal{O}} S^{k+2}\underline{T}^* \otimes_{\mathcal{O}} \underline{T})$$

is an invariant subsheaf. In fact, we shall prove that $\mathcal{A}d F$ commutes with $\underline{\delta}$. Let $G \in \Gamma_{k+2}X$ be a lifting of F and let

$$w \in \Lambda^{p-1}\underline{T}^* \otimes_{\mathcal{O}} S^{k+2}\underline{T}^* \otimes_{\mathcal{O}} \underline{T} \subset \Lambda^{p-1}\underline{T}^* \otimes_{\mathcal{O}} \tilde{\mathcal{J}}_{k+2} T .$$

Since $\mathcal{A}d G$ commutes with ρ_{k+1} and $\tilde{\rho}_{k+1}$,

$$\tilde{\rho}_{k+1}\, \mathcal{A}d G(w) = \mathcal{A}d F(\tilde{\rho}_{k+1} w) = 0 ,$$

hence

$$\mathcal{A}d G(w) \in \Lambda^{p-1}\underline{T}^* \otimes_{\mathcal{O}} S^{k+2}\underline{T}^* \otimes \underline{T} .$$

We infer that $\mathcal{A}d G$ preserves $\mathcal{ker}\, \tilde{\rho}_{k+1}$. Applying the formula (23.14) we obtain

$$\zeta_{p,k+1} \circ \underline{\delta}_{p-1,k+2}(\mathcal{A}d G(w)) = \tilde{D}(\mathcal{A}d G(w)) = \mathcal{A}d F(\tilde{D}w - [\![\tilde{\mathcal{D}}G^{-1}, \tilde{\rho}_{k+1} w]\!]) =$$
$$\mathcal{A}d F(\zeta_{p,k+1} \circ \underline{\delta}_{p-1,k+2}(w))$$

since $\tilde{\rho}_{k+1} w = 0$. This is the desired relation. It follows that $\mathcal{A}d$ factors to a contravariant action of $\Gamma_{k+1}X$ on the sheaf $\tilde{\mathcal{C}}^\bullet_{k+1} T$. If $\alpha \in \Lambda \underline{T}^*$ and $u \in \tilde{\mathcal{C}}^\bullet_{k+1} T$ then

$$\mathcal{A}d F(\alpha \wedge u) = \mathcal{A}d F(\alpha) \wedge \mathcal{A}d F(u)$$

where $\wedge$ is the left $\Lambda \underline{T}^*$-module structure on $\tilde{\mathcal{C}}^\bullet_{k+1} T$ induced by the corresponding structure on $\Lambda \underline{T}^* \otimes_{\mathcal{O}} \tilde{\mathcal{J}}_{k+1} T$. In particular, it follows that $\mathcal{A}d F$ is semilinear with respect to $f^*\colon \mathcal{O} \to \mathcal{O}$. Moreover (cf. Theorem 1, Section 23),

$$\mathcal{A}d F([\![u, v]\!]) = [\![\mathcal{A}d F(u), \mathcal{A}d F(v)]\!]$$

hence $F \mapsto \mathcal{A}d F^{-1}$ is a representation of $\Gamma_{k+1}X$ into the groupoid of germs of automorphisms of the Lie algebra sheaf $\tilde{\mathcal{C}}^\bullet_{k+1} T$.

We know that $(23.9)_{k+1}$ is a finite form of $(23.18)_{k+1}$. Since each step in the construction of $(25.3)_{k+1}$ has a corresponding step in the construction of

$$(25.7)_{k+1} \qquad 0 \longrightarrow \underline{T} \xrightarrow{\tilde{j}_{k+1}} \tilde{\mathcal{C}}^0_{k+1}T \xrightarrow{\tilde{D}'} \tilde{\mathcal{C}}^1_{k+1}T \xrightarrow{\tilde{D}'} \tilde{\mathcal{C}}^2_{k+1}T$$

and each step is an $\mathbb{R}$-linear factorization, it follows that $(25.3)_{k+1}$ is a finite form of $(25.7)_{k+1}$. Precisely, we have the relations

$$(25.8) \qquad \frac{d}{dt}\tilde{\mathfrak{D}}'F_t\big|_{t=t_o} = \mathfrak{Ad}F_{t_o}(\tilde{D}'\Xi)$$

where (F_t) is a differentiable 1-parameter family of sections in $\Gamma(U, \Gamma_{k+1}X)$ with F_{t_o} admissible, $\Xi = \frac{d}{dt}F_t|_{t=t_o} \in \Gamma(V, \tilde{\mathfrak{J}}_{k+1}T)$ and $V = f_{t_o}(U)$, and

$$(25.9) \qquad \frac{d}{dt}\tilde{\mathfrak{D}}'_1 v_t\big|_{t=t_o} = \tilde{D}'u - [u, v_{t_o}]$$

where (v_t) is a differentiable 1-parameter family of sections in $\Gamma(U, \tilde{\mathcal{C}}^1_{k+1}T)$ and $u = \frac{d}{dt}v_t|_{t=t_o}$.

The main identities for the non-linear operators $\tilde{\mathfrak{D}}$ and $\tilde{\mathfrak{D}}_1$ also hold for $\tilde{\mathfrak{D}}'$ and $\tilde{\mathfrak{D}}'_1$. The identity (23.11) yields

$$(25.10) \qquad ad(\tilde{\mathfrak{D}}'F) = d - \mathfrak{Ad}F \circ d \circ \mathfrak{Ad}F^{-1} \in \mathfrak{Der}\, \Lambda\underline{T}^*$$

with $F \in \Gamma_{k+1}X$, $k \geq 0$, since

$$ad\colon \Lambda\underline{T}^* \otimes_{\mathcal{O}} \tilde{\mathfrak{J}}_{k+1}T \to \mathfrak{Der}\, \Lambda\underline{T}^*$$

factors to $\tilde{\mathcal{C}}^\bullet_{k+1}T$ (cf. the theorem, Section 15). The identity (23.12) yields

$$(25.11) \qquad \tilde{\mathfrak{D}}'(F \bullet G) = \mathfrak{Ad}G(\tilde{\mathfrak{D}}'F) + \tilde{\mathfrak{D}}'G$$

with $F, G \in \Gamma_{k+1}X$. In particular,

$$(25.12) \qquad \tilde{\mathfrak{D}}'F^{-1} = -\mathfrak{Ad}F^{-1}(\tilde{\mathfrak{D}}'F)\,.$$

The relation (23.14) yields

$$\tilde{D}'[\mathfrak{A}d F(u)] = \mathfrak{A}d F(\tilde{D}'u - [\![\tilde{\mathfrak{D}}'F^{-1}, u]\!]) \in \tilde{\mathcal{C}}^{\bullet}_{k+1}T \tag{25.13}$$

where $F \in \Gamma_{k+1}X$, $u \in \tilde{\mathcal{C}}^{\bullet}_{k+1}T$ and $k \geq 0$.

Let $u \in \tilde{\mathcal{C}}^{1}_{k+1}T$, $F \in \Gamma_{k+1}X$ and define

$$u^F = \mathfrak{A}d F(u) + \tilde{\mathfrak{D}}'F \in \tilde{\mathcal{C}}^{1}_{k+1}T\ . \tag{25.14}$$

If $G \in \Gamma_{k+2}X$ is a lifting of F and $v \in \underline{T}^* \otimes_{\mathcal{O}} \tilde{\mathcal{J}}_{k+1}T$ is a representative of u then u^F is the quotient of v^G modulo $\zeta \circ \underline{\delta}(\ )$ where v^G is defined by (23.15). We have the relations

$$u^{F \bullet G} = (u^F)^G, \quad u^1 = u \quad \text{and} \quad u^{F^{-1}} = \mathfrak{A}d F^{-1}(u - \tilde{\mathfrak{D}}'F) \tag{25.15}$$

and

$$\tilde{\mathfrak{D}}'_1(u^F) = \mathfrak{A}d F(\tilde{\mathfrak{D}}'_1 u)\ . \tag{25.16}$$

The first formula (25.15) shows that $\Gamma_{k+1}X$ operates contravariantly on $\tilde{\mathcal{C}}^{1}_{k+1}T$ by

$$(u, F) \mapsto u^F$$

and this action is a representation of $\Gamma_{k+1}X$ into the groupoid of germs of affine motions of the sheaf $\tilde{\mathcal{C}}^{1}_{k+1}T$. These affine motions are the extensions, to germs, of affine maps of vector bundles; the sheaf $\tilde{\mathcal{C}}^{\bullet}_{k+1}T$ is locally free of finite rank for the left and right $\mathcal{O}$-module structures and its definition clearly indicates how to obtain it as the sheaf of germs of sections of a quotient vector bundle. The relation (25.16) shows that $\mathscr{ker}\,\tilde{\mathfrak{D}}'_1$ is invariant under this action.

Remark. Let $\mathcal{R}_k$ be a formally integrable linear Lie equation and $\Gamma(\mathcal{R}_k)$ the corresponding non-linear equation. Then the complexes $(25.3)_{\ell+1}$ also restrict to $\Gamma(\mathcal{R}_k)$ and its prolongations.

CHAPTER V

DERIVATIONS OF JET FORMS

The purpose of this chapter is to extend the Frölicher-Nijenhuis theory (Section 13) to the sheaf

$$\Lambda(\tilde{\mathcal{J}}_k T)^* \otimes_{\mathcal{O}} \mathcal{J}_k$$

and transcribe into this context some of the previous results, in particular, the linear and non-linear complexes.

26. *The sheaves*

Recall that $\mathcal{J}_k$ is a left and right $\mathcal{O}$-algebra and that $\tilde{\mathcal{J}}_k T$ is a module over the algebra $\mathcal{J}_k$ hence $\tilde{\mathcal{J}}_k T$ also carries a left and right $\mathcal{O}$-module structure. Let $(\tilde{\mathcal{J}}_k T)^*$ be the dual of $\tilde{\mathcal{J}}_k T$ with respect to the left $\mathcal{O}$-module structure. $(\tilde{\mathcal{J}}_k T)^*$ is the sheaf of germs of (left) $\mathcal{O}$-linear 1-forms $\tilde{\mathcal{J}}_k T \to \mathcal{O}$. Since $\tilde{\mathcal{J}}_k T$ is locally free of finite rank, the stalk of $(\tilde{\mathcal{J}}_k T)^*$ at $x \in X$ is equal to the $\mathcal{O}(x)$-dual of the stalk of $\tilde{\mathcal{J}}_k T$ at x. Moreover, $(\tilde{\mathcal{J}}_k T)^*$ is locally free of finite rank (equal to the rank of $\tilde{\mathcal{J}}_k T$) and identifies, via ε_k^*, with $(\mathcal{J}_k T)^*$ hence identifies with the sheaf of germs of sections of the dual vector bundle $(J_k T)^*$. The sheaf $\Lambda^r(\tilde{\mathcal{J}}_k T)^*$ identifies with the sheaf of germs of (left) $\mathcal{O}$-multilinear skew-symmetric r-forms on $\tilde{\mathcal{J}}_k T$. Let

$$\mathcal{A}\ell t^r_{\mathcal{O}}(\tilde{\mathcal{J}}_k T, \mathcal{J}_k)$$

be the sheaf of germs of $\mathcal{O}$-multilinear alternating (skew-symmetric) r-forms on $\tilde{\mathcal{J}}_k T$ with values in $\mathcal{J}_k$ (linearity with respect to the left $\mathcal{O}$-module structures). The stalk of $\mathcal{A}\ell t^r_{\mathcal{O}}(\tilde{\mathcal{J}}_k T, \mathcal{J}_k)$ at x is equal to

$$\mathrm{Alt}^r_{\mathcal{O}(x)}(\tilde{\mathcal{J}}_k T(x), \mathcal{J}_k(x))$$

and

$$\mathcal{A}\ell t^r_{\mathcal{O}}(\tilde{\mathcal{J}}_k T, \mathcal{J}_k) \simeq \mathcal{H}om_{\mathcal{O}}(\Lambda^r \tilde{\mathcal{J}}_k T, \mathcal{J}_k) \simeq \Lambda^r(\tilde{\mathcal{J}}_k T)^* \otimes_{\mathcal{O}} \mathcal{J}_k$$

(exterior products with respect to the left $\mathcal{O}$-module structure) since all the modules involved are locally free of finite rank. To simplify the notations we set

$$\mathcal{A}^r_k(X) = \Lambda^r(\mathcal{J}_k T)^*, \ \mathcal{A}_k(X) = \oplus \mathcal{A}^r_k(X) = \Lambda(\mathcal{J}_k T)^*$$

and

$$\mathcal{B}^r_k(X) = \mathcal{A}\ell t^r_{\mathcal{O}}(\tilde{\mathcal{J}}_k T, \mathcal{J}_k), \ \mathcal{B}_k(X) = \oplus \mathcal{B}^r_k(X) .$$

$\mathcal{B}_k(X)$ carries a natural graded $\mathcal{J}_k$-module structure defined by

$$(f\omega)(\xi_1, \ldots, \xi_r) = f[\omega(\xi_1, \ldots, \xi_r)]$$

where $f \in \mathcal{J}_k$, $\omega \in \mathcal{B}^r_k(X)$ and $\xi_i \in \tilde{\mathcal{J}}_k T$, hence natural left and right graded $\mathcal{O}$-module structures. The structure on $\mathcal{B}_k(X)$ identifies with the natural $\mathcal{J}_k$-module structure of $\Lambda(\tilde{\mathcal{J}}_k T)^* \otimes_{\mathcal{O}} \mathcal{J}_k$ obtained by extension of the ring of scalars $\mathcal{O}$ to $\mathcal{J}_k$ via the imbedding $i_k \colon \mathcal{O} \to \mathcal{J}_k$. The $\mathcal{J}_k$-module $\mathcal{B}^r_k(X)$ is canonically isomorphic to

$$\mathcal{A}\ell t^r_{\mathcal{J}_k}(\tilde{\mathcal{J}}_k T \otimes_{\mathcal{O}} \mathcal{J}_k, \mathcal{J}_k) \simeq \Lambda^r(\tilde{\mathcal{J}}_k T \otimes_{\mathcal{O}} \mathcal{J}_k)^*$$

(tensor product with respect to the left $\mathcal{O}$-module structure of $\tilde{\mathcal{J}}_k T$, duality and exterior product with respect to the $\mathcal{J}_k$-module structures).

$\mathcal{A}_k(X)$ carries a structure of graded $\mathcal{O}$-algebra defined by the exterior product. $\mathcal{B}_k(X)$ also carries a structure of graded $\mathcal{J}_k$-algebra defined by the exterior product of $\mathcal{J}_k$-valued forms. The algebra structure of

$$\mathcal{B}_k(X) \simeq \Lambda(\tilde{\mathcal{J}}_k T)^* \otimes_{\mathcal{O}} \mathcal{J}_k$$

is equal to the tensor product of the exterior algebra $\Lambda(\tilde{\mathcal{J}}_k T)^*$ with the commutative algebra $\mathcal{J}_k$. It is also canonically isomorphic to the exterior algebra $\Lambda(\tilde{\mathcal{J}}_k T \otimes_{\mathcal{O}} \mathcal{J}_k)^*$ (over the ring $\mathcal{J}_k$). Moreover, $\mathcal{B}_k(X)$ is a graded left module over the graded $\mathcal{O}$-algebra $\mathcal{A}_k(X)$ where the external operation is the exterior product $\omega \wedge \mu$ of $\omega \in \mathcal{A}_k(X)$ and $\mu \in \mathcal{B}_k(X)$ with respect to the pairing

$$(f, g) \in \mathcal{O} \times \mathcal{J}_k \longmapsto i_k(f) \bullet g \in \mathcal{J}_k .$$

If $k = 0$, all the previous structures collapse to the usual structure of $\Lambda \underline{T}^*$.

Since $i_k(\mathcal{O}) \subset \mathcal{J}_k$, $\mathcal{A}_k(X)$ can be considered as a subsheaf of $\mathcal{B}_k(X)$. This imbedding is also given by

$$\Lambda(\tilde{\mathcal{J}}_k T)^* \simeq \Lambda(\tilde{\mathcal{J}}_k T)^* \otimes_{\mathcal{O}} i_k(\mathcal{O}) \longrightarrow \Lambda(\tilde{\mathcal{J}}_k T)^* \otimes_{\mathcal{O}} \mathcal{J}_k .$$

$\mathcal{B}_k(X)$ is generated by $\mathcal{A}_k(X)$ over $\mathcal{J}_k$. In particular, any $\mathcal{O}$-basis of $\mathcal{A}_k(X)$ is a $\mathcal{J}_k$-basis of $\mathcal{B}_k(X)$. The mapping

$$\tilde{\beta}_k : \tilde{\mathcal{J}}_k T \to \underline{T}$$

induces an injective algebra morphism

$$\tilde{\beta}_k^* : \Lambda \underline{T}^* \to \mathcal{A}_k(X) \subset \mathcal{B}_k(X)$$

hence $\Lambda \underline{T}^*$ can be considered as a subsheaf of $\mathcal{B}_k(X)$.

Let

$$\mathcal{V}_k^r(X) = \mathcal{A}\ell t^r_{\mathcal{O}}(\tilde{\mathcal{J}}_k T, \tilde{\mathcal{J}}_k T) \simeq \Lambda^r(\tilde{\mathcal{J}}_k T)^* \otimes_{\mathcal{O}} \tilde{\mathcal{J}}_k T$$

be the sheaf of germs of (left) $\mathcal{O}$-multilinear alternating r-forms on $\tilde{\mathcal{J}}_k T$ with values in $\tilde{\mathcal{J}}_k T$. The elements of $\mathcal{V}_k(X) = \oplus \mathcal{V}_k^r(X)$ will be called vector forms. $\mathcal{V}_k(X)$ is a graded left module over the graded algebra $\mathcal{B}_k(X)$ via the exterior product of scalar forms with vector forms. In particular $\mathcal{V}_k(X)$ is a graded module over $\mathcal{J}_k$ hence a left and right graded $\mathcal{O}$-module. The stalk of $\mathcal{V}_k(X)$ at a point $x \in X$ is equal to the module

$$\oplus \mathrm{Alt}^r_{\mathcal{O}(x)}(\tilde{\mathcal{J}}_k T(x), \tilde{\mathcal{J}}_k T(x))$$

of vector forms on the stalk $\tilde{\mathcal{J}}_k T(x)$. If $k = 0$ then $\mathcal{O}_k(X)$ becomes $\wedge \underline{T}^* \otimes_{\mathcal{O}} \underline{T}$.

All the previous sheaves are locally free of finite rank for the $\mathcal{J}_k$-module structure and the left $\mathcal{O}$-module structure.

If we were to carry out the theory as in Section 13, we would proceed as follows. The set of global sections

$$\Gamma(X, \mathcal{B}_k(X)) \simeq \oplus \mathrm{Alt}^r_{\mathcal{O}}(\tilde{\mathcal{J}}_k T, \mathcal{J}_k)$$

is a graded algebra over the ring $\Gamma(X, \mathcal{J}_k)$ and a graded left module over the graded $\mathcal{F}(X)$-algebra $\Gamma(X, \mathcal{A}_k(X))$. A derivation of degree p of the graded R-algebra $\Gamma(X, \mathcal{B}_k(X))$ is an R-linear map

$$D: \Gamma(X, \mathcal{B}_k(X)) \longrightarrow \Gamma(X, \mathcal{B}_k(X))$$

which satisfies the two conditions given in Section 13 where $A(X)$ is replaced by $\Gamma(X, \mathcal{B}_k(X))$. The direct sum

$$\mathrm{Der}\, \Gamma(X, \mathcal{B}_k X) = \oplus \mathrm{Der}^p \Gamma(X, \mathcal{B}_k(X))$$

carries a natural graded left module structure over the graded algebra $\Gamma(X, \mathcal{B}_k X)$ and a structure of graded R-Lie algebra with respect to the (graded) bracket of derivations. A derivation D is, in virtue of condition (ii), Section 13, a local operator and, for any open set U in X, there is a graded Lie algebra morphism

$$\mathrm{Der}\, \Gamma(X, \mathcal{B}_k(X)) \longrightarrow \mathrm{Der}\, \Gamma(U, \mathcal{B}_k(U))$$

which is semilinear with respect to the algebra morphism

$$\Gamma(X, \mathcal{B}_k(X)) \longrightarrow \Gamma(U, \mathcal{B}_k(U)) .$$

A derivation $D \in \mathrm{Der}^p \Gamma(X, \mathcal{B}_k(X))$ is called *admissible* if

$$\text{(a)} \quad D(\Gamma(X, \mathcal{J}^0_k)) \subset \Gamma(X, \mathcal{J}^0_k \mathcal{B}^p_k(X)) \quad \text{and}$$

$$\text{(b)}\quad D[\Gamma(X, \mathcal{Q}_k(X))] \subset \Gamma(X, \mathcal{Q}_k(X))$$

where $\mathcal{J}_k^0 = \mathcal{J}/\mathcal{J}^{k+1}$ is the kernel of $\beta_k\colon \mathcal{J}_k \to \mathcal{J}_0 = \mathcal{O}$. The set

$$\mathrm{Der}_a\Gamma(X, \mathcal{B}_k(X)) = \oplus \mathrm{Der}_a^p\mathcal{B}_k(X)$$

of admissible derivations is a Lie subalgebra and a graded left $\Gamma(X, \mathcal{Q}_k(X))$ submodule of $\mathrm{Der}\,\Gamma(X, \mathcal{B}_k(X))$. The Frölicher-Nijenhuis theory can be extended to $\mathrm{Der}_a\Gamma(X, \mathcal{B}_k(X))$ and this extension reduces to the classical theory for $k = 0$. One can prove that $\mathrm{Der}_a\Gamma(X, \mathcal{B}_k(X))$ is the direct sum of the subalgebras of *interior* and *Lie* derivations (derivations of *type* i_* and d_* in the terminology of [3]). Moreover, each of these subalgebras is represented by the graded left $\Gamma(X, \mathcal{B}_k(X))$-module

$$\Gamma(X, \mathcal{V}_k(X)) \simeq \oplus \mathrm{Alt}^r_{\mathcal{O}}(\tilde{\mathcal{J}}_k T, \tilde{\mathcal{J}}_k T)\,.$$

The bracket of interior derivations transports to $\Gamma(X, \mathcal{V}_k(X))$ and defines a structure of graded $\Gamma(X, \mathcal{J}_k)$-Lie algebra. The transported bracket is simply given by the (graded) commutator of the $\barwedge$ operation. The bracket of Lie derivations transports to $\Gamma(X, \mathcal{V}_k(X))$ and defines a structure of graded $\mathbb{R}$-Lie algebra. The transported bracket is the Nijenhuis bracket of vector forms. All the results and formulas given in Section 13 transcribe into this context. Since derivations are local operators, we can define the sheaf

$$\mathcal{D}er_a\mathcal{B}_k(X)$$

of germs of admissible derivations of $\Gamma(X, \mathcal{B}_k(X))$. The global sections of this sheaf identify with the elements of $\mathrm{Der}_a\Gamma(X, \beta_k(X))$. The previous sheaf is a graded $\mathbb{R}$-Lie algebra and a graded left module over the graded $\mathcal{J}_k$-algebra $\mathcal{Q}_k(X)$. All the theory transports to the sheaf level. In particular the algebras of (germs of) interior and Lie derivations are represented by the sheaf of vector forms

$$\mathcal{V}_k(X) \simeq \Lambda(\tilde{\mathcal{J}}_k T)^* \otimes_{\mathcal{O}} \tilde{\mathcal{J}}_k T.$$

Using this fact one proves that the stalk of

$$\mathcal{D}er_a \mathcal{B}_k(X)$$

at $x \in X$ is equal to the algebra of admissible derivations of the stalk of

$$\mathcal{B}_k(X) \simeq \Lambda(\tilde{\mathcal{J}}_k T)^* \otimes_{\mathcal{O}} \mathcal{J}_k$$

at the point x.

Our main interest lies in the sheaf $\mathcal{D}er_a \mathcal{B}_k(X)$. In the next section we shall develop the theory by a slightly different, though equivalent, approach. Instead of considering derivations of $\Gamma(X, \mathcal{B}_k(X))$, we shall consider derivations of the sheaf $\mathcal{B}_k(X)$, i.e., $\mathbf{R}$-linear sheaf maps

$$D: \mathcal{B}_k(X) \longrightarrow \mathcal{B}_k(X)$$

which satisfy the two conditions for graded derivations. Once again, the theory can be entirely transcribed for the set of those derivations which are admissible. The direct sum decomposition into interior and Lie derivations implies that this algebra of admissible derivations identifies with

$$\mathrm{Der}_a \Gamma(X, \mathcal{B}_k(X)).$$

Furthermore, the sheaf

$$\mathcal{D}er_a \mathcal{B}_k(X)$$

identifies with the sheaf of germs of admissible derivations of the sheaf $\mathcal{B}_k(X)$ (germs taken in the sense of sheaf maps) and each such germ identifies with the admissible derivation it induces in the stalk of $\mathcal{B}_k(X)$.

27. *The derivations*

For any $\xi \in \tilde{\Theta}$, the Lie derivative $\mathcal{L}(\xi)$ operates as a (germ of) derivation on $\mathcal{O}_{X^2}$. Since ξ is diagonal, the derivation $\mathcal{L}(\xi)$ preserves the filtration $\mathcal{J}^k$ and since ξ is π_1-projectable, $\mathcal{L}(\xi)$ preserves the subsheaf $i(\mathcal{O}) \subset \mathcal{O}_{X^2}$. Let

$$\mathcal{D}er_{\mathbf{R},a} \mathcal{J}_k$$

be the sheaf of germs of derivations of $\mathcal{J}_k$ which are *admissible* in the sense that:

a) $D(\mathcal{J}_k^h) \subset \mathcal{J}_k^h$ for any $0 \leq h < k$ where $\mathcal{J}_k^h = \mathcal{J}^{h+1}/\mathcal{J}^{k+1}$ is the subsheaf of $\mathcal{J}_k$ composed of those elements which vanish to order $h+1$ on Δ. Since the ideal $\mathcal{J}_k^h$ is equal to $(\mathcal{J}_k^0)^{h+1}$, it is enough to assume that $D(\mathcal{J}_k^0) \subset \mathcal{J}_k^0$.

b) $D[i_k\mathcal{O}] \subset i_k\mathcal{O}$.

The Lie algebra representation

$$\mathcal{L}: \tilde{\Theta} \longrightarrow \mathcal{D}er_R \mathcal{O}_{X^2}$$

factors to a left $\mathcal{O}$-linear Lie algebra representation

$$\mathcal{L}: \tilde{\mathcal{J}}_k T \longrightarrow \mathcal{D}er_{R,a} \mathcal{J}_k \,.$$

LEMMA 1. *The representation*

$$\mathcal{L}: \tilde{\mathcal{J}}_k T \longrightarrow \mathcal{D}er_{R,a} \mathcal{J}_k$$

is an isomorphism.

Proof. We first prove that $\mathcal{L}$ is faithful. Let $\Xi \in \tilde{\mathcal{J}}_k T$ and assume that $\mathcal{L}(\Xi) = 0$. If $\xi \in \tilde{\Theta}$ represents Ξ, then $\mathcal{L}(\xi)f$ vanishes to order $k+1$ on Δ for any $f \in \mathcal{O}_{X^2}$ hence ξ vanishes to order $k+1$ on Δ, i.e., $\Xi = 0$. We prove next that $\mathcal{L}$ is surjective. Let

$$\mathfrak{D} \in \mathcal{D}er_{R,a} \mathcal{J}_k$$

be a germ of admissible derivation at the point $p \in X$ and let

$$D: \mathcal{J}_k \to \mathcal{J}_k$$

be a representative of $\mathfrak{D}$. Taking a coordinate system in a neighborhood U of p, any element in $\mathcal{J}_k|_U$ can be uniquely represented by a germ of polynomial

$$\sum_{|\alpha|\le k} a_\alpha(x)\frac{y^\alpha}{\alpha!}\ .$$

By assumption,

$$D(x_i) = f_i \in \mathcal{O}$$

and

$$D(y_i) = P_i = \sum_{|\alpha|\le k} a_{\alpha,i}(x)\frac{y^\alpha}{\alpha!}\ .$$

Consider

$$\xi = P_i\frac{\partial}{\partial y_i} + f_i\frac{\partial}{\partial x_i} \in \underline{TX^2}\ .$$

The field ξ is π_1-projectable by construction and, since $D(y_i - x_i) = P_i - f_i$ vanishes on Δ, it follows that $P_i(x,x) = f_i(x)$ hence ξ is admissible. Let

$$\Xi = \xi \bmod \mathcal{J}^{k+1}\Delta\tilde{\Theta}\ .$$

Then clearly $\mathfrak{D} = \mathfrak{L}(\Xi)$. For curiosity, we give a second proof of the surjectivity. Any element in $\mathcal{J}_k|_U$ can be uniquely represented by a germ of polynomial

$$\sum_{|\alpha|\le k} a_\alpha(x)\frac{(y-x)^\alpha}{\alpha!}\ .$$

Take the new coordinates $z_i = y_i - x_i$ and $w_i = x_i$. Then

$$\frac{\partial}{\partial z_i} = \frac{\partial}{\partial y_i},\ \frac{\partial}{\partial w_i} = \frac{\partial}{\partial x_i} + \frac{\partial}{\partial y_i},\ D(x_i) = f_i \in \mathcal{O}$$

and

$$D(y_i - x_i) = Q_i = \sum_{1\le|\alpha|\le k} a_{\alpha,i}\frac{(y-x)^\alpha}{\alpha!}$$

since Q_i must vanish on Δ. It is clear that

$$\xi = Q_i \frac{\partial}{\partial z_i} + f_i \frac{\partial}{\partial w_i} = (Q_i + f_i) \frac{\partial}{\partial y_i} + f_i \frac{\partial}{\partial x_i}$$

is admissible and $\mathcal{D} = \mathcal{L}(\Xi)$.

We infer from this lemma that the algebra

$$\mathrm{Der}_{\mathbf{R},a} \mathcal{J}_k$$

of admissible derivations of $\mathcal{J}_k$ is isomorphic with the algebra $\Gamma(X, \tilde{\mathcal{J}}_k T)$. Observe next that any germ of derivation

$$\mathcal{D} \in \mathcal{D}er_{\mathbf{R}} \mathcal{J}_k$$

at a point x induces a derivation of the stalk $\mathcal{J}_k(x)$. The argument given in the proof of the lemma can be rewritten for a stalk of $\mathcal{J}_k$ thus proving that any admissible derivation of the stalk $\mathcal{J}_k(x)$ is induced by a germ of derivation $\mathcal{L}(\Xi)$ with $\Xi \in \tilde{\mathcal{J}}_k T$. We infer that

$$\mathcal{D}er_{\mathbf{R},a} \mathcal{J}_k = \bigcup_{x \in X} \mathrm{Der}_{\mathbf{R},a} \mathcal{J}_k(x) ,$$

that is, a germ of admissible derivation identifies with an admissible derivation of the stalk.

Define the exterior differentiation

$$d\colon \mathcal{B}_k(X) \longrightarrow \mathcal{B}_k(X)$$

by the usual formula

$$(27.1)\quad \langle \xi_1 \wedge \ldots \wedge \xi_{r+1}, d\omega \rangle = \sum_{1 \le i \le r+1} (-1)^{i+1} \mathcal{L}(\xi_i) \langle \xi_1 \wedge \ldots \wedge \hat{\xi}_i \wedge \ldots \wedge \xi_{r+1}, \omega \rangle +$$

$$\sum_{i<j} (-1)^{i+j} \langle [\![\xi_i, \xi_j]\!] \wedge \xi_1 \wedge \ldots \wedge \hat{\xi}_i \wedge \ldots \wedge \hat{\xi}_j \wedge \ldots \wedge \xi_{r+1}, \omega \rangle$$

where $\omega \in \mathcal{B}_k^r(X)$, $\xi_i \in \tilde{\mathcal{J}}_k T$ and $[\![\ ,\ ]\!]$ is the bracket in $\tilde{\mathcal{J}}_k T$ induced by the Lie bracket of admissible vector fields in X^2. An easy computation shows that $d\omega$ is left $\mathcal{O}$-multilinear, i.e., $d\omega \in \mathcal{B}_k^{r+1}(X)$. If $f \in \mathcal{B}_k^0(X) = \mathcal{J}_k$ then

$$df(\xi) = \mathcal{L}(\xi) f .$$

If $k = 0$, then d is equal to the usual exterior differentiation in $\Lambda \underline{T}^*$ since $[\![\, , \,]\!]$ reduces to the usual bracket of vector fields in $\underline{T}$. The subalgebra $\mathcal{A}_k(X)$ is clearly invariant by d and, since

$$\tilde{\beta}_k \colon \tilde{\mathcal{J}}_k T \to \underline{T}$$

is a Lie algebra morphism for the bracket $[\![\, , \,]\!]$ on $\tilde{\mathcal{J}}_k T$ and the usual bracket $[\, , \,]$ on $\underline{T}$, it follows that $\Lambda \underline{T}^*$, considered as a subsheaf of $\mathcal{B}_k(X)$ (via $\tilde{\beta}_k^*$), is also invariant by d.

A derivation of degree p of the R-algebra $\mathcal{B}_k(X)$ is an R-linear map

$$D \colon \mathcal{B}_k(X) \to \mathcal{B}_k(X)$$

which satisfies the conditions (i) and (ii) of Section 13 with $A(X)$ replaced by $\mathcal{B}_k(X)$. Let

$$\mathrm{Der}\, \mathcal{B}_k(X) = \oplus_{\mathbf{R}} \mathrm{Der}^p\, \mathcal{B}_k(X) .$$

The graded bracket of derivations defines on $\mathrm{Der}\, \mathcal{B}_k(X)$ a structure of graded R-Lie algebra. Furthermore, $\mathrm{Der}\, \mathcal{B}_k(X)$ carries a natural structure of graded left module over the graded $\mathcal{J}_k$-algebra $\mathcal{B}_k(X)$, in particular it is a graded module over $\mathcal{B}_k^0(X) = \mathcal{J}_k$. For any open set $U \subset X$, the restriction map

$$\mathrm{Der}\, \mathcal{B}_k(X) \longrightarrow \mathrm{Der}[\mathcal{B}_k(X)|_U]$$

is a left $\mathcal{B}_k(X)$-linear (hence also $\mathcal{J}_k$-linear) graded Lie algebra morphism. Let $\mathcal{D}er\, \mathcal{B}_k(X)$ be the sheaf of germs of derivations of $\mathcal{B}_k(X)$ (germs taken in the sense of sheaf maps). This sheaf inherits the graded R-Lie algebra structure and the graded left $\mathcal{B}_k(X)$-module structure of $\mathrm{Der}\, \mathcal{B}_k(X)$.

A derivation $D \in \mathrm{Der}^p \mathcal{B}_k(X)$ is *admissible* if

a) $D(\mathcal{J}_k^0) \subset \mathcal{J}_k^0 \mathcal{B}_k^p(X)$,

b) $D[\mathcal{Q}_k(X)] \subset \mathcal{Q}_k(X)$.

LEMMA 2. *If* D *is admissible then*

$$D[\mathcal{J}_k^h \mathcal{B}_k(X)] \subset \mathcal{J}_k^h \mathcal{B}_k(X), \quad 0 \leq h < k .$$

A proof of this assertion follows from condition (ii) and the fact that the ideal $\mathcal{J}_k^h$ of $\mathcal{J}_k$ is equal to $(\mathcal{J}_k^0)^{h+1}$. Let

$$\mathrm{Der}_a^p \mathcal{B}_k(X)$$

be the set of admissible derivations of degree p. Then

$$\mathrm{Der}_a \mathcal{B}_k(X) = \oplus \mathrm{Der}_a^p \mathcal{B}_k(X)$$

is a graded Lie subalgebra and a homogeneous graded left $\mathcal{Q}_k(X)$-submodule of $\mathrm{Der}\, \mathcal{B}_k(X)$. Denote by

$$\mathcal{D}er_a \mathcal{B}_k(X)$$

the sheaf of germs of admissible graded derivations. Then

$$\mathcal{D}er_a \mathcal{B}_k(X) = \oplus \mathcal{D}er_a^p \mathcal{B}_k(X)$$

is a graded R-Lie algebra for the graded bracket of derivations and a graded left $\mathcal{Q}_k(X)$-module. Unless mentioned otherwise, all the derivations considered in the sequel are assumed to be admissible.

We now examine some special derivations:

(I) A derivation

$$D \in \mathrm{Der}_a^p \mathcal{B}_k(X)$$

is called an *interior* derivation if D vanishes on $\mathcal{B}_k^0(X) = \mathcal{J}_k$, i.e., $D(\mathcal{J}_k) = 0$. The condition (ii) in the definition of graded derivations implies that the interior derivations are the $\mathcal{J}_k$-linear derivations. The theory of interior derivations is purely algebraic and its properties can be proved by methods analogous to those used in [3] and [4] where the interior derivations are called derivations of type i_*. It is clear that the interior derivations are closed under the bracket and the $\mathcal{A}_k(X)$-module operations. Moreover, the bracket operation restricted to interior derivations is $\mathcal{O}$-bilinear. The sheaf of germs of interior derivations is a graded Lie subalgebra and a homogeneous graded $\mathcal{A}_k(X)$-submodule of $\mathcal{D}er_a\mathcal{B}_k(X)$. The induced bracket on this sheaf defines a structure of graded Lie algebra over $\mathcal{O}$.

Any $u \in \mathcal{V}_k^r(X)$ defines the germ of interior derivation

$$i(u) \in \mathcal{D}er_a^{r-1}\mathcal{B}_k(X)$$

given by $i(u)\omega = u \barwedge \omega$ where $\omega \in \mathcal{B}_k(X)$. More precisely, one should first extend u to a section σ of $\mathcal{V}_k^r(X)$, define the interior derivation

$$i(\sigma) = \sigma \barwedge \bullet : \mathcal{B}_k(X) \to \mathcal{B}_k(X)$$

and then take its germ. This germ induces in the stalk the interior derivation $i(u)$. The map

$$i: \mathcal{V}_k(X) \longrightarrow \mathcal{D}er_a\mathcal{B}_k(X)$$

is a homogeneous left $\mathcal{A}_k(X)$-linear injective map of degree -1 whose image is the graded Lie subalgebra and graded $\mathcal{A}_k(X)$-submodule of interior derivations. If $\mathcal{D}$ is a germ of interior derivation of degree r then $\mathcal{D} = i(u)$ where

$$u = \sum \mathcal{D}\omega^a \otimes \xi_a \in \Lambda^{r+1}(\tilde{\mathcal{J}}_k T)^* \otimes_{\mathcal{O}} \tilde{\mathcal{J}}_k T \simeq \mathcal{V}_k^{r+1}(X)\,,$$

$\{\xi_a\}$ is an $\mathcal{O}$-basis of $\tilde{\mathcal{J}}_k T$ (in one stalk) and $\{\omega^a\}$ is the $\mathcal{O}$-dual basis of $(\tilde{\mathcal{J}}_k T)^*$ (in the corresponding stalk). The graded Lie algebra structure of the image transports to $\mathcal{V}_k(X)$ and is given, for homogeneous elements, by the commutator

$$[u, v] = u \barwedge v - (-1)^{rs} v \barwedge u$$

where $deg\, u = r+1$ and $deg\, v = s+1$. This bracket on $\mathcal{O}_k(X)$ is obviously $\mathcal{J}_k$-bilinear.

From the previous discussion we infer the

PROPOSITION 1. *An interior derivation is uniquely determined by its action on* $\mathcal{C}_k^1(X)$. *Conversely, any* $\mathcal{O}$*-linear map*

$$\delta\colon \mathcal{C}_k^1(X) \to \mathcal{C}_k^{r+1}(X)$$

which satisfies condition (ii) *for the elements in* $\mathcal{C}_k^1(X)$ *can be extended uniquely to an admissible interior derivation of degree* r *of* $\mathcal{B}_k(X)$. *A similar property holds for germs of interior derivations and germs of* $\mathcal{O}$*-linear maps* δ.

The extension D can be defined, as previously, by $D = i(u)$ where

$$u = \sum \delta\omega^\alpha \otimes \xi_\alpha .$$

One should remark that the last formula defines u in each stalk of $\mathcal{O}_k^{r+1}(X)$, the definition being independent of the choice of $\{\xi_\alpha\}$. Another method of extension is the following. Any $\omega \in \mathcal{B}_k^p(X)$ can be written uniquely as

$$\omega = \sum_{\alpha_1 < \cdots < \alpha_p} f_{\alpha_1 \cdots \alpha_p}\, \omega^{\alpha_1} \wedge \ldots \wedge \omega^{\alpha_p}$$

where $\{\omega^\alpha\}$ is an $\mathcal{O}$-basis of $(\tilde{\mathcal{J}}_k T)^*$ and the indices α are given an ordering. The extension D is given by

$$D\omega = \sum_{\alpha_1 < \cdots < \alpha_p} \sum_{1 \le j \le p} (-1)^{r(j+1)} f_{\alpha_1 \cdots \alpha_p}\, \omega^{\alpha_1} \wedge \ldots \wedge \delta\omega^{\alpha_j} \wedge \ldots \wedge \omega^{\alpha_p}$$

for $p \geq 1$ and $D(\mathcal{J}_k) = 0$. Similarly, one proves the following

PROPOSITION 2. *Any* $\mathcal{J}_k$*-linear map*

$$\delta\colon \mathcal{B}_k^1(X) \to \mathcal{B}_k^{r+1}(X)$$

which is admissible, i.e.,

$$\delta[\mathcal{J}_k^0\mathcal{B}_k^1(X)] \subset \mathcal{J}_k^0\mathcal{B}_k^{r+1}(X) \quad \textit{and} \quad \delta[\mathcal{A}_k^1(X)] \subset \mathcal{A}_k^{r+1}(X)$$

and which satisfies condition (ii) *for the elements in* $\mathcal{B}_k^1(X)$ *can be extended uniquely to an admissible interior derivation of degree* r *of* $\mathcal{B}_k(X)$. *A similar statement holds for germs.*

The extension can be defined in exactly the same way as in the case of Proposition 1. It is interesting, however, to compare the maps δ given in the two propositions and reduce the second to the first. An admissible $\mathcal{J}_k$-linear map δ, as in Proposition 2, restricts obviously to an $\mathcal{O}$-linear map δ_1 as in Proposition 1. Conversely, any $\mathcal{O}$-linear map δ_1 satisfying the conditions of Proposition 1, can be extended uniquely to an admissible $\mathcal{J}_k$-linear map satisfying the conditions of Proposition 2 since $\mathcal{B}_k^1(X)$ is $\mathcal{J}_k$-generated by $\mathcal{A}_k^1(X)$ and any $\mathcal{O}$-basis of $\mathcal{A}_k^1(X)$ is a $\mathcal{J}_k$-basis of $\mathcal{B}_k^1(X)$. Another way of stating this amounts to observing that $\mathcal{B}_k^1(X)$ is obtained from $\mathcal{A}_k^1(X)$ by extension of the ring of scalars to $\mathcal{J}_k$.

(II) We shall now introduce *Lie* derivations. First observe that d is an admissible derivation of degree 1 and that $d^2 = d \circ d = 0$. We infer that $[d, d] = 2d^2 = 0$. A derivation

$$D \in \mathrm{Der}_a^p\mathcal{B}_k(X)$$

is called a *Lie* derivation if $[D, d] = 0$. The graded Jacobi identity implies that the Lie derivations are closed under the bracket. If $\omega \in \mathcal{A}_k(X)$ and D is a Lie derivation then $\omega \wedge D$ is not necessarily a Lie derivation unless ω is closed, i.e., $d\omega = 0$. The sheaf of germs of Lie derivations is a homogeneous graded R-Lie subalgebra of $\mathcal{D}er_a\mathcal{B}_k(X)$.

With every $u \in \mathcal{O}_k^r(X)$ is associated the germ of Lie derivation

$$\mathcal{L}(u) = [i(u), d] \in \mathcal{D}er_a^r\mathcal{B}_k(X) .$$

The remark concerning the definition of $i(u)$ applies as well for $\mathcal{L}(u)$. The map

$$\mathcal{L}: \mathcal{O}_k(X) \longrightarrow \mathcal{D}er_a\mathcal{B}_k(X)$$

is a homogeneous $\mathbf{R}$-linear map of degree 0. We shall prove that $\mathcal{L}$ is injective and that its image is equal to the graded Lie algebra of germs of admissible Lie derivations.

LEMMA 3. $\mathcal{L}$ *is injective.*

Proof. Let $u \in \mathcal{O}_k^r(X)$ and assume that $\mathcal{L}(u) = 0$. For any $f \in \mathcal{J}_k$ we have

$$0 = \mathcal{L}(u)f = i(u)df = (df) \circ u$$

hence

$$df \circ u(\xi_1, \ldots, \xi_r) = \mathcal{L}[u(\xi_1, \ldots, \xi_r)]f = 0$$

where $\xi_i \in \tilde{\mathcal{J}}_k T$. Lemma 1 implies that $u(\xi_1, \ldots, \xi_r) = 0$ hence $u = 0$.

LEMMA 4. *The image of* $\mathcal{L}$ *is equal to the algebra of germs of admissible Lie derivations.*

Proof. We shall adapt to the present context the proof of Proposition (4.5) in [3] where Lie derivations are called derivations of type d_*. Let $\mathfrak{D}$ be a germ of admissible Lie derivation of degree r and let D be a representative of $\mathfrak{D}$. We shall construct a vector form $u \in \mathcal{O}_k^r(X)$ such that $\mathfrak{D} = \mathcal{L}(u)$. This construction is, actually, independent of the fact that D is a Lie derivation, i.e., applies to any admissible derivation D. Let $\xi_1, \ldots, \xi_r \in \tilde{\mathcal{J}}_k T$ and consider the map

$$\lambda_\xi: f \in \mathcal{J}_k \longmapsto Df(\xi_1, \ldots, \xi_r) \in \mathcal{J}_k .$$

Since D is a graded derivation, we infer that λ_ξ is a derivation of $\mathcal{J}_k$. Moreover, since D is admissible, then λ_ξ is also admissible hence, by Lemma 1, $\lambda_\xi = \mathcal{L}(\eta)$ with $\eta \in \tilde{\mathcal{J}}_k T$. Define u by the relation

$$u(\xi_1, \ldots, \xi_r) = \eta .$$

Since Df is a left $\mathcal{O}$-multilinear r-form it follows that u is also a left $\mathcal{O}$-multilinear vector form. Let us prove that $\mathfrak{D} = \mathcal{L}(u)$. By construction we have $Df = i(u)df$ hence $Df = \mathcal{L}(u)f$ for any $f \in \mathcal{J}_k$. Consider the derivation $D' = D - \mathcal{L}(u)$. Then D' vanishes on $\mathcal{J}_k$ hence it is an interior derivation and $D' = i(v)$ with $v \in \mathcal{O}_k^{r+1}(X)$. Moreover, since D' is also a Lie derivation, it follows that

$$\mathcal{L}(v) = [i(v), d] = [D', d] = 0$$

hence, by Lemma 3, $v = 0$ and $D' = 0$.

The graded Lie algebra structure of the image of $\mathcal{L}$ transports to $\mathcal{O}_k(X)$ and defines a bracket $[u, v]$ which is characterized by the property

$$\mathcal{L}([u, v]) = [\mathcal{L}(u), \mathcal{L}(v)] .$$

This bracket is only R-bilinear and will be referred to as the Nijenhuis bracket of $\mathcal{O}_k(X)$.

From the previous discussion we infer the

PROPOSITION 3. *A Lie derivation is uniquely determined by its action on* $\mathcal{B}_k^0(X) = \mathcal{J}_k$. *Conversely, any* R*-linear map*

$$\delta\colon \mathcal{J}_k \longrightarrow \mathcal{B}_k^r(X)$$

which is admissible, i.e., $\delta(\mathcal{J}_k^0) \subset \mathcal{J}_k^0 \mathcal{B}_k^r$ *and* $\delta(i_k \mathcal{O}) \subset \mathcal{A}_k^r(X)$, *and satisfies*

$$\delta(fg) = f\delta(g) + g\delta(f)$$

(i.e., condition (ii) *for the elements of* $\mathcal{J}_k$*) can be extended uniquely to an admissible Lie derivation of degree* r *of* $\mathcal{B}_k(X)$. *A similar statement holds for germs.*

The extension of δ is given by $\mathcal{L}(u)$ where $u \in \mathcal{O}_k^r(X)$ is defined in the same way as in the proof of Lemma 4. We observe that the knowledge of δ is sufficient to define the derivation λ_ξ.

THEOREM. $\mathcal{D}er_a \mathcal{B}_k(X) = \mathcal{L}[\mathcal{O}_k(X)] \oplus_{\mathbf{R}} i[\mathcal{O}_k(X)]$.

Proof. Let $\mathfrak{D} \in \mathcal{D}er_a^r \mathcal{B}_k(X)$ and let $u \in \mathcal{O}_k^r(X)$ be the vector form constructed in the proof of Lemma 4 using the derivation $\mathfrak{D}$. The derivation

$$\mathfrak{D}' = \mathfrak{D} - \mathcal{L}(u)$$

vanishes on $\mathcal{J}_k$ hence $\mathfrak{D}' = i(v)$ with $v \in \mathcal{O}_k^{r+1}(X)$ and $\mathfrak{D} = \mathcal{L}(u) + i(v)$. Let us prove that the sum is direct. Assume that $\mathcal{L}(u) = i(v)$ with $u \in \mathcal{O}_k^r(X)$ and $v \in \mathcal{O}_k^{r+1}(X)$. Then

$$\mathcal{L}(v) = [i(v), d] = [\mathcal{L}(u), d] = 0$$

implies, by Lemma 3, that $v = 0$ hence $\mathcal{L}(u) = i(v) = 0$.

COROLLARY. *An admissible derivation is determined uniquely by its action on* $\mathcal{J}_k \oplus \mathcal{C}_k^1(X)$. *Conversely, any* $\mathbf{R}$*-linear map*

$$\delta\colon \mathcal{J}_k \oplus \mathcal{C}_k^1(X) \longrightarrow \mathcal{B}_k(X)$$

which is admissible and satisfies conditions (i) *and* (ii) *for the elements in* $\mathcal{J}_k \oplus \mathcal{C}_k^1(X)$ *and a certain integer* p, *can be extended uniquely to an admissible derivation of degree* p *of* $\mathcal{B}_k(X)$. *A similar statement holds for germs.*

Given δ, we first define the Lie component of the extension D as being the derivation $\mathcal{L}(u)$ which is the extension of

$$\delta\colon \mathcal{J}_k \longrightarrow \mathcal{B}_k^p(X)\,.$$

The interior component of D is the derivation $i(v)$ which is the extension of the map

$$\delta - \mathcal{L}(u)\colon \mathfrak{A}_k^1(X) \longrightarrow \mathfrak{A}_k^{p+1}(X) .$$

We could replace in the above corollary $\mathfrak{A}_k^1(X)$ by $\mathcal{B}_k^1(X)$.

We now give some standard formulas which are replicas of corresponding formulas in Section 13. Let

$$1 \in \mathcal{O}_k^1(X) \simeq (\tilde{\mathcal{J}}_k T)^* \otimes_{\mathcal{O}} \tilde{\mathcal{J}}_k T$$

be the identity map of $\tilde{\mathcal{J}}_k T$. Then

$$\mathcal{L}(1) = d, \quad [u, 1] = 0 \quad \text{and} \quad i(1) = \oplus\, r\, \mathrm{Id}_r \tag{27.2}$$

where $u \in \mathcal{O}_k(X)$ and Id_r is the identity map of $\mathcal{B}_k^r(X)$.

If $\xi \in \tilde{\mathcal{J}}_k T = \mathcal{O}_k^0(X)$, $\omega \in \mathcal{B}_k(X)$ and $u \in \mathcal{O}_k(X)$ then $\mathcal{L}(\xi)\omega$ is the Lie derivative, in the usual sense, of the scalar form ω along ξ and $[\xi, u]$ is the usual Lie derivative $\mathcal{L}(\xi)u$ of the vector form u along ξ. To see this, recall (cf. Section 20) that the groupoid $\Gamma_k X$ operates covariantly (to the left) on the sheaf $\tilde{\mathcal{J}}_k T$ hence by transposition it operates contravariantly on

$$\Lambda(\tilde{\mathcal{J}}_k T)^* \simeq \mathfrak{A}_k(X) .$$

This action, combined with the contravariant action of $\Gamma_k X$ on $\mathcal{J}_k$, gives a contravariant action of $\Gamma_k X$ on

$$\mathcal{B}_k(X) \simeq \Lambda(\tilde{\mathcal{J}}_k T)^* \otimes_{\mathcal{O}} \mathcal{J}_k .$$

Furthermore, the same action combined with the contravariant action of $\Gamma_k X$ on $\tilde{\mathcal{J}}_k T$ (obtained via the inversion map of the groupoid) yields a contravariant action of $\Gamma_k X$ on

$$\mathcal{O}_k(X) \simeq \Lambda(\tilde{\mathcal{J}}_k T)^* \otimes_{\mathcal{O}} \tilde{\mathcal{J}}_k T .$$

If $u \in \mathcal{O}_k^r(X)$ and $F \in \Gamma_k X$ then the latter is defined by

$$(F^* \bullet u)(\xi_1, \ldots, \xi_r) = F^{-1} \bullet u(F \bullet \xi_1, \ldots, F \bullet \xi_r)$$

or, in tensor product notation,

$$F^* \bullet (\alpha \otimes \xi) = (F^* \bullet \alpha) \otimes (F^{-1} \bullet \xi) .$$

The former action is given by

$$(F^* \bullet \nu)(\xi_1, \ldots, \xi_r) = F^{-1} \bullet \nu(F \bullet \xi_1, \ldots, F \bullet \xi_r)$$

or

$$F^* \bullet (\alpha \otimes f) = (F^* \bullet \alpha) \otimes (F^{-1} \bullet f)$$

where $\nu \in \mathcal{B}_k^r(X)$, $\alpha \in \Lambda(\tilde{\mathfrak{J}}_k T)^*$, $f \in \mathfrak{J}_k$ and where $F^{-1} \bullet f \in \mathfrak{J}_k$ is the quotient of $\varphi \circ F \in \mathcal{O}_{X^2}$ with $\varphi \in \mathcal{O}_{X^2}$ a representative of f and $F \in \widetilde{\mathfrak{Aut}}\, X^2$ a representative of F. These actions extend to contravariant actions of admissible section of $\Gamma_k X$ onto sections of $\mathcal{B}_k(X)$ and $\mathcal{V}_k(X)$. With this in mind, let Ξ be a local section of $\tilde{\mathfrak{J}}_k T$ which extends the germ ξ, Ω a local section of $\mathcal{B}_k(X)$ which extends ω and $\mathfrak{U}$ a local section of $\mathcal{V}_k(X)$ which extends u. Let (F_t), $|t| < \varepsilon$, be a differentiable 1-parameter family of local sections in $\Gamma(U, \Gamma_k X)$ with F_0 the identity section and such that $\frac{d}{dt} F_t|_{t=0} = \Xi$ (cf. Section 19). Then

$$\mathfrak{L}(\Xi)\Omega = \frac{d}{dt}(F_t^* \bullet \Omega)\,|_{t=0} \tag{27.3}$$

and

$$[\Xi, \mathfrak{U}] = \mathfrak{L}(\Xi)\mathfrak{U} = \frac{d}{dt}(F_t^* \bullet \mathfrak{U})\,|_{t=0} \tag{27.4}$$

The proof of (27.3) is similar to the usual proof of the Cartan formula for the Lie derivative of a scalar form along a vector field. The formula (27.4) is a consequence of (27.3) and the formula

$$[\![\Xi, \Sigma]\!] = \mathfrak{L}(\Xi)\Sigma = \frac{d}{dt}(F_t^{-1} \bullet \Sigma)\,|_{t=0} \tag{27.5}$$

where Σ is another section of $\tilde{\mathfrak{J}}_k T$. Moreover (27.5) is a special case of (27.4) since $[\![\Xi, \Sigma]\!]$ (bracket in $\tilde{\mathfrak{J}}_k T$) is equal to $[\Xi, \Sigma]$ (Nijenhuis bracket of vector forms of degree zero, i.e., in $\mathcal{V}_k^0(X)$) and is a consequence of the corresponding formula for admissible vector fields in X^2.

From the previous formulas we infer that $\mathcal{L}(\xi)\omega$, $[\xi, u]$ and $[\![\xi, \sigma]\!] =$ $[\xi, \sigma]$ are simply the germs induced by the right hand sides of formulas (27.3-5), i.e., germs of Lie derivatives. In particular, we can replace F_t by the 1-parameter local group $\operatorname{Exp} t\Xi$.

For decomposable elements we have the following relations. Let $\omega \in \mathcal{B}_k(X)$, $u = \alpha \otimes \xi \in \Lambda^r(\tilde{\mathcal{J}}_k T)^* \otimes_{\mathcal{O}} \tilde{\mathcal{J}}_k T$, $v = \beta \otimes \eta \in \Lambda^s(\tilde{\mathcal{J}}_k T)^* \otimes_{\mathcal{O}} \tilde{\mathcal{J}}_k T$ and $w \in \mathcal{B}_k^r(X)$. Then

$$(27.6) \qquad i(u)\omega = \alpha \wedge i(\xi)\omega, \mathcal{L}(u)\omega = \alpha \wedge \mathcal{L}(\xi)\omega + (-1)^r d\alpha \wedge i(\xi)\omega$$

$$(27.7) \quad [w, \beta \otimes \eta] = (\mathcal{L}(w)\beta) \otimes \eta - (-1)^{rs} \beta \wedge \mathcal{L}(\eta) w - (-1)^{rs+s} (d\beta \otimes \eta) \barwedge w$$

where $\mathcal{L}(\eta)w = [\eta, w]$,

$$(27.8) \qquad [u,v] = (\mathcal{L}(u)\beta) \otimes \eta - (-1)^{rs}(\mathcal{L}(v)\alpha) \otimes \xi + (\alpha \wedge \beta) \otimes [\![\xi, \eta]\!] .$$

For notational convenience, let us introduce the map

$$\mathcal{K} \colon \mathcal{O}_k(X) \longrightarrow \mathcal{D}er_a \mathcal{B}_k(X)$$

which on the homogeneous component $\mathcal{O}_k^{r+1}(X)$ is equal to $\mathcal{K} = (-1)^r i$. More generally, we extend this notation to the $\barwedge$ operation on vector forms, i.e.,

$$\mathcal{K}(u)v = (-1)^r u \barwedge v$$

where $u \in \mathcal{O}_k^{r+1}(X)$ and $v \in \mathcal{O}_k(X)$. For decomposable elements we find

$$\mathcal{K}(\alpha \otimes \xi)(\beta \otimes \eta) = (\mathcal{K}(\alpha \otimes \xi)\beta) \otimes \eta .$$

The direct sum decomposition theorem for $\mathcal{D}er_a \mathcal{B}_k(X)$ gives rise to the R-linear isomorphism

$$\mathcal{L} + \mathcal{K} \colon \mathcal{O}_k(X) \times_X \mathcal{O}_k(X) \longrightarrow \mathcal{D}er_a \mathcal{B}_k(X)$$

which transforms $\mathcal{O}_k^r(X) \times_X \mathcal{O}_k^{r+1}(X)$ onto $\mathcal{D}er_a^r \mathcal{B}_k(X)$. The graded R-Lie algebra structure of $\mathcal{D}er_a \mathcal{B}_k(X)$ transports to $\mathcal{O}_k(X) \times_X \mathcal{O}_k(X)$. Let us write explicitely the transported bracket. Take $u_1 \in \mathcal{O}_k^r(X)$, $u_2 \in \mathcal{O}_k^s(X)$,

$v_1 \in \mathcal{V}_k^{r+1}(X)$ and $v_2 \in \mathcal{V}_k^{s+1}(X)$. Then

$$w_1 = \mathcal{L}(u_1) + \mathcal{K}(v_1) \in \mathcal{D}er_a^r \mathcal{B}_k(X)$$

and

$$w_2 = \mathcal{L}(u_2) + \mathcal{K}(v_2) \in \mathcal{D}er_a^s \mathcal{B}_k(X) .$$

By definition

$$\begin{aligned}(\mathcal{L}+\mathcal{K})[(u_1,v_1),(u_2,v_2)] = [w_1,w_2] = [\mathcal{L}(u_1) + \mathcal{K}(v_1), \mathcal{L}(u_2) + \mathcal{K}(v_2)] = \\ [\mathcal{L}(u_1), \mathcal{L}(u_2)] + \{[\mathcal{K}(v_1), \mathcal{L}(u_2)] + \\ (-1)^{rs+1}[\mathcal{K}(v_2), \mathcal{L}(u_1)]\} + [\mathcal{K}(v_1), \mathcal{K}(v_2)]\end{aligned}$$

and the various terms are given by the formulas (for the second formula see [3])

(27.9) $$[\mathcal{L}(u_1), \mathcal{L}(u_2)] = \mathcal{L}([u_1, u_2])$$

(27.10) $$[\mathcal{K}(v_1), \mathcal{L}(u_2)] = (-1)^r \mathcal{L}(v_1 \barwedge u_2) + \mathcal{K}([v_1, u_2])$$

(27.11) $$[\mathcal{K}(v_1), \mathcal{K}(v_2)] = \mathcal{K}(v_1 \barwedge v_2 - (-1)^{rs} v_2 \barwedge v_1) .$$

We infer that

$$[(u_1, v_1), (u_2, v_2)] = (u, v) \in \mathcal{V}_k^{r+s}(X) \times_X \mathcal{V}_k^{r+s+1}(X)$$

where

(27.12) $$u = [u_1,u_2] + (-1)^r v_1 \barwedge u_2 + (-1)^{rs+s+1} v_2 \barwedge u_1$$

and

(27.13) $$v = [v_1,u_2] + (-1)^{rs+1}[v_2,u_1] + v_1 \barwedge v_2 - (-1)^{rs} v_2 \barwedge v_1 .$$

The definitions and methods of proof suggest quite clearly that we have been essentially working in each stalk. In fact, the whole theory of admissible derivations can be transcribed for each stalk and the direct sum decomposition theorem implies that

$$\mathcal{D}er_a \mathcal{B}_k(X) \simeq \bigcup_{x \in X} \mathcal{D}er_a[\mathcal{B}_k(X)(x)]$$

where $\mathcal{B}_k(X)(x)$ is the stalk of $\mathcal{B}_k(X)$ at the point x. For $k = 0$, the theory reduces to the classical Frölicher-Nijenhuis theory for $\Lambda \underline{T}^*$.

28. *The* $\hat{D}$-*complex*

Let

$$\hat{D}: \mathcal{D}er_a \mathcal{B}_k(X) \longrightarrow \mathcal{D}er_a \mathcal{B}_k(X)$$

be the differential operator of order 1 defined by

$$\hat{D}w = [d, w], \quad w \in \mathcal{D}er_a \mathcal{B}_k(X) . \tag{28.1}$$

Here $\hat{D}$ is a homogeneous $\mathbf{R}$-linear map of degree 1, i.e.,

$$\hat{D}: \mathcal{D}er^r_a \mathcal{B}_k(X) \longrightarrow \mathcal{D}er^{r+1}_a \mathcal{B}_k(X) .$$

From the graded Jacobi identity and the identity $[d,d] = 0$ we infer that

$$\hat{D}^2 = \hat{D} \circ \hat{D} = 0 .$$

Moreover, we also obtain, from the Jacobi identity, the formula

$$\hat{D}[w_1, w_2] = [\hat{D}w_1, w_2] + (-1)^r [w_1, \hat{D}w_2] \tag{28.2}$$

where $w_1 \in \mathcal{D}er^r_a \mathcal{B}_k(X)$ and $w_2 \in \mathcal{D}er_a \mathcal{B}_k(X)$. This formula says that $\hat{D}$ is a graded derivation of degree 1 of the graded $\mathbf{R}$-Lie algebra

$$\mathcal{D}er_a \mathcal{B}_k(X) .$$

$\hat{D}$ is an *interior* derivation in the sense of Lie algebras since $\hat{D} = ad\ d =$ the adjoint derivation of d with respect to the Nijenhuis bracket [,]. If $u \in \mathcal{O}^r_k(X)$ then

$$\mathcal{L}(u) = [i(u),d] = i(u) \circ d + (-1)^r d \circ i(u) = -(d \circ \mathcal{K}(u) + (-1)^r \mathcal{K}(u) \circ d) = -[d, \mathcal{K}(u)]$$

hence

(28.3) $$\mathcal{L} = -\hat{D}\circ\mathcal{K}$$

From the second formula (27.2) we also derive the identity

(28.4) $$rw = [\mathcal{K}(1), w], \ w \in \mathcal{D}er_a^r\mathcal{B}_k(X) .$$

In fact, let $w = \mathcal{L}(u) + \mathcal{K}(v)$ with $u \in \mathcal{O}_k^r(X)$ and $v \in \mathcal{O}_k^{r+1}(X)$. Since $\mathcal{K}(1)$ is a derivation of degree 0, we have by (27.10-11)

$$[\mathcal{K}(1), w] = [\mathcal{K}(1), \mathcal{L}(u)] + [\mathcal{K}(1), \mathcal{K}(v)] =$$
$$\mathcal{L}(1 \barwedge u) + \mathcal{K}([1, u]) + \mathcal{K}(1 \barwedge v - v \barwedge 1) .$$

But $[1, u] = 0$, $1 \barwedge u = ru$ and $1 \barwedge v - v \barwedge 1 = (r+1)v - v = rv$ hence

$$[\mathcal{K}(1), w] = r\mathcal{L}(u) + r\mathcal{K}(v) = rw .$$

Remark. It is natural to define a derivation of degree p of the graded R-Lie algebra $\mathcal{D}er_a\mathcal{B}_k(X)$ to be an R-linear sheaf map

$$W\colon \mathcal{D}er_a\mathcal{B}_k(X) \to \mathcal{D}er_a\mathcal{B}_k(X)$$

which satisfies the conditions

i) $$W[\mathcal{D}er_a^r\mathcal{B}_k(X)] \subset \mathcal{D}er_a^{r+p}\mathcal{B}_k(X) .$$

ii) $W[w_1,w_2] = [W(w_1),w_2] + (-1)^{pr}[w_1,W(w_2)],\ w_1 \in \mathcal{D}er_a^r\mathcal{B}_k(X)$.

We have seen already that $\hat{D}$ is a derivation of degree 1. Let W be a derivation of degree p of $\mathcal{D}er_a\mathcal{B}_k(X)$. Applying W to (28.4) we obtain

$$rW(w) = [W\mathcal{K}(1), w] + [\mathcal{K}(1), W(w)] = [W\mathcal{K}(1), w] + (r+p)W(w)$$

hence

(28.5) $$pW(w) = -[W\mathcal{K}(1), w] .$$

This formula generalizes the definition (28.1) of $\hat{D}$ since

$$\hat{D}\mathcal{K}(1) = -\mathcal{L}(1) = -d .$$

The following complex is exact, where $\mathcal{D}er^r_{a,k}$ denotes $\mathcal{D}er^r_a\mathcal{B}_k(X)$ and where N denotes the (local) rank of $\tilde{\mathcal{J}}_kT$ over $\mathcal{O}$.

$$(28.6)_k \qquad 0 \longrightarrow \tilde{\mathcal{J}}_kT \xrightarrow{\mathcal{L}} \mathcal{D}er^0_{a,k} \xrightarrow{\hat{D}} \mathcal{D}er^1_{a,k} \xrightarrow{\hat{D}} \cdots \xrightarrow{\hat{D}} \mathcal{D}er^N_{a,k} \longrightarrow 0 \ .$$

In fact, if $w \in \mathcal{D}er^r_{a,k}$ and $\hat{D}w = [d, w] = 0$ then w is a Lie derivation, hence $w = \mathcal{L}(u)$ with

$$u \in \mathcal{O}^r_k(X) \simeq \Lambda^r(\tilde{\mathcal{J}}_kT)^* \otimes_{\mathcal{O}} \tilde{\mathcal{J}}_kT \ .$$

If $r = 0$ then $u \in \tilde{\mathcal{J}}_kT$ and if $r > 0$ then, by the formula (28.3),

$$w = -\hat{D}\mathcal{K}(u) \quad \text{with} \quad \mathcal{K}(u) \in \mathcal{D}er^{r-1}_{a,k} \ .$$

The complex ends at N steps since $\Lambda^{N+1}(\tilde{\mathcal{J}}_kT)^* = 0$, so that

$$\mathcal{D}er^N_{a,k} = \mathcal{L}[\mathcal{O}^N_k(X)] = \hat{D}[\mathcal{K}\mathcal{O}^N_k(X)] \quad \text{and} \quad \mathcal{D}er^r_{a,k} = 0 \quad \text{for} \quad r > N \ .$$

In the sequence $(28.6)_k$ the term $\tilde{\mathcal{J}}_kT$ can be replaced by

$$\mathcal{D}er^{-1}_{a,k} = \mathcal{K}(\tilde{\mathcal{J}}_kT)$$

and $\mathcal{L}$ is then replaced by $\hat{D}$. We obtain the exact complex

$$(28.7)_k \qquad 0 \longrightarrow \mathcal{D}er^{-1}_{a,k} \xrightarrow{\hat{D}} \mathcal{D}er^0_{a,k} \xrightarrow{\hat{D}} \cdots \xrightarrow{\hat{D}} \mathcal{D}er^N_{a,k} \longrightarrow 0 \ .$$

This second way of writing (28.6) is equivalent to the following. An element $w \in \mathcal{D}er_a\mathcal{B}_k(X)$ is of the form $w = \hat{D}\mu$ with $\mu \in \mathcal{D}er_a\mathcal{B}_k(X)$ if and only if $\hat{D}w = 0$. Alternatively, the first order differential equation $\hat{D}\mu = w$ with unknown μ is solvable for a given w if and only if the compatibility condition $\hat{D}w = 0$ is satisfied.

29. *The restricted* $\hat{D}$*-complex*

Recall that $\Lambda\underline{T}^*$ can be considered as an $\mathcal{O}$-subalgebra of $\mathcal{B}_k(X)$ via the imbedding

$$\tilde{\beta}_k^*: \Lambda\underline{T}^* \to \mathcal{A}_k(X) \subset \mathcal{B}_k(X) \simeq \Lambda(\tilde{\mathcal{J}}_k T)^* \otimes_{\mathcal{O}} \mathcal{J}_k .$$

Similarly,

$$\Lambda\underline{T}^* \otimes_{\mathcal{O}} \mathcal{J}_k \quad \text{and} \quad \Lambda\underline{T}^* \otimes_{\mathcal{O}} \tilde{\mathcal{J}}_k T$$

can be considered as subsheaves of

$$\mathcal{B}_k(X) \quad \text{and} \quad \mathcal{V}_k(X) \simeq \Lambda(\tilde{\mathcal{J}}_k T)^* \otimes_{\mathcal{O}} \tilde{\mathcal{J}}_k T$$

respectively. The former is a homogeneous graded $\mathcal{J}_k$-subalgebra of $\mathcal{B}_k(X)$ and a graded left submodule of $\mathcal{B}_k(X)$ over the graded $\mathcal{O}$-algebra $\Lambda\underline{T}^*$. The latter is a homogeneous graded left submodule of $\mathcal{V}_k(X)$ over the graded $\mathcal{J}_k$-algebra $\Lambda\underline{T}^* \otimes_{\mathcal{O}} \mathcal{J}_k$. Let

$$\tilde{\mathcal{J}}_k^0 T = \mathcal{J}_k^0 \,\Delta\, \tilde{\mathcal{J}}_k T = \mathit{ker}(\tilde{\beta}_k : \tilde{\mathcal{J}}_k T \to \underline{T}) .$$

The sheaf

$$\Lambda^r \underline{T}^* \otimes_{\mathcal{O}} \tilde{\mathcal{J}}_k T \simeq \mathcal{A}\ell t^r_{\mathcal{O}}(\underline{T}, \tilde{\mathcal{J}}_k T)$$

is isomorphic to the subsheaf of $\mathcal{V}_k^r(X)$ composed of those germs of alternating r-forms u on $\tilde{\mathcal{J}}_k T$ with values in $\tilde{\mathcal{J}}_k T$ that satisfy the condition

(29.1) $\qquad u(\xi_1, \ldots, \xi_r) = 0$ whenever some ξ_i belongs to $\tilde{\mathcal{J}}_k^0 T$.

A similar description holds for $\Lambda\underline{T}^*$ and $\Lambda\underline{T}^* \otimes_{\mathcal{O}} \mathcal{J}_k$ considered as subsheaves of $\mathcal{A}_k(X)$ and $\mathcal{B}_k(X)$ respectively.

An admissible derivation $D \in \mathrm{Der}_a \mathcal{B}_k(X)$ is called *strongly admissible* if it satisfies the further conditions:

c) $D(\Lambda\underline{T}^* \otimes_{\mathcal{O}} \mathcal{J}_k) \subset \Lambda\underline{T}^* \otimes_{\mathcal{O}} \mathcal{J}_k$

d) $D(\Lambda\underline{T}^*) \subset \Lambda\underline{T}^*$.

LEMMA 1. *An admissible Lie derivation* D *of degree* p *is strongly admissible if and only if*

$$D(\mathcal{J}_k) \subset \Lambda^p\underline{T}^* \otimes_{\mathcal{O}} \mathcal{J}_k \quad \textit{and} \quad D(\mathcal{O}) \subset \Lambda^p\underline{T} \quad (\mathcal{O} \simeq i_k\mathcal{O}) .$$

Proof. The condition is obviously necessary. It is also sufficient since $\Lambda\underline{T}^*\otimes_{\mathcal{O}}\mathcal{J}_k$ is $\mathcal{J}_k$-generated by $\{\mathcal{O}, d\mathcal{O}\}$, $\Lambda\underline{T}^*$ is $\mathcal{O}$-generated by $\{\mathcal{O}, d\mathcal{O}\}$ and D commutes with d.

Remark. The exterior differentiation d is not strongly admissible except for $k = 0$ in which case all the admissibility notions vanish. This can be checked directly on an example. A formal justification is given by Theorem 1 of this section since 1, the identity mapping of $\tilde{\mathcal{J}}_k T$, does not belong to $\Lambda\underline{T}^*\otimes_{\mathcal{O}}\tilde{\mathcal{J}}_k T$ except in the case $k = 0$. The sheaf of germs of strongly admissible derivations is a homogeneous graded R-Lie subalgebra of $\mathcal{D}er_a\mathcal{B}_k(X)$. If $k = 0$, this sheaf reduces to $\mathcal{D}er\, \Lambda\underline{T}^*$.

LEMMA 2. *For any* $u \in \Lambda\underline{T}^*\otimes_{\mathcal{O}}\tilde{\mathcal{J}}_k T$ *the germs of derivations* $i(u)$, *hence* $\mathcal{K}(u)$, *and* $\mathcal{L}(u)$ *are strongly admissible.*

Proof. It is trivial to verify that $i(u)$ is strongly admissible. Assume that $u \in \Lambda^r\underline{T}^*\otimes_{\mathcal{O}}\tilde{\mathcal{J}}_k T$. Then

$$\mathcal{L}(u) = [i(u), d] = i(u)\circ d + (-1)^r d\circ i(u) .$$

If $f \in \mathcal{O}$

$$\mathcal{L}(u)f = i(u)df = (df)\circ u \in \Lambda^r\underline{T}^*$$

and if $f \in \mathcal{J}_k$ then

$$\mathcal{L}(u)f = (df)\circ u \in \Lambda^r\underline{T}^*\otimes_{\mathcal{O}}\mathcal{J}_k .$$

Denote by $\mathcal{D}er_\sigma\mathcal{B}_k(X)$ the sheaf of (germs of) strongly admissible Lie derivations. Since the conditions (c) and (d) are obviously invariant by the commutator of derivations it follows that $\mathcal{D}er_\sigma\mathcal{B}_k(X)$ is a homogeneous R-Lie subalgebra of $\mathcal{D}er_a\mathcal{B}_k(X)$.

LEMMA 3. $\mathcal{L}(\Lambda\underline{T}^*\otimes_{\mathcal{O}}\tilde{\mathcal{J}}_k T) = \mathcal{D}er_\sigma\mathcal{B}_k(X)$.

Proof. Let D be a strongly admissible Lie derivation of degree r. Then $D = \mathcal{L}(u)$ with $u \in \mathcal{O}_k^r(X)$ and we have to prove that $u(\xi_1, \ldots, \xi_r) = 0$ whenever some ξ_i belongs to $\tilde{\mathcal{J}}_k^0 T$. Recall that $u(\xi_1, \ldots, \xi_r) = \lambda_\xi$ where $\lambda_\xi f = Df(\xi_1, \ldots, \xi_r)$ for any $f \in \mathcal{J}_k$. Since

$$D(\mathcal{J}_k) \in \Lambda^r \underline{T}^* \otimes_{\mathcal{O}} \mathcal{J}_k$$

it follows that $Df(\xi_1, \ldots, \xi_r) = 0$ hence $\lambda_\xi = 0$.

The map

$$\mathcal{L}: \Lambda \underline{T}^* \otimes_{\mathcal{O}} \tilde{\mathcal{J}}_k T \longrightarrow \mathcal{D}er_\sigma \mathcal{B}_k(X)$$

being bijective and $\mathcal{D}er_\sigma \mathcal{B}_k(X)$ being a Lie subalgebra of $\mathcal{D}er_a \mathcal{B}_k(X)$, we infer that $\Lambda \underline{T}^* \otimes_{\mathcal{O}} \tilde{\mathcal{J}}_k T$ is a homogeneous graded R-Lie subalgebra of

$$\Lambda(\tilde{\mathcal{J}}_k T)^* \otimes_{\mathcal{O}} \tilde{\mathcal{J}}_k T \simeq \mathcal{O}_k(X)$$

for the Nijenhuis bracket.

Remark. The last property can also be verified directly using formulas (27.6) and (27.8). In fact, if $u, v \in \Lambda \underline{T}^* \otimes_{\mathcal{O}} \tilde{\mathcal{J}}_k T$ are decomposable then Lemma 2 and (27.6) imply that $[u, v] \in \Lambda \underline{T}^* \otimes_{\mathcal{O}} \tilde{\mathcal{J}}_k T$.

From (27.3), the formula

$$b_k \circ \operatorname{Exp} t\Xi = \exp t(\tilde{\beta}_k \circ \Xi)$$

(cf. the last formula of (b)(2), Section 19) and the commutativity relation

$$F^* \bullet \tilde{\beta}_k^* \omega = \tilde{\beta}_k^* \circ f^* \omega$$

where $\omega \in \Lambda \underline{T}^*$, $F \in \Gamma(U, \Gamma_k X)$ and $f = b_k \circ F$, we infer that the term $\mathcal{L}(\xi)\omega$, $\omega \in \Lambda \underline{T}^*$, in (27.6) is equal to the Lie derivative $\mathcal{L}(\xi)\omega$ where the last $\mathcal{L}$ is the representation

$$\mathcal{L}: \Lambda \underline{T}^* \otimes_{\mathcal{O}} \tilde{\mathcal{J}}_k T \to \mathcal{D}er\, \Lambda \underline{T}^*$$

defined in the beginning of Section 15. It follows from (27.6) that this representation $\mathcal{L}$ is equal to the restriction of the representation

$$\mathcal{L}: \Lambda\underline{T}^* \otimes_{\mathcal{O}} \tilde{\mathcal{J}}_k T \longrightarrow \mathcal{D}er_a \mathcal{B}_k(X)$$

to the subsheaf $\Lambda\underline{T}^*$ imbedded in $\mathcal{B}_k(X)$, i.e.,

$$\mathcal{L}: u \in \Lambda\underline{T}^* \otimes_{\mathcal{O}} \tilde{\mathcal{J}}_k T \longmapsto \mathcal{L}(u) \mid_{\Lambda\underline{T}^*} \in \mathcal{D}er\, \Lambda\underline{T}^*.$$

If $k = 0$ then

$$\tilde{\mathcal{J}}_0 T \simeq \underline{T},\ \mathcal{D}er_a \mathcal{B}_0(X) = \mathcal{D}er\, \Lambda\underline{T}^*$$

and the two representations coincide. We infer from this and the formula (27.8) that the Nijenhuis bracket on $\Lambda\underline{T}^* \otimes_{\mathcal{O}} \tilde{\mathcal{J}}_k T$ as defined above is equal to the Nijenhuis bracket as defined in Section 14 (by quotienting the Nijenhuis bracket of $\tilde{\Theta}$-valued horizontal forms on X^2).

THEOREM 1. *The map*

$$\mathcal{L}: \Lambda\underline{T}^* \otimes_{\mathcal{O}} \tilde{\mathcal{J}}_k T \longrightarrow \mathcal{D}er_a \mathcal{B}_k(X)$$

is a graded R*-Lie algebra isomorphism when the first sheaf is given the Nijenhuis bracket.*

Remark. A similar property does not hold for interior derivations. In fact, the sheaf of strongly admissible interior derivations is much larger than $\mathcal{K}(\Lambda\underline{T}^* \otimes_{\mathcal{O}} \tilde{\mathcal{J}}_k T)$. If $u \in \mathcal{O}_k(X)$ takes values in $\tilde{\mathcal{J}}_k^0 T$, i.e., $u \in \Lambda(\tilde{\mathcal{J}}_k T)^* \otimes_{\mathcal{O}} \tilde{\mathcal{J}}_k^0 T$ then $\mathcal{K}(u)$ vanishes on $\Lambda\underline{T}^* \otimes_{\mathcal{O}} \tilde{\mathcal{J}}_k$.

We now define the adjoint representation

$$ad: \Lambda\underline{T}^* \otimes_{\mathcal{O}} \tilde{\mathcal{J}}_{k+1} T \longrightarrow \mathcal{D}er_a \mathcal{B}_k(X)$$

by

$$ad(u) = \mathcal{L}(\tilde{\rho}_k u) - \mathcal{K}(\tilde{D}u)$$

where $\tilde{D}$ is the operator defined in Section 14(*). Observe that $ad(u)$ is a

(*) (Added in proof) The map ad is $\Lambda\underline{T}^*$-linear and, in particular, $ad: \tilde{\mathcal{J}}_{k+1} T \to \mathcal{D}er_a \mathcal{B}_k(X)$ is the left $\mathcal{O}$-linear sheaf map which corresponds to the (germ of) strongly admissible derivation and, in particular, $\mathcal{ad}(u)$ preserves the subsheaf $\wedge\underline{T}^*$ imbedded in $\mathcal{B}_k(X)$. From the previous considerations concerning the representation $\mathcal{L}$ we infer that the restriction

$$\mathcal{ad}\colon u \in \wedge\underline{T}^* \otimes_{\mathcal{O}} \tilde{\mathcal{J}}_{k+1}T \longmapsto \mathcal{ad}(u)\,|\,\wedge\underline{T}^* \in \mathcal{D}er\ \wedge\underline{T}^*$$

is equal to the representation $\mathcal{ad}$ defined in Section 15. Moreover, if $k = 0$, the two representations coincide.

Denote by $\mathcal{D}er_\Sigma\mathcal{B}_k(X)$ the image of $\mathcal{ad}$ in $\mathcal{D}er_a\mathcal{B}_k(X)$.

LEMMA 4. *The sheaf* $\mathcal{D}er_\Sigma\mathcal{B}_k(X)$ *is a homogeneous graded* R*-Lie subalgebra of* $\mathcal{D}er_a\mathcal{B}_k(X)$. *Each element of* $\mathcal{D}er_\Sigma\mathcal{B}_k(X)$ *is strongly admissible.*

Proof. If $k = 0$ then

$$\mathcal{D}er_\Sigma\mathcal{B}_0(X) = \bigoplus_{r\geq 0} \mathcal{D}er_a^r\mathcal{B}_0(X) = \bigoplus_{r\geq 0} \mathcal{D}er^r\ \wedge\underline{T}^*$$

(cf. the Theorem of Section 15). Assume that $k > 0$ and let

$$u \in \wedge^r\underline{T}^* \otimes_{\mathcal{O}} \tilde{\mathcal{J}}_{k+1}T \quad \text{and} \quad v \in \wedge^s\underline{T}^* \otimes_{\mathcal{O}} \tilde{\mathcal{J}}_{k+1}T\,.$$

We shall prove that

$$\mathcal{ad}([\![u, v]\!]) = [\mathcal{ad}u, \mathcal{ad}v]$$

where $[\![\ ,\ ]\!]$ is the twisted bracket on $\wedge\underline{T}^* \otimes_{\mathcal{O}} \tilde{\mathcal{J}}_{k+1}T$ as defined in Section 14. Now,

$$\begin{aligned}[\mathcal{ad}u, \mathcal{ad}v] &= [\mathcal{L}(\tilde{\rho}_k u) - \mathcal{K}(\tilde{D}u), \mathcal{L}(\tilde{\rho}_k v) - \mathcal{K}(\tilde{D}v)] = \\ &[\mathcal{L}(\tilde{\rho}_k u) + \mathcal{K}(-\tilde{D}u), \mathcal{L}(\tilde{\rho}_k v) + \mathcal{K}(-\tilde{D}v)] = \mathcal{L}(U) + \mathcal{K}(V)\end{aligned}$$

where

$$(U, V) \in \mathcal{V}_k^{r+s}(X) \times_X \mathcal{V}_k^{r+s+1}(X)$$

and we have to compute

(k+1)-st order differential operator $\mathcal{L}\circ\tilde{j}_k$ where $\tilde{j}_k\colon \underline{T} \to \tilde{\mathcal{J}}_k T$ is of order k and $\mathcal{L}\colon \mathcal{J}_k T \to \mathcal{D}er_a\mathcal{B}_k(X)$ is of order 1. This remark simplifies the proof of Lemma 4 (compare the footnote on pages 118, 119).

$$(U,V) = [(\tilde{\rho}_k u, -\tilde{D}u), (\tilde{\rho}_k v, -\tilde{D}v)]$$

where the last bracket is the graded R-Lie algebra bracket in $\mathcal{O}_k(X) \times_X \mathcal{O}_k(X)$ transported from $\mathcal{D}er_a\mathcal{B}_k(X)$ via the map $\mathcal{L} + \mathcal{K}$. The formulas (27.12-13) give

$$U = [\tilde{\rho}_k u, \tilde{\rho}_k v] + (-1)^r(-\tilde{D}u) \barwedge \tilde{\rho}_k v + (-1)^{rs+s+1}(-\tilde{D}v) \barwedge \tilde{\rho}_k u =$$
$$\tilde{\rho}_k[u,v] + \tilde{\rho}_k\{(-1)^{r+1}\tilde{D}u \barwedge v + (-1)^{rs+s}\tilde{D}v \barwedge u\} = \tilde{\rho}_k[\![u,v]\!]$$

and

$$V = -[\tilde{D}u, \tilde{\rho}_k v] + (-1)^{rs}[\tilde{D}v, \tilde{\rho}_k u] + \tilde{D}u \barwedge \tilde{D}v + (-1)^{rs+1}\tilde{D}v \barwedge \tilde{D}u \ .$$

On the other hand

$$ad([\![u,v]\!]) = \mathcal{L}(\tilde{\rho}_k[\![u,v]\!]) - \mathcal{K}(\tilde{D}[\![u,v]\!]) = \mathcal{L}(\tilde{\rho}_k[\![u,v]\!]) + \mathcal{K}(-\tilde{D}[\![u,v]\!]) \ .$$

It is therefore enough to check that $\tilde{\rho}_k[\![u,v]\!] = U$, which has already been done, and that $-\tilde{D}[\![u,v]\!] = V$. Since $k+1 \geq 2$, Lemma 6 of Section 16 yields

$$-\tilde{D}[\![u,v]\!] = -[\![\tilde{D}u, \tilde{\rho}_k v]\!] + (-1)^{r+1}[\![\tilde{\rho}_k u, \tilde{D}v]\!] =$$
$$-[\tilde{D}u, \tilde{\rho}_k v] + (-1)^{r+1}[\tilde{\rho}_k u, \tilde{D}v] + (-1)^{(r+1)s+s+1}\tilde{D}(\tilde{\rho}_k v) \barwedge \tilde{D}u +$$
$$(-1)^{r+1}(-1)^{r+1}\tilde{D}\tilde{\rho}_k u \barwedge \tilde{D}v \ .$$

A simple computation shows that the second term is equal to $(-1)^{rs}[\tilde{D}v, \tilde{\rho}_k u]$, the third term is equal to

$$(-1)^{rs+1}(\tilde{\rho}_k\tilde{D}v) \barwedge \tilde{D}u = (-1)^{rs+1}\tilde{D}v \barwedge \tilde{D}u$$

since $w \barwedge \tilde{D}u = (\tilde{\rho}_0 w) \barwedge \tilde{D}u$ and finally the last term is equal to $\tilde{D}u \barwedge \tilde{D}v$. This completes the proof since, by Lemma 2, the last assertion is obvious.

THEOREM 2. *The mapping*

$$ad\colon \Lambda\underline{T}^* \otimes_{\mathcal{O}} \tilde{\mathcal{J}}_{k+1}T \longrightarrow \mathcal{D}er_{\Sigma}\mathcal{B}_k(X)$$

is a surjective homogeneous graded R*-Lie algebra morphism of degree zero with respect to the bracket* $[\![\ ,\ ]\!]$ *on* $\Lambda\underline{T}^*\otimes_{\mathcal{O}}\tilde{\mathcal{J}}_{k+1}T$ *and the derivations bracket on* $\mathcal{D}er_{\Sigma}\mathcal{B}_k(X)$.

Proof. If $k > 0$, the result follows from the proof of the Lemma 4. If $k = 0$, the result is a special case of the theorem in Section 15.

This theorem is the birthcradle of the twisted bracket $[\![\ ,\ ]\!]$. In his papers [13 (a)], Spencer defined this bracket as the transport of a derivation bracket.

LEMMA 5. $\mathcal{K}er\ ad = \underline{\delta}(\Lambda\underline{T}^*\otimes_{\mathcal{O}} S^{k+2}\underline{T}^*\otimes_{\mathcal{O}}\underline{T})$.

Proof. Let $u \in \Lambda\underline{T}^*\otimes_{\mathcal{O}}\tilde{\mathcal{J}}_{k+1}T$ and assume that $ad(u) = 0$. Since $\mathcal{D}er_a\mathcal{B}_k(X)$ is the direct sum of $\mathcal{L}[\mathcal{O}_k(X)]$ and $\mathcal{K}[\mathcal{O}_k(X)]$ it follows that

(a) $\mathcal{L}(\tilde{\rho}_k u) = 0$ hence $\tilde{\rho}_k u = 0$, i.e., $u \in \Lambda\underline{T}^*\otimes_{\mathcal{O}} S^{k+1}\underline{T}^*\otimes_{\mathcal{O}}\underline{T}$ and, by the exactness of the $\underline{\delta}$-complex,

(b) $\mathcal{K}(\tilde{D}u) = 0$ hence $\tilde{D}u = \underline{\delta}u = 0$, i.e., $u = \underline{\delta}v$ with $v \in \Lambda\underline{T}^*\otimes_{\mathcal{O}} S^{k+2}\underline{T}^*\otimes_{\mathcal{O}}\underline{T}$.

The converse statement is obvious since

$$ad(\underline{\delta}v) = \mathcal{L}(\tilde{\rho}_k\underline{\delta}v) - \mathcal{K}(\underline{\delta}\circ\underline{\delta}v) = \mathcal{L}(0) - \mathcal{K}(0)\ .$$

THEOREM 3. *The graded* R*-Lie algebra*

$$\tilde{\mathcal{C}}^{\bullet}_{k+1}T = \Lambda\underline{T}^*\otimes_{\mathcal{O}}\tilde{\mathcal{J}}_{k+1}T\,/\,\underline{\delta}(\Lambda\underline{T}^*\otimes_{\mathcal{O}} S^{k+2}\underline{T}^*\otimes_{\mathcal{O}}\underline{T})$$

with the quotient bracket $[\![\ ,\ ]\!]$ *is canonically isomorphic, via the quotient of* ad, *to the algebra* $\mathcal{D}er_{\Sigma}\mathcal{B}_k(X)$ *(cf. Section* 25).

We shall now prove that the complexes $(28.6)_k$ and $(28.7)_k$ restrict to $\mathcal{D}er_{\Sigma}\mathcal{B}_k(X)$. For convenience, let us write

$$\mathcal{D}er^r_{\Sigma,k} = \mathcal{D}er^r_{\Sigma}\mathcal{B}_k(X) = ad(\Lambda^r\underline{T}^*\otimes_{\mathcal{O}}\tilde{\mathcal{J}}_{k+1}T)$$

for $r \geq 0$ and

$$\underline{T} \simeq \mathcal{D}er^{-1}_{\Sigma,k} = \mathcal{K}[\tilde{j}_k(\underline{T})]$$

where $\tilde{j}_k(\underline{T})$ is considered as a Lie subalgebra of $\tilde{\mathcal{J}}_k T$. Denote by ad the isomorphism of Theorem 3 which is the quotient of the representation ad. The following diagram is commutative

$$\begin{array}{ccc}
\underline{\delta}(\Lambda^{r-1}\underline{T}^* \otimes_{\mathcal{O}} S^{k+2}\underline{T}^* \otimes_{\mathcal{O}} \underline{T}) & & \\
\Big\downarrow \zeta_{k+1} & & \\
\Lambda^r \underline{T}^* \otimes_{\mathcal{O}} \tilde{\mathcal{J}}_{k+1} T & \xrightarrow{\quad ad \quad} & \mathcal{D}er^r_{\Sigma,k} \\
\Big\downarrow & \nearrow_{ad} & \\
\tilde{\mathcal{C}}^r_{k+1} T & &
\end{array} \tag{29.2}$$

and in particular

$$\begin{array}{ccc}
\tilde{\mathcal{J}}_{k+1} T & \xrightarrow[\simeq]{\quad ad \quad} & \mathcal{D}er^0_{\Sigma,k} \\
\simeq \Big\downarrow & \nearrow^{ad}_{\simeq} & \\
\tilde{\mathcal{C}}^0_{k+1} T & &
\end{array} \tag{29.3}$$

Let $u \in \Lambda^r \underline{T}^* \otimes_{\mathcal{O}} \tilde{\mathcal{J}}_{k+1} T$ and consider $ad\, u \in \mathcal{D}er^r_{\Sigma,k}$. The relation (28.3) yields

$$\hat{D}(ad\, u) = \hat{D}[\mathcal{L}(\tilde{\rho}_k u)] - \hat{D} \circ \mathcal{K}(\tilde{D}u) = [d, \mathcal{L}(\tilde{\rho}_k u)] + \mathcal{L}(\tilde{D}u) = \mathcal{L}(\tilde{D}u) .$$

Let $v \in \Lambda^r \underline{T}^* \otimes_{\mathcal{O}} \tilde{\mathcal{J}}_{k+2} T$ be a lifting of u, i.e., $\tilde{\rho}_{k+1} v = u$. Then

$$\tilde{D}v \in \Lambda^{r+1} \underline{T}^* \otimes_{\mathcal{O}} \tilde{\mathcal{J}}_{k+1} T$$

and

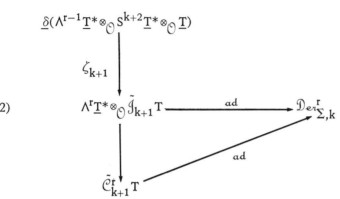

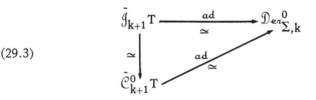

$$\mathit{ad}(\tilde{D}v) = \mathfrak{L}(\tilde{\rho}_k \tilde{D}v) - \mathcal{K}(\tilde{D}\circ\tilde{D}v) = \mathfrak{L}(\tilde{D}\tilde{\rho}_{k+1}v) = \mathfrak{L}(\tilde{D}u)$$

hence

$$\hat{D}(\mathit{ad}u) = \mathit{ad}(\tilde{D}v) .$$

The following diagram is commutative, where $\tilde{D}^1_{r,k+2}$ stands for the operator $D^1_{r,k+2}$ defined in Section 7.

$$\text{(29.4)} \qquad \begin{array}{ccccc}
\Lambda^r\underline{T}^*\otimes_{\mathcal{O}}\tilde{\mathcal{J}}_{k+2}T & \xrightarrow{\tilde{D}} & & \Lambda^{r+1}\underline{T}^*\otimes_{\mathcal{O}}\tilde{\mathcal{J}}_{k+1}T & \\
\downarrow{\scriptstyle \tilde{\rho}_{k+1}} & & \swarrow & & \\
\Lambda^r\underline{T}^*\otimes_{\mathcal{O}}\tilde{\mathcal{J}}_{k+1}T & \xrightarrow{\tilde{D}^1_{r,k+2}} & \tilde{\mathcal{C}}^{r+1}_{k+1}T & \downarrow{\scriptstyle \mathit{ad}} & \\
\downarrow{\scriptstyle \mathit{ad}} & & \searrow{\scriptstyle \mathit{ad}} & & \\
\mathcal{D}er^r_{\Sigma,k} & \xrightarrow{\hat{D}} & & \mathcal{D}er^{r+1}_{\Sigma,k} &
\end{array}$$

From Lemma 5 we infer that the lower square of (29.4) factors to the commutative diagram

$$\text{(29.5)} \qquad \begin{array}{ccc}
\tilde{\mathcal{C}}^r_{k+1}T & \xrightarrow{\tilde{D}'} & \tilde{\mathcal{C}}^{r+1}_{k+1}T \\
\downarrow{\scriptstyle \mathit{ad}} & & \downarrow{\scriptstyle \mathit{ad}} \\
\mathcal{D}er^r_{\Sigma,k} & \xrightarrow{\hat{D}} & \mathcal{D}er^{r+1}_{\Sigma,k}
\end{array}$$

where $\tilde{D}'$ is the operator in the complex $(25.1)_{k+1}$. Observe now that the restriction

$$\mathit{ad}\colon \tilde{\mathcal{J}}_{k+1}T \longrightarrow \mathcal{D}er^0_{\Sigma,k}$$

(diagram (29.3)) is an isomorphism since

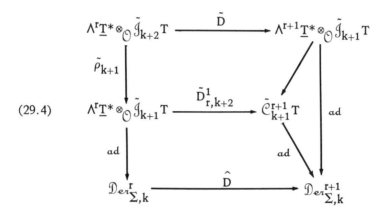

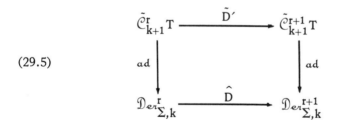

$$0 \longrightarrow S^{k+1}\underline{T}^* \otimes_{\mathcal{O}} \underline{T} \xrightarrow{\delta} \underline{T}^* \otimes_{\mathcal{O}} S^k\underline{T}^* \otimes_{\mathcal{O}} \underline{T}$$

is exact. If $\tilde{\rho}_k\xi = 0$, $\xi \in \tilde{\mathcal{J}}_{k+1}T$, then

$$\mathit{ad}\xi = -\mathcal{K}(\underline{\delta}\xi) .$$

Let

$$\mathit{ad}\xi \in \mathcal{D}er^0_{\Sigma,k},\ \xi \in \tilde{\mathcal{J}}_{k+1}T ,$$

and assume that $\hat{D}(\mathit{ad}\xi) = 0$. This condition is equivalent to

$$\hat{D}(\mathcal{L}(\tilde{\rho}_k\xi) - \mathcal{K}(\tilde{D}\xi)) = \mathcal{L}(\tilde{D}\xi) = 0$$

and, by the injectivity of $\mathcal{L}$, the latter is equivalent to $\tilde{D}\xi = 0$, i.e., $\xi = \tilde{j}_{k+1}\theta$ with $\theta \in \underline{T}$. We infer that $\hat{D}(\mathit{ad}\xi) = 0$ is equivalent to

$$\mathit{ad}\xi = \mathcal{L}(\tilde{j}_k\theta) = -\hat{D} \circ \mathcal{K}(\tilde{j}_k\theta)$$

with $\theta \in \underline{T}$.

THEOREM 4. *The complexes* $(28.6)_k$ *and* $(28.7)_k$ *restricted to* $\mathcal{D}er_\Sigma\mathcal{B}_k(X)$ *are equivalent to the complex* $(25.1)_{k+1}$ *and the natural equivalences are given by the commutative diagram, where* $n = \dim X$,

$(29.6)_k$

$$\begin{array}{ccccccccccccc}
0 & \longrightarrow & \mathcal{D}er^{-1}_{\Sigma,k} & \xrightarrow{\hat{D}} & \mathcal{D}er^{0}_{\Sigma,k} & \xrightarrow{\hat{D}} & \mathcal{D}er^{1}_{\Sigma,k} & \xrightarrow{\hat{D}} & \cdots & \xrightarrow{\hat{D}} & \mathcal{D}er^{n}_{\Sigma,k} & \longrightarrow & 0 \\
 & & \uparrow{\scriptstyle -\mathcal{K}\circ\tilde{j}_k} & & \| & & \| & & & & \| & & \\
0 & \longrightarrow & \underline{T} & \xrightarrow{\mathcal{L}\circ\tilde{j}_k} & \mathcal{D}er^{0}_{\Sigma,k} & \xrightarrow{\hat{D}} & \mathcal{D}er^{1}_{\Sigma,k} & \xrightarrow{\hat{D}} & \cdots & \xrightarrow{\hat{D}} & \mathcal{D}er^{n}_{\Sigma,k} & \longrightarrow & 0 \\
 & & \| & & \uparrow{\scriptstyle \mathit{ad}} & & \uparrow{\scriptstyle \mathit{ad}} & & & & \uparrow{\scriptstyle \mathit{ad}} & & \\
0 & \longrightarrow & \underline{T} & \xrightarrow{\tilde{j}_{k+1}} & \tilde{\mathcal{C}}^0_{k+1}T = \tilde{\mathcal{J}}_{k+1}T & \xrightarrow{\tilde{D}'} & \tilde{\mathcal{C}}^1_{k+1}T & \xrightarrow{\tilde{D}'} & \cdots & \xrightarrow{\tilde{D}'} & \tilde{\mathcal{C}}^n_{k+1}T & \longrightarrow & 0
\end{array}$$

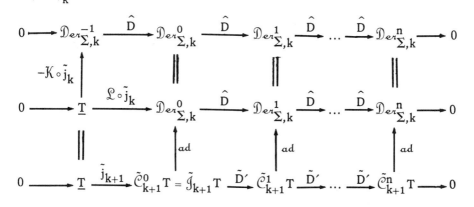

Here again, the diagram (29.6) is the origin of the operator $\tilde{D}'$. In his papers [13(a)], Spencer defined $(25.1)_{k+1}$ as the transported complex of the derivations complex $\hat{D}$. The first or second line of (29.6) will be referred to as the *restricted* $\hat{D}$-*complex*.

COROLLARY. *If* $k = 0$, *then the restricted* $\hat{D}$-*complex is equal to the Frölicher-Nijenhuis resolution of* $\underline{T}$ *namely*

$$0 \longrightarrow \underline{T} \xrightarrow{\mathcal{L}} \mathcal{D}er^0 \wedge \underline{T}^* \xrightarrow{[d,\ ]} \mathcal{D}er^1 \wedge \underline{T}^* \xrightarrow{[d,\ ]} \cdots \xrightarrow{[d,\ ]} \mathcal{D}er^n \wedge \underline{T}^* \longrightarrow 0$$

or equivalently

$$0 \longrightarrow \mathcal{D}er^{-1} \wedge \underline{T}^* \xrightarrow{[d,\ ]} \mathcal{D}er^0 \wedge \underline{T}^* \xrightarrow{[d,\ ]} \cdots \xrightarrow{[d,\ ]} \mathcal{D}er^n \wedge \underline{T}^* \longrightarrow 0$$

since

$$\mathcal{D}er^{-1} \wedge \underline{T}^* = \mathcal{K}(\underline{T}) .$$

In fact, if $k = 0$, then

$$ad: \wedge \underline{T}^* \otimes_{\mathcal{O}} \tilde{\mathcal{J}}_1 T \to \bigoplus_{r \geq 0} \mathcal{D}er^r \wedge \underline{T}^*$$

is surjective. The Frölicher-Nijenhuis resolution is equivalent to the resolution $(25.1)_1$. If we write

$$\mathcal{D}er^r \wedge \underline{T}^* = (\mathcal{L} + \mathcal{K})(\wedge^r \underline{T}^* \otimes_{\mathcal{O}} \underline{T} \times_X \wedge^{r+1} \underline{T}^* \otimes_{\mathcal{O}} \underline{T})$$

then the Frölicher-Nijenhuis resolution can be written in terms of vector forms and the operator $[d,\ ]$ becomes $[1,\ ]$ where 1 is the identity map of $\underline{T}$ and $[\ ,\ ]$ is the Nijenhuis bracket of vector forms.

30. *The* $\hat{\mathcal{D}}$-*complex and the twisting of* d

As in the case of the restricted $\hat{D}$-complex, we can also transport, via the isomorphism ad, the non-linear complex $(25.3)_{k+1}$ to a non-linear complex where the objects are sheaves of (germs of) strongly admissible derivations. More precisely, we define the first order differential operators $\hat{\mathfrak{D}}$ and $\hat{\mathfrak{D}}_1$ of the *fourth non-linear complex* by the commutativity of the diagram

$$\begin{array}{llccccccccc}
(30.1)_{k+1} & 1 \longrightarrow & \mathcal{A}ut\, X & \xrightarrow{\tilde{j}_{k+1}} & \Gamma_{k+1}X & \xrightarrow{\hat{\mathfrak{D}}} & \mathcal{D}er^1_{\Sigma,k} & \xrightarrow{\hat{\mathfrak{D}}_1} & \mathcal{D}er^2_{\Sigma,k} \\
 & & \| & & \| & & \uparrow ad & & \uparrow ad \\
(25.3)_{k+1} & 1 \longrightarrow & \mathcal{A}ut\, X & \xrightarrow{\tilde{j}_{k+1}} & \Gamma_{k+1}X & \xrightarrow{\tilde{\mathfrak{D}}'} & \tilde{\mathcal{C}}^1_{k+1}T & \xrightarrow{\tilde{\mathfrak{D}}'_1} & \tilde{\mathcal{C}}^2_{k+1}T
\end{array}$$

By Theorems 3 and 4 we infer that

$$(30.2) \qquad \hat{\mathfrak{D}}_1 w = \hat{D}w - \tfrac{1}{2}[w, w] = [d, w] - \tfrac{1}{2}[w, w]$$

where $w \in \mathcal{D}er^1_{\Sigma,k}$ and [,] is the derivations bracket. Let us now compute $\hat{\mathfrak{D}}$. Take $F \in \Gamma_{k+1}X$ and let $G \in \Gamma_{k+2}X$ be a lifting of F, i.e., $F = \rho_{k+1}G$. Then

$$ad \circ \tilde{\mathfrak{D}}'(F) = ad \circ \tilde{\mathfrak{D}}(G) = \mathfrak{L}(\tilde{\rho}_k \tilde{\mathfrak{D}}G) - \mathcal{K}(\tilde{D}\tilde{\mathfrak{D}}G) =$$
$$\mathfrak{L}(\tilde{\mathfrak{D}}F) - \tfrac{1}{2}\mathcal{K}(\tilde{\rho}_k [\![\tilde{\mathfrak{D}}G, \tilde{\mathfrak{D}}G]\!]) = -\hat{D} \circ \mathcal{K}(\tilde{\mathfrak{D}}F) - \tfrac{1}{2}\mathcal{K}(\tilde{\rho}_k [\![\tilde{\mathfrak{D}}G, \tilde{\mathfrak{D}}G]\!])$$

where the third equality is a consequence of the structure equation (23.8) and the last equality follows from (28.3). We infer that

$$(30.3) \quad \hat{\mathfrak{D}}(F) = -\hat{D} \circ \mathcal{K}(\tilde{\mathfrak{D}}F) - \tfrac{1}{2}\mathcal{K}(\tilde{\rho}_k [\![\tilde{\mathfrak{D}}G, \tilde{\mathfrak{D}}G]\!]) = -[d, \mathcal{K}(\tilde{\mathfrak{D}}F)] - \tfrac{1}{2}\mathcal{K}(\tilde{\rho}_k [\![\tilde{\mathfrak{D}}G, \tilde{\mathfrak{D}}G]\!]).$$

If $k > 0$ then

$$\tilde{\rho}_k [\![\tilde{\mathfrak{D}}G, \tilde{\mathfrak{D}}G]\!] = [\![\tilde{\rho}_k \tilde{\mathfrak{D}}G, \tilde{\rho}_k \tilde{\mathfrak{D}}G]\!]$$

hence (30.3) becomes

$$(30.4) \quad \hat{\mathfrak{D}}(F) = \mathfrak{L}(\tilde{\mathfrak{D}}F) - \tfrac{1}{2}\mathcal{K}([\![\tilde{\mathfrak{D}}F, \tilde{\mathfrak{D}}F]\!]) = -[d, \mathcal{K}(\tilde{\mathfrak{D}}F)] - \tfrac{1}{2}\mathcal{K}([\![\tilde{\mathfrak{D}}F, \tilde{\mathfrak{D}}F]\!])$$

All the properties of $(25.3)_{k+1}$ transcribe into corresponding properties for $(30.1)_{k+1}$. This complex is exact at $\Gamma_{k+1}X$, verifies the *structure equation*

$$(30.5) \qquad \hat{D}\hat{\mathfrak{D}}F - \tfrac{1}{2}[\hat{\mathfrak{D}}F, \hat{\mathfrak{D}}F] = 0$$

and is partially exact at $\mathfrak{D}er^{1}_{\Sigma,k}$ in a sense analogous to $(25.3)_{k+1}$. Moreover, the main identities for the operators $\tilde{\mathfrak{D}}'$ and $\tilde{\mathfrak{D}}'_1$ yield corresponding identities for $\hat{\mathfrak{D}}$ and $\hat{\mathfrak{D}}_1$. The complex $(30.1)_{k+1}$ is a finite form of the initial portion of the restricted $\hat{D}$-complex (first or second line of diagram (29.6)).

In [13(c)], Spencer defined the operator $\hat{\mathfrak{D}}$ by the *twisting* of the derivation d. We terminate this section by giving a brief outline of the twisting procedure.

Recall that the groupoid $\Gamma_{k+1}X$ operates covariantly on $\check{\mathcal{J}}_k T$ and, in particular, it operates on the subsheaves $\mathcal{J}_k T$ and $\tilde{\mathcal{J}}_k T$ (cf. Section 20). Recall also that

$$\varepsilon_k\colon \tilde{\mathcal{J}}_k T \to \mathcal{J}_k T$$

is a diagonal $\mathcal{J}_k$-linear isomorphism and, in particular, a left $\mathcal{O}$-linear isomorphism. If $F \in \Gamma_{k+1}X$, we define the action $\mathcal{A}d F$ on $\tilde{\mathcal{J}}_k T$ as the transported standard action F on $\mathcal{J}_k T$ via the isomorphism ε_k, i.e.,

$$\mathcal{A}d F\colon \Xi \in \tilde{\mathcal{J}}_k T \mapsto \varepsilon_k^{-1}[F \bullet \varepsilon_k(\Xi)] \in \tilde{\mathcal{J}}_k T\ . \tag{30.6}$$

$\Gamma_{k+1}X$ operates covariantly on $\tilde{\mathcal{J}}_k T$ via the action $\mathcal{A}d$. If $x = a_{k+1}F =$ source of F and $y = b_{k+1}F =$ target of F, then

$$\mathcal{A}d F\colon \tilde{\mathcal{J}}_k T(x) \to \tilde{\mathcal{J}}_k T(y)$$

and this mapping is semilinear with respect to

$$f^{-1*}\colon g \in \mathcal{O}(x) \mapsto g \circ f^{-1} \in \mathcal{O}(y)$$

where $f = \beta_{k+1} \circ F \in \mathcal{A}ut\, X$. It follows that $\mathcal{A}d$ transposes to a contravariant action of $\Gamma_{k+1}X$ on $\mathcal{A}_k(X) \simeq \Lambda(\tilde{\mathcal{J}}_k T)^*$, namely

$$\mathcal{A}d F(\omega)(\xi_1, \ldots, \xi_r) = f^*[\omega(\mathcal{A}d F \xi_1, \ldots, \mathcal{A}d F \xi_r)] = \omega(\mathcal{A}d F\xi_1, \ldots, \mathcal{A}d F \xi_r) \circ f$$

where $\xi_i \in \tilde{\mathcal{J}}_k T(x)$, $\omega \in \mathcal{A}^r_k(X)(y)$ and $\mathcal{A}d F(\omega) \in \mathcal{A}^r_k(X)(x)$. Denote by F^* the contravariant action of $F \in \Gamma_{k+1}X$ on $\mathcal{J}_k$, i.e.,

$$F^* \bullet ([\varphi]_{(y,y)} \bmod \mathcal{J}^{k+1}) = [\varphi \circ F]_{(x,x)} \bmod \mathcal{J}^{k+1}$$

where $F \in \widetilde{\mathfrak{Aut}}\, X^2$ is a representative of F and

$$[\varphi]_{(y,y)} \in \mathcal{O}_{X^2}(y,y)$$

(this action on $\mathcal{J}_k$ is actually defined for $F \in \Gamma_k X$). With this definition, $\mathfrak{Ad}$ extends to a contravariant action of $\Gamma_{k+1}X$ on the sheaf

$$\mathcal{B}_k(X) \simeq \Lambda(\tilde{\mathcal{J}}_k T)^* \otimes_{\mathcal{O}} \mathcal{J}_k$$

by

$$\mathfrak{Ad}F(\omega)(\xi_1, \ldots, \xi_r) = F^* \bullet [\omega(\mathfrak{Ad}F\xi_1, \ldots, \mathfrak{Ad}F\xi_r)]$$

or, in terms of decomposable elements,

$$\mathfrak{Ad}F(\nu \otimes \varphi) = \mathfrak{Ad}F(\nu) \otimes F^* \bullet \varphi \,.$$

It can be verified that the action $\mathfrak{Ad}$ preserves the subsheaf $\Lambda \underline{T}^*$ imbedded in $\mathcal{B}_k(X)$ and that the restriction of $\mathfrak{Ad}$ to $\Lambda \underline{T}^*$ is equal to the representation $\mathfrak{Ad}$ defined in the beginning of Section 22. Moreover, if $\varphi \in \mathcal{B}_k^0(X) \simeq \mathcal{J}_k$ and $F \in \Gamma_{k+1}X$ then $\mathfrak{Ad}F(\varphi) = F^* \bullet \varphi$ and if $\omega \in \mathcal{B}_k^1(X)$ then

$$\mathfrak{Ad}F(\omega) = F^* \bullet \omega - i(\tilde{\mathfrak{D}}F)(F^* \bullet \omega) \tag{30.7}$$

where $F^* \bullet \omega$ represents the action on $\mathcal{B}_k(X)$ transposed to the standard action of $\Gamma_{k+1}X$ on $\tilde{\mathcal{J}}_k T$ (Section 20). Since the standard action is actually defined for $\Gamma_k X$ then $F^* \bullet \varphi = G^* \bullet \varphi$ and (30.7) can be rewritten by

$$\mathfrak{Ad}F(\omega) = G^* \bullet \omega - i(\tilde{\mathfrak{D}}F)(G^* \bullet \omega) \tag{30.8}$$

where $G = \rho_k F \in \Gamma_k X$. The above relations generalize formula (23.10) and reduce to it when $\omega \in \underline{T}^*$. We infer that the action of $\mathfrak{Ad}F$ on $\mathcal{B}_k(X)$ is equal to the extension

$$(G^* \bullet \; - i(\tilde{\mathfrak{D}}F) \circ G^* \bullet \;)^\wedge \,, \tag{30.8$'$}$$

this formula generalizing (23.10′).

We shall now extend the action $\mathcal{A}d$ of $\Gamma_{k+1}X$ to $\mathcal{D}er_a\mathcal{B}_k(X)$. If $w \in \mathcal{D}er_a\mathcal{B}_k(X)$ and $F \in \Gamma_{k+1}X$ with w in the stalk above the point $y = b_{k+1}F$, we define

$$\mathcal{A}dF(w) = \mathcal{A}dF \circ w \circ \mathcal{A}dF^{-1} . \tag{30.9}$$

$\mathcal{A}dF(w)$ is an admissible (germ of) derivation since $\mathcal{A}dF$ preserves the subsheaves which enter in the definition of admissibility. If $w \in \mathcal{D}er^r_a\mathcal{B}_k(X)(y)$ then $\mathcal{A}dF(w) = \mathcal{D}er^r_a\mathcal{B}_k(X)(x)$, $x = a_{k+1}F$, and $\mathcal{A}dF$ preserves the bracket of derivations hence this extension is an isomorphism of graded R-Lie algebras which is homogeneous of degree zero and semilinear with respect to f^*. If $\omega \in \Lambda\underline{T}^*$ and $w \in \mathcal{D}er_a\mathcal{B}_k(X)$, then

$$\mathcal{A}dF(\omega \wedge w) = \mathcal{A}dF(\omega) \wedge \mathcal{A}dF(w) .$$

The extension $\mathcal{A}d$ is a contravariant action of $\Gamma_{k+1}X$ on $\mathcal{D}er_a\mathcal{B}_k(X)$ which preserves the subsheaf of strongly admissible derivations. Moreover, this action also preserves the subsheaf $\mathcal{D}er_{\Sigma,k}$ and the induced representation is equal to the representation $\mathcal{A}d$ of $\Gamma_{k+1}X$ on $\tilde{\mathcal{C}}^\bullet_{k+1}T$, as defined in the end of Section 25, transported by the isomorphism

$$ad: \tilde{\mathcal{C}}^\bullet_{k+1}T \to \mathcal{D}er_{\Sigma,k} .$$

The last assertion means simply that $\mathcal{A}d$ commutes with ad and generalizes Lemma 1 of Section 23. In fact, it is a restatement of the Theorem 1 of Section 23, factored to $\tilde{\mathcal{C}}^\bullet_{k+1}T$ (cf. end of Section 25).

THEOREM. $\hat{\mathcal{D}}F = d - \mathcal{A}dF(d)$ *where* $F \in \Gamma_{k+1}X$ *and* $d \in \mathcal{D}er_a\mathcal{B}_k(X)$ *is the exterior differentiation.*

It is easy to check that $d - \mathcal{A}dF(d)$ is strongly admissible (d is not strongly admissible). A proof of the theorem could be carried out by straightforward computation in local coordinates, using the formulas (22.11), (30.3) and (30.8). A better approach would consist in extending the representation

$$\mathcal{ad}\colon \wedge \underline{T}^* \otimes_{\mathcal{O}} \tilde{\mathfrak{J}}_{k+1} T \to \mathcal{D}er_{\Sigma,k}$$

to a larger sheaf which contains the element χ_{k+1} (for example $\wedge \underline{T}^* \otimes_{\mathcal{O}} \check{\mathfrak{J}}_{k+1} T$). One would then prove that

$$\mathcal{ad}(\chi_{k+1}) = d$$

and that $\mathcal{ad}$ commutes with $\mathfrak{A}d$ for elements of a subsheaf which itself would contain χ_{k+1} (essentially $(\wedge \underline{T}^* \otimes_{\mathcal{O}} \tilde{\mathfrak{J}}_{k+1} T) \oplus_{R} R\chi_{k+1}$) and these relations would trivially yield

$$\mathcal{ad} \circ \tilde{\mathfrak{D}}'(F) = d - \mathfrak{A}dF(d)$$

since

$$\mathfrak{D}G = \chi_{k+1} - \mathfrak{A}dG(\chi_{k+1}),\ G \in \Gamma_{k+2} X \quad \text{and} \quad \rho_{k+1} G = F\ .$$

Such an extension can be found in [13 (c)] and in (unpublished) joint work of Malgrange and Spencer. It can also be found in a transposed version in [9 (c)]. We do not give here a detailed proof of the theorem since, for one thing, it would be excessively long and, for another, a substantial effort has been made in Chapters III and IV precisely to avoid such an extension which resorts to the derivation d′ (cf. [9(c)]). The expression

$$d - \mathfrak{A}dF(d)$$

is the *twisting* of d.

Remark. The definition and the properties of the complex $(30.1)_{k+1}$ have been derived from the complex $(25.3)_{k+1}$ and the results of Chapter IV. However, it is worthwhile to note that the third non-linear complex can be studied independently of $(25.3)_{k+1}$. In fact, one defines the operator $\hat{\mathfrak{D}}$ by the twisting of d and the operator $\hat{\mathfrak{D}}_1$ by the formula (30.2). It turns out that the proof of some properties becomes easier, in some instances trivial (*e.g.*, the invariance of the derivations bracket by $\mathfrak{A}dF$).

For later reference we list below the main identities for the operators $\hat{\mathfrak{D}}$ and $\hat{\mathfrak{D}}_1$. Let $F, G \in \Gamma_{k+1}X$; then

(30.10) $$ad(\tilde{\mathfrak{D}}'F) = d - \mathcal{A}dF(d) \quad \text{(last theorem)},$$

(30.11) $$\hat{\mathfrak{D}}(F \bullet G) = \mathcal{A}dG(\hat{\mathfrak{D}}F) + \hat{\mathfrak{D}}G\,,$$

(30.12) $$\hat{\mathfrak{D}}F^{-1} = -\,\mathcal{A}dF^{-1}(\hat{\mathfrak{D}}F)\,,$$

(30.13) $$\hat{D}[\mathcal{A}dF(w)] = \mathcal{A}dF(\hat{D}w - [\hat{\mathfrak{D}}F^{-1}, w]) \in \mathcal{D}er^{\bullet}_{\Sigma,k}$$

where $w \in \mathcal{D}er^{\bullet}_{\Sigma,k}$,

(30.14) $$w^F \overset{def.}{=\!=} \mathcal{A}dF(w) + \hat{\mathfrak{D}}F \in \mathcal{D}er^1_{\Sigma,k}$$

where $w \in \mathcal{D}er^1_{\Sigma,k}$,

(30.15) $$w^{F \bullet G} = (w^F)^G, \quad w^1 = w, \quad w^{F^{-1}} = \mathcal{A}dF^{-1}(w - \hat{\mathfrak{D}}F)\,,$$

(30.16) $$\hat{\mathfrak{D}}_1(w^F) = \mathcal{A}dF(\hat{\mathfrak{D}}_1 w)\,.$$

31. *The projective limits*

The isomorphism

$$ad\colon \tilde{\mathcal{C}}^{\bullet}_{k+1}T \longrightarrow \mathcal{D}er_{\Sigma}\mathcal{B}_k(X) = \mathcal{D}er_{\Sigma,k}$$

was obtained by factoring the surjective graded Lie algebra morphism

(31.1) $$ad\colon \Lambda\underline{T}^* \otimes_{\mathcal{O}} \tilde{\mathcal{J}}_{k+1}T \longrightarrow \mathcal{D}er_{\Sigma,k}$$

modulo the kernel. We shall prove that (31.1) induces an isomorphism on the projective limits. Let

$$\tilde{\mathcal{J}}_{\infty}T = proj\ lim\,(\tilde{\mathcal{J}}_k T, \tilde{\rho}_k)$$

and

$$\mathcal{D}er_{\Sigma,\infty} = proj\ lim\,(\mathcal{D}er_{\Sigma,k}, \rho_k)$$

where

$$\rho_k\colon \mathcal{D}er_{\Sigma,k+h} \longrightarrow \mathcal{D}er_{\Sigma,k}$$

is the surjective morphism induced by $\tilde{\rho}_k$ via ad. It is clear that

$$\Lambda\underline{T}^* \otimes_{\mathcal{O}} \tilde{\mathcal{J}}_\infty T = proj\ lim\,(\Lambda\underline{T}^* \otimes_{\mathcal{O}} \tilde{\mathcal{J}}_k T, \tilde{\rho}_k)\ .$$

Moreover, all the structures given on the component spaces extend to corresponding structures on the projective limit. In fact, $\Lambda\underline{T}^* \otimes_{\mathcal{O}} \tilde{\mathcal{J}}_\infty T$ is a sheaf of graded R-Lie algebras with respect to the Nijenhuis bracket $[\ ,\ ]_\infty$ and the twisted bracket $[\![\ ,\ ]\!]_\infty$, the grading being given by the subsheaves $\Lambda^r\underline{T}^* \otimes_{\mathcal{O}} \tilde{\mathcal{J}}_\infty T$. It is also a sheaf of modules over the ring

$$\mathcal{J}_\infty = proj\ lim\,(\mathcal{J}_k, \rho_k)$$

hence a left and right $\mathcal{O}$-module. $\mathcal{D}er_{\Sigma,\infty}$ is a sheaf of graded R-Lie algebras, the grading being given by

$$\mathcal{D}er^r_{\Sigma,\infty} = proj\ lim\,(\mathcal{D}er^r_{\Sigma,k}, \rho_k)\ .$$

The representation (31.1) induces a surjective graded Lie algebra morphism

$$ad_\infty\colon \Lambda\underline{T}^* \otimes_{\mathcal{O}} \tilde{\mathcal{J}}_\infty T \longrightarrow \mathcal{D}er_{\Sigma,\infty} \tag{31.2}$$

with respect to the bracket $[\![\ ,\ ]\!]_\infty$. We claim that this morphism is also injective. Let $u = (u_k)_{k\geq 0}$ be an element of $\Lambda\underline{T}^* \otimes_{\mathcal{O}} \tilde{\mathcal{J}}_\infty T$. If $ad_\infty(u) = 0$ then $ad(u_{k+1}) = 0$ for all k hence $u_{k+2} \in \underline{\delta}(\Lambda\underline{T}^* \otimes_{\mathcal{O}} S^{k+3}\underline{T}^* \otimes_{\mathcal{O}} \underline{T})$, $u_{k+2} = \underline{\delta}(v)$, and $u_{k+1} = \rho_{k+1} u_{k+2} = \rho_{k+1} \circ \underline{\delta}(v) = 0$. It follows that $ker\ ad_\infty = 0$.

PROPOSITION. *ad_∞ is a graded R-Lie algebra isomorphism.*

We can extend the operators $\tilde{D}$ and $\hat{D}$ to $\Lambda\underline{T}^* \otimes_{\mathcal{O}} \tilde{\mathcal{J}}_\infty T$ and $\mathcal{D}er_{\Sigma,\infty}$ respectively. The extended operator

$$\tilde{D}_\infty = proj\ lim\ \tilde{D}_{k+1}$$

is a derivation of degree 1 of the graded algebra $\Lambda\underline{T}^* \otimes_{\mathcal{O}} \tilde{\mathcal{J}}_\infty T$ with the bracket $[\![\ ,\ ]\!]_\infty$ (the $\tilde{\rho}_{k-1}$ of Lemma 6, Section 16, disappears in the projective limit) and

$$\hat{D}_\infty = proj\ lim\ \hat{D}_k$$

is a derivation of degree 1 of $\mathcal{D}er_{\Sigma,\infty}$. These extended operators yield the equivalent exact *linear complexes*

$$\begin{array}{ll}
(31.3) & 0 \longrightarrow \mathcal{D}er^{-1}_{\Sigma,\infty} \xrightarrow{\hat{D}_\infty} \mathcal{D}er^{0}_{\Sigma,\infty} \xrightarrow{\hat{D}_\infty} \mathcal{D}er^{1}_{\Sigma,\infty} \xrightarrow{\hat{D}_\infty} \cdots \xrightarrow{\hat{D}_\infty} \mathcal{D}er^{n}_{\Sigma,\infty} \longrightarrow 0 \\
& \quad\ \uparrow{\scriptstyle -\mathcal{K}\circ\tilde{j}_\infty} \qquad\quad \uparrow{\scriptstyle ad_\infty} \qquad\qquad \uparrow{\scriptstyle ad_\infty} \qquad\qquad\qquad \uparrow{\scriptstyle ad_\infty} \\
(31.4) & 0 \longrightarrow \underline{T} \xrightarrow{\tilde{j}_\infty} \tilde{\mathcal{J}}_\infty T \xrightarrow{\tilde{D}_\infty} \underline{T}^*\otimes_{\mathcal{O}}\tilde{\mathcal{J}}_\infty T \xrightarrow{\tilde{D}_\infty} \cdots \xrightarrow{\tilde{D}_\infty} \Lambda^n\underline{T}^*\otimes_{\mathcal{O}}\tilde{\mathcal{J}}_\infty T \longrightarrow 0
\end{array}$$

where $\tilde{j}_\infty = proj\ lim\ \tilde{j}_k$, $\mathcal{K}\circ\tilde{j}_\infty = proj\ lim\ \mathcal{K}\circ\tilde{j}_k$ and $\mathcal{D}er^{-1}_{\Sigma,\infty} \simeq \underline{T}$.

Let

$$\mathcal{X}_\infty = (\mathcal{X}_k)_{k\geq 0} = proj\ lim\ \mathcal{X}_k$$

and

$$d_\infty = (d_k)_{k\geq 0} = proj\ lim\ d_k$$

where d_k is the exterior differentiation d on $\mathcal{B}_k(X)$. For a fixed k and $u \in \Lambda\underline{T}^*\otimes_{\mathcal{O}}\tilde{\mathcal{J}}_{k+1}T$ we have

$$[\![\mathcal{X}_{k+1}, u]\!] = \tilde{D}u \in \Lambda\underline{T}^*\otimes_{\mathcal{O}}\tilde{\mathcal{J}}_k T$$

(Theorem 2, Section 16) and

$$[d_k, ad\,u] = \hat{D}(ad\,u)\ .$$

From the remark following the proof of Lemma 6, Section 16, we infer that $\Lambda\underline{T}^*\otimes_{\mathcal{O}}\tilde{\mathcal{J}}_\infty T$ and $\mathcal{X}_\infty$ generate a graded R-Lie algebra

$$\Xi_\infty = (\Lambda\underline{T}^*\otimes_{\mathcal{O}}\tilde{\mathcal{J}}_\infty T) \oplus_R R\mathcal{X}_\infty$$

where the homogeneous component of degree 1 is equal to

$$(\underline{T}^* \otimes_{\mathcal{O}} \tilde{\mathcal{J}}_\infty T) \oplus_R R\mathcal{X}_\infty ,$$

the subsheaf $\Lambda \underline{T}^* \otimes_{\mathcal{O}} \tilde{\mathcal{J}}_\infty T$ is an ideal and

$$[\![\mathcal{X}_\infty, u]\!]_\infty = \tilde{D}_\infty u .$$

Similarly, $\mathcal{D}er_{\Sigma,\infty}$ and d_∞ generate a graded R-Lie algebra

$$\Sigma_\infty = \mathcal{D}er_{\Sigma,\infty} \oplus_R Rd_\infty$$

where $\mathcal{D}er_{\Sigma,\infty}$ is an ideal and

$$[d_\infty, u]_\infty = \hat{D}_\infty u .$$

The map ad_∞ extends to these graded algebras by setting $ad_\infty(\mathcal{X}_\infty) = d_\infty$ or $ad(\mathcal{X}_{k+1}) = d_k$.

THEOREM. *The map* $ad_\infty \colon \Xi_\infty \to \Sigma_\infty$ *is a graded* R*-Lie algebra isomorphism.*

The extended operators $\tilde{D}_\infty$ and $\hat{D}_\infty$ are given by the adjoints $[\![\mathcal{X}_\infty, \;]\!]_\infty$ and $[d_\infty, \;]_\infty$ with respect to the Lie algebra structures of Ξ_∞ and Σ_∞.

The non-linear operators $\tilde{\mathcal{D}}$, $\tilde{\mathcal{D}}_1$ and $\hat{\mathcal{D}}$, $\hat{\mathcal{D}}_1$ can also be extended to the projective limits and yield the equivalent *non-linear complexes*

$$\begin{array}{llllllllll}
(31.5) & 1 \longrightarrow & \mathcal{A}ut\, X & \xrightarrow{\tilde{j}_\infty} & \Gamma_\infty X & \xrightarrow{\hat{\mathcal{D}}_\infty} & \mathcal{D}er^1_{\Sigma,\infty} & \xrightarrow{\hat{\mathcal{D}}_{1,\infty}} & \mathcal{D}er^2_{\Sigma,\infty} \\
& & \| & & \| & & \uparrow ad_\infty & & \uparrow ad_\infty \\
(31.6) & 1 \longrightarrow & \mathcal{A}ut\, X & \xrightarrow{\tilde{j}_\infty} & \Gamma_\infty X & \xrightarrow{\tilde{\mathcal{D}}_\infty} & \underline{T}^* \otimes_{\mathcal{O}} \tilde{\mathcal{J}}_\infty T & \xrightarrow{\tilde{\mathcal{D}}_{1,\infty}} & \Lambda^2 \underline{T}^* \otimes_{\mathcal{O}} \tilde{\mathcal{J}}_\infty T
\end{array}$$

where $\Gamma_\infty X = proj\ lim\,(\Gamma_k X, \rho_k)$, $\tilde{j}_\infty = proj\ lim\ \tilde{j}_k$. The $\mathcal{A}d$ representations extend to contravariant representations $\mathcal{A}d_\infty$ of the groupoid $\Gamma_\infty X$ on the sheaves Ξ_∞ and Σ_∞ and each $\mathcal{A}d_\infty F$, $F \in \Gamma_\infty X$, is a graded Lie algebra isomorphism (of the corresponding stalks). Moreover, the two actions $\mathcal{A}d_\infty F$ commute with the infinitesimal representation ad_∞ of the theorem, i.e.,

$$ad_\infty \circ \mathcal{A}d_\infty F = \mathcal{A}d_\infty F \circ ad_\infty$$

where $F \in \Gamma_\infty X$. The non-linear operators $\tilde{\mathfrak{D}}_\infty$ and $\hat{\mathfrak{D}}_\infty$ are given by the twisting of $\mathcal{X}_\infty$ and d_∞ respectively i.e.,

$$\tilde{\mathfrak{D}}_\infty F = \mathcal{X}_\infty - \mathcal{A}d_\infty F(\mathcal{X}_\infty)$$

and

$$\hat{\mathfrak{D}}_\infty F = d_\infty - \mathcal{A}d_\infty F(\mathcal{X}_\infty) .$$

These relations follow from the corresponding relations at each level k. The operators satisfy the *structure equations*

$$\text{(31.7)} \qquad \tilde{D}_\infty \tilde{\mathfrak{D}}_\infty F - \tfrac{1}{2}[\![\tilde{\mathfrak{D}}_\infty F, \tilde{\mathfrak{D}}_\infty F]\!]_\infty = 0$$

$$\text{(31.8)} \qquad \hat{D}_\infty \hat{\mathfrak{D}}_\infty F - \tfrac{1}{2}[\hat{\mathfrak{D}}_\infty F, \hat{\mathfrak{D}}_\infty F]_\infty = 0$$

which are consequences of the corresponding equations at each level k or the fact that $\mathcal{A}d_\infty F$ commutes with the brackets in Ξ_∞ and Σ_∞ and that

$$[\![\mathcal{X}_\infty, \mathcal{X}_\infty]\!]_\infty = [d_\infty, d_\infty]_\infty = 0 .$$

The operators $\tilde{\mathfrak{D}}_{1,\infty}$ and $\hat{\mathfrak{D}}_{1,\infty}$ are given by

$$\tilde{\mathfrak{D}}_{1,\infty} = \tilde{D}_\infty - \tfrac{1}{2}[\![\ ,\ ]\!]_\infty$$

and

$$\hat{\mathfrak{D}}_{1,\infty} = \hat{D}_\infty - \tfrac{1}{2}[\ ,\ ]_\infty .$$

The complexes (31.5) and (31.6) are exact at $\Gamma_\infty X$, partially exact (in a sense similar to $(25.3)_{k+1}$) at $\underline{T}^* \otimes_{\mathcal{O}} \mathcal{J}_\infty T$ and $\mathcal{D}er^1_{\Sigma,\infty}$ respectively, and are finite forms of the initial portion of (31.3) and (31.4). All the main identities for the non-linear operators transcribe into this context.

APPENDIX

LIE GROUPOIDS

We shall review some differential-geometric notions introduced by Ehresmann [2 (a), (b), (c)] and relate them to some notions discussed in Chapter IV.

32. *Definitions*

In these notes we consider only categories and groupoids that are small, i.e., their underlying class is a set.

A partially defined internal operation in a set A is a mapping $B \to A$ where $B \subset A \times A$. As usual, we shall indicate B by $A * A$ and say that a is composable with b (a, b in this order) if $(a, b) \in A * A$. An element $e \in A$ is called a unit if $ae = a$ for any $a \in A$ that is composable with e and $eb = b$ for any $b \in A$ such that e is composable with b.

DEFINITION. A groupoid is a non-empty set G together with a partially defined internal operation $G * G \to G$ satisfying the following properties:

1. (associativity) If $(g, h), (h, k) \in G * G$ then $(gh, k), (g, hk) \in G * G$ and $(gh)k = g(hk)$.
2. (existence of units) For any $g \in G$ there exists a unit e which is right composable with g (right unit) and a unit e' which is left composable with g (left unit), i.e., $ge = e'g = g$.
3. (existence of inverses) For any $g \in G$ there exists an element g' which is left and right composable with g and such that $gg' = e'$ and $g'g = e$.

From (1) and (3) we infer that $(gh, k) \in G*G$ if and only if $(h,k) \in G*G$. A simple computation shows that the left and right units associated to each $g \in G$ as well as the element g' are unique. We denote g' by g^{-1} and define the *source* map α: $G \to G$, $\alpha(g) = e$, and the *target* map β: $G \to G$, $\beta(g) = e'$. The maps α and β are retractions $(\alpha^2 = \alpha, \beta^2 = \beta)$ whose common image is the set X of units of G. The element g is composable with h if and only if $\alpha(g) = \beta(h)$. The inverse map φ: $g \in G \to g^{-1} \in G$ is an involution $(\varphi^2 = \mathrm{Id})$. There are several equivalent ways of stating the groupoid axioms, the associativity law being usually expressed by:

$$(g, h), (gh, k) \in G*G \iff (h, k), (g, hk) \in G*G$$

and, this being the case, the identity $(gh)k = g(hk)$ holds.

A differentiable groupoid is a set G together with a differentiable structure and a groupoid structure satisfying the following compatibility conditions:

1. The set X of unit elements of G is a submanifold of G.
2. The source and target maps α, β: $G \to X$ are differentiable and transversal.
3. The map $(g, h) \in G \times_X G \mapsto gh \in G$ is differentiable, where $G \times_X G = G*G$ is the fibre product of α and β.
4. The map $g \in G \to g^{-1} \in G$ is differentiable.

The natural inclusion ι: $X \to G$ is a right inverse to α: $G \to X$, i.e., $\alpha \circ \iota = \mathrm{Id}_X$; hence α is of maximal rank at every point of X and therefore also in a neighborhood U of X which we can choose to be symmetrical. A simple argument shows that α is also of maximal rank at each point of the subgroupoid $G(U)$ of G generated by U (all finite products of elements in U) which in fact will be an open subgroupoid of G. It follows that, for any differentiable groupoid G, the map α (and, by inversion, β) is of maximal rank on the α-connected component of X (union of the connected components, in the α-fibres, of the points of X) which is an open subgroupoid of G. Since α is a retraction onto X we also infer that X is a regularly imbedded submanifold of G.

A Lie groupoid is a differentiable groupoid for which the map $\alpha: G \to X$ is a fibration (surjective submersion). By inversion, β is equally a fibration and we infer that, for Lie groupoids, the condition (2) of the above definition is a consequence of the hypothesis on α. Let G be a Lie groupoid and denote by G_x the fibre of $\alpha: G \to X$ over the point $x \in X$. Any $h \in G$ defines a diffeomorphism

$$\phi_h: g \in G_y \to gh \in G_x$$

where $x = \alpha(h)$ and $y = \beta(h)$. A local vector field ξ on G is right-invariant if it is α-vertical (i.e., $\alpha_* \xi = 0$) and if $\phi_{h_*}\xi = \xi$ for any $h \in G$ (whenever this relation has a meaning). If ξ is right-invariant and defined in the open set U, it can be extended to a right-invariant vector field ξ' defined on the open set $V = \beta^{-1}(\beta U)$. In fact, if $h \in V$ and $g \in U$ with $\beta(h) = \beta(g)$, we set $\xi'_h = \phi_{\ell_*}\xi_g$ where $\ell = g^{-1}h$. By restriction of ξ', the field ξ determines an α-vertical vector field ξ_0 along $U = \beta(U) \subset X$ and

$$(\xi_0)_y = \phi_{g^{-1}*}\xi_g \in T_y\alpha^{-1}(y)$$

for any $g \in U$ and $y = \beta(g)$. Conversely, if ξ_0 is an α-vertical vector field defined along the open set U of X then there exists a unique right-invariant vector field ξ defined on $\beta^{-1}(U)$ which induces ξ_0. We infer that the natural domains of definition of right-invariant vector fields on G are the β-saturated open sets of G. Denote by VG the subbundle of TG composed of α-vertical vectors (i.e., tangent to the α-fibres). The previous discussion shows that the presheaf of right-invariant vector fields defined on β-saturated open sets of G is canonically isomorphic to the presheaf of local sections of the vector bundle $VG|_X$, restriction of VG to the submanifold X. It is obvious that the bracket of two right-invariant vector fields is again right-invariant, hence, by transportation, a structure of R-Lie algebra on the presheaf $\Gamma_{loc}(VG|_X)$. Passing to germs, we obtain a sheaf of R-Lie algebras $\mathfrak{g}$ of base space X which, by definition, is called the Lie (pseudo-)algebra of G. For reasons which later will become apparent, we denote the bracket of $\mathfrak{g}$ by $[\![\ ,\ ]\!]$. This bracket satisfies the identity

$$(32.1)\qquad [\![f\mu, g\eta]\!] = fg[\![\mu, \eta]\!] + f[\mathcal{L}(\beta_*\mu)g]\eta - g[\mathcal{L}(\beta_*\eta)f]\mu$$

where $\mu, \eta \in \mathfrak{g}$, $f, g \in \mathcal{O}_X$ and $\beta_*\colon \mathfrak{g} \to \underline{TX}$ is the extension, to germs, of $\beta_*\colon VG|_X \to TX$. For any $x \in X$, let

$$G_{(x,x)} = \{g \in G \mid \alpha(g) = \beta(g) = x\}.$$

$G_{(x,x)}$ is a group called the isotropy group of G at the point x. The isotropy group has some differentiable properties though, in general, it does not possess any suitable Lie group structure. $G_{(x,x)}$ operates to the right on the fibre $\alpha^{-1}(x)$, the operation being induced by the product in G. This right action is of principal type with respect to the map $\beta\colon \alpha^{-1}(x) \to X$, i.e., the action is simply transitive on each β-fibre. For any $g \in G_{(x,x)}$, the right translation τ_g on $\alpha^{-1}(x)$ is a diffeomorphism that preserves (leaves invariant) the β-fibres. Observe that $\beta\colon \alpha^{-1}(x) \to X$ is not necessarily of locally constant rank and that $\beta(\alpha^{-1}(x))$ need not have any suitable structure of differentiable submanifold of X (in such a way, for example, that $\beta\colon \alpha^{-1}(x) \to \beta(\alpha^{-1}(x))$ be a fibration).

Many of the elementary properties of Lie groups transcribe for Lie groupoids. For example, any (local or global) differentiable morphism of Lie groupoids $\varphi\colon G \to H$, which restricted to the units is a diffeomorphism, induces an infinitesimal morphism $\varphi_*\colon \mathfrak{g} \to \mathfrak{h}$ of Lie algebras. Conversely, any infinitesimal morphism of Lie algebras $\lambda\colon \mathfrak{g} \to \mathfrak{h}$ generates a local Lie groupoid morphism $\varphi\colon U \to H$, where U is an open neighborhood of X in G, such that $\varphi_* = \lambda$. A proof of this statement, in a more restricted situation, can be found in [10(b)]. Globalizing such a local morphism φ and associating to a given (pseudo-)algebra $\mathfrak{g}$ on X a Lie groupoid G such that $\mathfrak{g}(G) = \mathfrak{g}$ are more difficult problems which were studied by Pradines [11]. We shall confine our attention to the more elementary aspects of the theory which are relevant to Chapter IV.

33. *The Exponential map*

Let G be a Lie groupoid and $\mathfrak{g}$ its Lie algebra. Since $\mathfrak{g}$ is the sheaf of germs of local sections of $VG|_X$ then $\Gamma(U, \mathfrak{g}) = \Gamma(U, VG|_X)$. A local section $\sigma \in \Gamma(U, G)$, with respect to α, is called admissible if $V = \beta \circ \sigma(U)$ is open in X and $\beta \circ \sigma : U \to V$ is a diffeomorphism. The groupoid structure of G extends to the presheaf $\Gamma_{a,loc}(X, G)$ of admissible local sections. In fact, if τ is defined on the open set V and σ on the open set U with $\beta \circ \sigma(U) \subset V$, we define $\tau \bullet \sigma$ by

$$(\tau \bullet \sigma)(x) = [\tau \circ \beta \circ \sigma(x)] \cdot \sigma(x)$$

where the right hand product is the composition in G.

Let $\Xi \in \Gamma(X, \mathfrak{g})$ be a section of $\mathfrak{g}$, ξ the corresponding right-invariant vector field on G and (φ_t), $\varphi_t = \exp t\xi$, the local 1-parameter group of transformations generated by ξ. If we set

$$\operatorname{Exp} t\Xi = \varphi_t \circ \iota : X \to G$$

with $\iota : X \to G$ the inclusion, then, for any $a \in X$, there is an open neighborhood U_a of a and $\varepsilon_a > 0$ such that $\operatorname{Exp} t\Xi : U_a \to G$ is defined on U_a for $|t| < \varepsilon_a$. Since ξ is tangent to the α-fibres of G, its integral curves are always contained in the α-fibres, hence (φ_t) preserves the α-fibration. It follows that $\operatorname{Exp} t\Xi : U_a \to G$ is an α-section. Moreover, since ξ is β-projectable, we infer that

$$\beta \circ \operatorname{Exp} t\Xi|_{U_a} = \exp t(\beta_* \xi)|_{U_a} ,$$

hence $\operatorname{Exp} t\Xi|_{U_a}$ is an admissible local section of G.

The family $(\operatorname{Exp} t\Xi)$ is a local differentiable 1-parameter subgroup of the groupoid $\Gamma_{a,loc}(X, G)$, generated by Ξ in the sense that

i) each local family $(\operatorname{Exp} t\Xi|_{U_a})$, $|t| < \varepsilon_a$, depends differentiably on the parameter t and two such local families agree on the overlap of their domains and for common values of t,

ii) $\frac{d}{dt} \operatorname{Exp} t\Xi|_{t=0} = \Xi$,

iii) (a) $\operatorname{Exp} 0\Xi = \iota$, the unit section,

(b) $\operatorname{Exp}(t+u)\Xi(x) = \operatorname{Exp} t\Xi(y) \cdot \operatorname{Exp} u\Xi(x)$, $y = \beta \circ \operatorname{Exp} u\Xi(x)$, whenever both members are defined.

We wish to find conditions on Ξ under which the Exponential map is global, i.e., for any $t \in R$, $\operatorname{Exp} t\Xi$ is an admissible section defined on X. Before stating the next lemma, we observe that any right-invariant vector field on a Lie groupoid G is always β-projectable. We say that a vector field θ on a manifold M is global if it generates a global 1-parameter group, i.e., $\exp t\theta$ is defined on M for any $t \in R$. θ is global if and only if its integral curves can be extended to infinity both ways.

LEMMA. *Let* G *be a Lie groupoid,* ξ *a right-invariant vector field on* G *and* $\theta = \beta_*\xi$ *the projected vector field on* X. ξ *is global if and only if* θ *is global.*

Proof. If ξ is global then so is θ since the integral curves of θ are the β-projections of the integral curves of ξ. For each fixed t, the transformation $\exp t\xi$ is β-projectable and its projection in X is equal to $\exp t\theta$. Assume conversely that θ is global. Since ξ is α-vertical, any integral curve of ξ is contained in an α-fibre. Moreover, since ξ is right-invariant, the integral curves of ξ are permuted by the right translations of G. We infer that, for any $h \in G$ with $\beta(h) = y \in X$ and any integral curve $\gamma(t)$ of ξ with initial data $\gamma(0) = y$, the right translated curve $\rho(t) = \gamma(t) \cdot h$ is an integral curve of ξ with initial data $\rho(0) = h$. To prove that ξ is global it is enough to show that, for any unit $x \in X$, there is a global integral curve $\gamma\colon R \to \alpha^{-1}(x)$ of ξ with initial data $\gamma(0) = x$. We distinguish two cases.

(a) $\theta(x) = 0$.

Observe that any integral curve $\gamma(t)$ of ξ with initial value $\gamma(0) \in G_{(x,x)}$ (the isotropy group at x) will be entirely contained in $G_{(x,x)}$ since $\rho(t) = \beta \circ \gamma(t)$ is an integral curve of θ, with $\rho(0) = x$, and $\theta(x) = 0$. Hence

$\rho(t) = x$ for all t. Let $\gamma\colon]-\varepsilon,\varepsilon[\to G_{(x,x)}$ be an integral curve of ξ with $\gamma(0) = e$, the unit element, and take any $h \in G_{(x,x)}$. Then $\rho(t) = \gamma(t)\cdot h$ is an integral curve of ξ, with $\rho(0) = h$, defined on the same interval $]-\varepsilon,\varepsilon[$. A standard continuation argument shows that γ can be extended to a global integral curve $\mathbf{R} \to G_{(x,x)}$ of ξ.

(b) $\theta(x) \neq 0$.

Let $\rho\colon \mathbf{R} \to X$ be the global integral curve of θ with initial data $\rho(0) = x$. $\theta(\rho(t)) = \frac{d\rho}{dt} \neq 0$ for any $t \in \mathbf{R}$, hence ρ is an immersion and the image of ρ is a one dimensional submanifold of X. Let $\gamma\colon]a,b[\to \alpha^{-1}(x)$ be the maximal integral curve of ξ with initial value $\gamma(0) = x$. We want to show that $]a,b[= \mathbf{R}$. Assume, for example, that $b < +\infty$. $\beta\circ\gamma$ is equal to the integral curve $\rho\colon]a,b[\to X$. Take a point $h \in \alpha^{-1}(x)$ such that $\beta(h) = \rho(b)$ and let $\delta_h\colon]b-\varepsilon, b+\varepsilon[\to \alpha^{-1}(x)$ be an integral curve of ξ with initial data $\delta_h(b) = h$. Since $\beta\circ\delta_h = \rho\colon]b-\varepsilon, b+\varepsilon[\to X$ and since $\theta(\rho(b)) \neq 0$, we can assume, by reducing ε if necessary, that $\rho\colon]b-\varepsilon, b+\varepsilon[\to X$ is injective. Moreover, since ξ is right-invariant, the right action of $G_{(x,x)}$ on $\alpha^{-1}(x)$ permutes the integral curves of ξ. Hence, for any $g \in G_{(x,x)}$, the curve

$$\delta_{hg}\colon]b-\varepsilon, b+\varepsilon[\to \alpha^{-1}(x), \quad \delta_{hg}(t) = \delta_h(t)\cdot g\,,$$

is the integral curve of ξ with initial data $\delta_{hg}(b) = hg$ and $\beta\circ\delta_{hg} = \rho\colon]b-\varepsilon, b+\varepsilon[\to X$. All the curves δ_{hg} are injective since their projections are injective. $M = \rho(]b-\varepsilon, b+\varepsilon[)$ is a one dimensional submanifold of X and each $M_g = \delta_{hg}(]b-\varepsilon, b+\varepsilon[)$ is a one dimensional submanifold of $\alpha^{-1}(x)$. If $g, g' \in G_{(x,x)}$ and $g \neq g'$ then $M_g \cap M_{g'} = \emptyset$. In fact, if $w = \delta_{hg}(t) = \delta_{hg'}(u)$, then $\beta(w) = \rho(t) = \rho(u)$ hence, by injectivity, $t = u$ and $\delta_{hg}(t) = \delta_{hg'}(t)$. The local uniqueness of integral curves implies that $\delta_{hg} = \delta_{hg'}$, hence $hg = \delta_{hg}(b) = \delta_{hg'}(b) = hg'$, i.e., $g = g'$. We infer that the subset

$$A = \{w \in \alpha^{-1}(x) \mid \beta(w) \in M\}$$

is the union of the disjoint submanifolds $M_g, g \in G_{(x,x)}$. If $b-\varepsilon < c < b$,

$w = \gamma(c) \in A$ and there exists a unique $g \in G_{(x,x)}$ such that $\gamma(c) = \delta_{hg}(d)$ with $b-\varepsilon < d < b+\varepsilon$. Moreover, $\rho(c) = \beta\circ\gamma(c) = \beta\circ\delta_{hg}(d) = \rho(d)$, hence $c = d$ and the two integral curves γ and δ_{hg} have the same value at the point c. It follows that $\gamma|]b-\varepsilon, b[= \delta_{hg}|]b-\varepsilon, b[$, hence δ_{hg} is a continuation of γ which therefore is not maximal.

Remark. The method of proof extends to the following more general situation. Let ξ and θ be as in the lemma. Then, for any $g \in G$, the maximal integral curve γ of ξ with initial data $\gamma(0) = g$ projects, by β, onto the maximal integral curve ρ of θ with initial data $\rho(0) = \beta(g)$.

The map $\beta: G \to X$ induces, by differentiation, a linear map $\beta_*: VG|_X \to TX$, hence an $\mathcal{O}$-linear sheaf map $\beta_*: \mathfrak{g} \to \underline{TX}$.

COROLLARY 1. *Let* G *be a Lie groupoid with Lie algebra* $\mathfrak{g}$ *and let* $\Xi \in \Gamma(X, \mathfrak{g})$. *Then* Ξ *is global (i.e.,* $\operatorname{Exp} t\Xi$ *is global) if and only if the same is true for the vector field* $\beta_*\circ\Xi$ *(compare with the lemma of Section* 19; *uniform is equivalent to global).*

Proof. If Ξ is global then so is $\beta_*\circ\Xi$ since $\exp t(\beta_*\circ\Xi) = \beta\circ\operatorname{Exp} t\Xi$. Conversely, if $\beta_*\circ\Xi$ is global, the right-invariant vector field ξ associated to Ξ is global, hence $\operatorname{Exp} t\Xi = (\exp t\xi)\circ\iota$ is global.

Let $\Gamma_c(X, \mathfrak{g})$ be the set of all sections Ξ of $\mathfrak{g}$ such that $\beta_*\circ\Xi$ has compact support. The Lie algebra structure of $\mathfrak{g}$ extends to $\Gamma_c(X, \mathfrak{g})$.

COROLLARY 2. *Any* $\Xi \in \Gamma_c(X, \mathfrak{g})$ *is global.*

Proof. $\theta = \beta_*\circ\Xi$ has compact support hence is global. Observe that the right-invariant vector field ξ associated to Ξ need not be compactly supported.

Let $\Gamma_a(X, G)$ be the set of admissible sections of G, i.e., sections σ, with respect to the fibration α, such that $\beta\circ\sigma$ is a diffeomorphism of X. The groupoid structure of G (or $\Gamma_{a,loc}(X, G)$) extends to a group structure on $\Gamma_a(X, G)$. We define the Exponential map

$$\text{(33.1)} \qquad \text{Exp}: \Gamma_c(X, \mathfrak{g}) \to \Gamma_a(X, G)$$

by $\text{Exp}\,\Xi = (\exp\xi)\circ\iota$ where ξ is the right-invariant vector field associated to Ξ and $\exp\xi = \exp 1\xi$. It is clear that $\text{Exp}(t\Xi) = \text{Exp}\,t(\Xi)$ since $\exp t\xi = \exp 1\eta$ with $\eta = t\xi$. This map is differentiable and satisfies the following properties:

1) For each $\Xi \in \Gamma_c(X, \mathfrak{g})$, the map $t \in \mathbf{R} \to \text{Exp}\,t\Xi \in \Gamma_a(X, G)$ is a differentiable 1-parameter subgroup of $\Gamma_a(X, G)$ generated by Ξ, i.e.,

$$\text{Exp}(t+u)\Xi = (\text{Exp}\,t\Xi)\bullet(\text{Exp}\,u\Xi),$$

$$\text{Exp}\,0 = \iota,$$

$$\text{Exp}(-t\Xi) = (\text{Exp}\,t\Xi)^{-1},$$

$$\frac{d}{dt}\,\text{Exp}\,t\Xi|_{t=0} = \Xi.$$

2) $\beta\circ\text{Exp}\,t\Xi = \exp t(\beta_*\circ\Xi)$.

3) Exp is determined uniquely by the property (1).

The differentiability means that Exp transforms differentiable families of sections of $\mathfrak{g}$ into differentiable families of sections of G. The uniqueness property is a consequence of local uniqueness for solutions of ordinary differential equations. Taking appropriate local coordinates, one can construct (locally) a system of ordinary differential equations whose solutions, with certain initial data, are (fragments of) the curves $\text{Exp}\,t\Xi(x)$. The above mentioned properties are also trivial consequences of the corresponding properties of $\exp t\xi$ where ξ is right-invariant on G.

If G is a Lie group, it is clear that Exp is the ordinary exponential map $|\mathfrak{g}| \to G$.

We end this section by examining some further analogies between Lie groups and Lie groupoids.

Let G be a Lie groupoid. A diffeomorphism $\phi: G \to G$ is called a right translation of G if:

(a) The map ϕ preserves the fibration $\alpha: G \to X$, hence induces a diffeomorphism $f: X \to X$ such that the diagram

$$\begin{array}{ccc} G & \xrightarrow{\ \phi\ } & G \\ {\scriptstyle\alpha}\downarrow & & \downarrow{\scriptstyle\alpha} \\ X & \xrightarrow{\ f\ } & X \end{array}$$

is commutative,

(b) For any $x \in X$, the restriction $\phi: G_x \to G_y$, $y = f(x)$, is a right translation, i.e., there exists an element $h(x) \in \alpha^{-1}(y)$ such that

$$\phi(g) = \phi_{h(x)}(g) = g \cdot h(x)$$

where $g \in G_x$. This implies that ϕ commutes with β.

Given a right translation $\phi: G \to G$ then, for any $x \in X$, the element $h(x)$ of condition (b) is unique, $h(x) = \phi(x)$ and $\beta[h(x)] = x$. The mapping $h: x \in X \mapsto h(x) \in G$ is a section of β which is α-admissible since $\alpha \circ h = f$ is a diffeomorphism of X. Equivalently, the map $y \in X \mapsto h \circ f^{-1}(y) \in G$ is a β-admissible section of α which, by inversion, yields the β-admissible section $\sigma: x \in X \mapsto h(x)^{-1} \in G$ of the projection α, where $h(x)^{-1}$ is the inverse of $h(x)$ in G. The right translation ϕ is given by

$$\phi(g) = g \cdot \sigma(x)^{-1}, \quad x = \alpha(g) . \tag{33.2}$$

Conversely, if $\sigma \in \Gamma_\alpha(X\ G)$, then (33.2) defines a right translation which induces σ. We infer that the group $\Gamma_\alpha(X, G)$ identifies canonically with the group of right translations of G and this identification is a group isomorphism.

Interchanging α with β, we can define left translations of G. These are β-preserving diffeomorphisms, i.e., they induce a commutative diagram

$$\begin{array}{ccc} G & \xrightarrow{\ \psi\ } & G \\ {\scriptstyle\beta}\downarrow & & \downarrow{\scriptstyle\beta} \\ X & \xrightarrow{\ f\ } & X \end{array}$$

and commute with α.

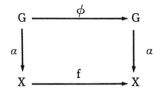

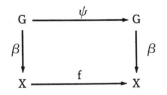

The group of left translations of G identifies canonically with the group $\Gamma_a(X, G)$ and the correspondence is given by

$$\psi(g) = \sigma(y) \cdot g, \quad y = \beta(g) . \tag{33.3}$$

We can also define local right translations whose domains are α-saturated open sets of G and local left translations whose domains are β-saturated open sets. The formula (33.2) (resp. (33.3)) establishes an isomorphism of the groupoid $\Gamma_{a,loc}(X, G)$ with the groupoid of local right (resp. left) translations of G.

If (σ_t), $|t-t_0| < \varepsilon$, is a local 1-parameter family of sections $\sigma_t \in \Gamma(U,G)$, with σ_{t_0} admissible, then $\phi_*\left[\frac{d\sigma_t}{dt}\Big|_{t=t_0}\right] \in \Gamma(f(U), \mathfrak{g})$ where $f = \beta \circ \sigma_{t_0}$ and ϕ is the right translation defined on $\alpha^{-1}(U)$ by $\sigma = \sigma_{t_0}$ (cf. (33.2)). We shall denote this section of $\mathfrak{g}$ by $\frac{d\sigma_t}{dt}\Big|_{t=t_0}$. In particular, the property (1) of the Exponential map implies that

$$\frac{d}{dt} \operatorname{Exp} t\Xi \Big|_{t=t_0} = \Xi .$$

Let $\Xi \in \Gamma(X, \mathfrak{g})$ be global and let ξ be the corresponding right-invariant vector field on G. Then $\operatorname{Exp} t\Xi = (\exp t\xi) \circ \iota$ and, since ξ is right-invariant, it follows that

$$\exp t\xi(g) = [\operatorname{Exp} t\Xi(y)] \cdot g$$

where $y = \beta(g)$. We infer that $(\exp t\xi)$ is the 1-parameter group of left translations of G associated to the 1-parameter group of sections $(\operatorname{Exp} t\Xi)$ in $\Gamma_a(X, G)$. Each of these left translations preserves the right-invariant vector field ξ. The uniqueness of the Exponential map is then a consequence of the uniqueness of the 1-parameter group associated to ξ.

If $[\![\Xi, \Sigma]\!] = 0$ then $[\xi, \eta] = 0$ and the 1-parameter groups of left translations commute. It follows that

$$\operatorname{Exp} t\Xi \bullet \operatorname{Exp} u\Sigma = \operatorname{Exp} u\Sigma \bullet \operatorname{Exp} t\Xi = \operatorname{Exp}(t\Xi + u\Sigma) .$$

More generally, one can also prove a *Campbell-Hausdorff formula* which gives a second order approximation of $\operatorname{Exp} t\Xi \bullet \operatorname{Exp} t\Sigma$ in terms of $\Xi + \Sigma$ and $[\![\Xi, \Sigma]\!]$.

We can also consider the 1-parameter group (ϕ_t) of right translations associated to $\operatorname{Exp} t\Xi$, namely,

$$\phi_t(g) = g \cdot [\operatorname{Exp} t\Xi(x)]^{-1}, \quad x = \alpha(g) .$$

The vector field η associated to (ϕ_t) is the left-invariant vector field of G obtained from ξ by the inversion $g \mapsto g^{-1}$ of G. If Ξ is not global, similar results hold locally.

It is not true, in general, that any admissible section $\sigma \in \Gamma_a(X, G)$, which is close to the unit section ι (in any reasonable topology), is of the form $\sigma = \operatorname{Exp} t\Xi$ with $\Xi \in \Gamma_c(X, \mathfrak{g})$. The result is not even true locally, i.e., in the neighborhood of a point $x \in X$. In fact, if $\sigma = \operatorname{Exp} t\Xi$, the diffeomorphism $\beta \circ \sigma$ is equal to $\exp t(\beta_* \circ \Xi)$. There are (even local) diffeomorphisms which are arbitrarily close to the identity map and, nevertheless, are not the flow of any vector field. One can however prove, quite easily, that the set

$$\{\operatorname{Exp} t\Xi(x) \mid \Xi \in \Gamma_c(X, \mathfrak{g}),\ x \in X\}$$

is an open neighborhood of X in G. Furthermore, taking a local basis of $\mathfrak{g}$ in the neighborhood of a point $x \in X$, one can define exponential coordinates in the neighborhood of x in $\alpha^{-1}(x)$.

Observe that if $\Xi \in \Gamma_c(X, \mathfrak{g}) = \Gamma_c(X, VG|_X)$ takes its values in the Lie algebras of the isotropy groups of G (i.e., in $T_{(x,x)}G_{(x,x)}$, assuming that each $G_{(x,x)}$ is a Lie group), then $\operatorname{Exp} t\Xi$ takes its values in the isotropy groups of G and

$$\operatorname{Exp} t\Xi(x) = \exp t[\Xi(x)]$$

where the right hand term is the ordinary exponential map $\mathfrak{g}_{(x,x)} \to G_{(x,x)}$, $\mathfrak{g}_{(x,x)}$ being the Lie algebra of $G_{(x,x)}$.

Let $\sigma \in \Gamma_a(X, G)$ be an admissible section of the Lie groupoid G, let ϕ be the right translation defined by σ (cf. (33.2)) and ψ the left translation defined by σ (cf. (33.3)). The composite

$$\psi \circ \phi = \phi \circ \psi : G \to G, \quad g \mapsto \sigma(y) \cdot g \cdot \sigma(x)^{-1}, \quad x = \alpha(g) \quad \text{and} \quad y = \beta(g),$$

is an automorphism of G which induces the identity on X. The maps

$$\sigma \in \Gamma_a(X, G) \mapsto \phi_\sigma \circ \psi_\sigma \in \text{Aut } G$$

and

$$\mathcal{A}d : \sigma \in \Gamma_a(X, G) \mapsto (\phi_\sigma \circ \psi_\sigma)_* \in \text{Aut } \mathfrak{g}$$

are group morphisms. If $\Xi, \Sigma \in \Gamma(X, \mathfrak{g})$, one also proves the formula formula

$$ad\Xi(\Sigma) \overset{\text{def.}}{=\!=} \frac{d}{dt} \mathcal{A}d(\text{Exp } t\Xi)\Sigma|_{t=0} = - [\Xi, \Sigma] .$$

We could also define a Lie algebra sheaf $\mathfrak{h}$ by taking the vector bundle HG composed of β-vertical vectors and the bracket of left-invariant vector fields on G. There would correspond an Exponential map $\Gamma_c(X, \mathfrak{h}) \to \Gamma_b(X, G)$ where $\Gamma_b(X, G)$ is the group of α-admissible sections of the fibration β. The inversion map $g \mapsto g^{-1}$ of G and its differential transport one Exponential map into the other.

34. *Prolongation of Lie groupoids*

Let G be a Lie groupoid and denote by J_kG the manifold of k-jets of local sections with respect to the fibration $\alpha: G \to X$. Then J_kG admits a natural structure of differentiable category: $j_k\tau(y)$ is composable with $j_k\sigma(x)$ if and only if $y = \beta \circ \sigma(x)$ and the composite is given by

$$[j_k\tau(y)] \cdot [j_k\sigma(x)] = j_k s(x)$$

where $s(z) = \tau(\beta \circ \sigma(z)) \cdot \sigma(z)$ and the right hand term is the composition in G. The units of $J_k G$ are the k-jets $j_k \iota(x)$, $x \in X$, of the unit section $\iota: X \to G$, hence can be identified with the elements of X. The groupoid G_k of invertible elements of $J_k G$ will be called, according to Ehresmann, the k-th prolongation of G. It is an open subset of $J_k G$ and, with the induced differentiable structure, becomes a Lie groupoid with X the submanifold of units. One checks easily that G_k is the set of jets $j_k \sigma(x)$ such that $j_1(\beta \circ \sigma)(x)$ is invertible, or equivalently, such that σ is an admissible local section in the neighborhood of x. It follows that the groupoid structure of G_k is equal to the extension, to k-jets, of the groupoid structure of $\Gamma_{a,loc}(X, G)$, or equivalently, the extension to k-jets, with source in X, of the groupoid structure of the set of local right (or left) translations of G. We denote by $\mathfrak{g}_k$ the Lie algebra of G_k. If $k = 0$ then $G_0 = G$. Let $\underline{G}$ be the sheaf of germs of local admissible sections of $\alpha: G \to X$. Then $j_k: \underline{G} \to \underline{G_k}$ is an injective groupoid morphism.

Let $\mathfrak{g}$ be the Lie algebra of G. $\mathfrak{g}$ is the sheaf of germs of local sections of the vector bundle $VG|_X$. Denote by

$$\underline{J_k(VG|_X)} \simeq \mathcal{J}_k \mathfrak{g}$$

the sheaf of germs of local sections of the bundle $J_k(VG|_X)$.

PROPOSITION. *There is a unique structure of Lie algebra on the sheaf* $\mathcal{J}_k \mathfrak{g}$ *which satisfies the relation*

$$[\![g j_k \mu, f j_k \eta]\!] = g f j_k [\![\mu, \eta]\!] + g[\mathcal{L}(\beta_* \circ \mu) f] j_k \eta - f[\mathcal{L}(\beta_* \circ \eta) g] j_k \mu \ ,$$

where $\mu, \eta \in \mathfrak{g}$, $g, f \in \mathcal{O}_X$; $\beta_* \circ \eta$, $\beta_* \circ \mu \in \underline{TX}$ *and* $[\![\mu, \eta]\!]$ *is the bracket in* $\mathfrak{g}$. *In particular,*

$$j_k: \mathfrak{g} \to \mathcal{J}_k \mathfrak{g}$$

is an injective Lie algebra morphism.

The uniqueness is obvious since any element of $\mathcal{J}_k\mathfrak{g}$ is a linear combination of (right) holonomic k-jets with coefficients in $\mathcal{O}_X$ (left structure). The existence can be proved by transporting the bracket of $\mathfrak{g}_k$ via the isomorphism of the next theorem. It can also be proved by extending a method used in a special case by Quê [10(a)]. This method consists in defining the bracket of two linear combinations by respecting additivity and the formula of the proposition. One checks, using the linear operator D, that the definition is independent from the choice of linear combinations.

THEOREM. *$\mathfrak{g}_k$ is canonically isomorphic to $\mathcal{J}_k\mathfrak{g}$.*

For a proof, we refer the reader to [7]. In the examples to follow, we shall carry out the proof in the special case where $G_k = \Pi_k X$.

If $h \leq k$, then $\rho_h\colon G_k \to G_h$ is a surjective morphism of Lie groupoids which is the identity on X and induces a surjective Lie algebra morphism $\rho_{h*}\colon \mathfrak{g}_k \to \mathfrak{g}_h$. The latter identifies with the standard projection $\rho_h\colon \mathcal{J}_k\mathfrak{g} \to \mathcal{J}_h\mathfrak{g}$. The morphisms ρ_h commute with the Exponential maps, i.e.,

$$\rho_h \circ \operatorname{Exp} t\Xi = \operatorname{Exp} t\rho_{h*}\Xi \tag{34.1}$$

where $\Xi \in \Gamma_c(X, \mathfrak{g}_k)$. The map $\rho_0 = \beta_k\colon G_k \to G$ verifies the same properties.

35. *Examples*

a) *Lie groups*

Let G be a Lie group. Then $\mathfrak{g}$ is the usual Lie algebra of G and, for any k, $G_k = G$. The unique isotropy group is $G_{(e,e)} = G$.

b) *The groupoids X^2 and $\Pi_k X$*

Let X be a manifold and consider the product manifold $Y \times X$ where Y is another copy of X (i.e., $Y = X$). Contrary to accepted norm, we shall think of X as the horizontal component and Y as the vertical component of the product. This could be avoided if, instead of the α-fibration, we gave preference to the β-fibration and, instead of right-invariant vector fields, we gave preference to left-invariant vector fields. The diagonal $\Delta \subset Y \times X$ identifies with X via the diagonal inclusion $x \mapsto (x, x)$. Define a groupoid structure on $Y \times X$ by the following rule: the element (y, x) is composable with (z, w) if and only if $x = z$ and, this being the case, the composite is defined by $(y, x)(x, w) = (y, w)$. It is trivial to check that $G = Y \times X$, together with this law of composition, is a Lie groupoid. The set of units of G is the submanifold Δ which identifies with X. The source map α is simply the second (horizontal) projection $(y, x) \mapsto x$ onto X and the target map β is the first (vertical) projection $(y, x) \mapsto y$ onto $Y = X$. This groupoid also satisfies the following transitivity condition: the map $\beta \times \alpha \colon G \to X \times X$ is a surjective map of maximal rank (surjective submersion). In fact $\beta \times \alpha = \mathrm{Id}$. If $x \in X$, the fibre $\alpha^{-1}(x)$ is equal to the vertical submanifold $\{(y, x) \mid y \in Y\} \simeq Y$. VG is equal to the sub-bundle of $T(Y \times X)$ composed of all vertical vectors, i.e., annihilated by π_{X*}. Hence $VG|_X = VG|_\Delta$ is canonically isomorphic, via π_{Y*}, to $TY = TX$. It follows that $\mathfrak{g} \simeq \underline{TX}$ as an $\mathcal{O}_X$-linear sheaf. Given any two points $x, z \in X$, there exists a unique right translation $G_x \to G_z$ of the α-fibres, namely the one produced by the element (x, z) and given by $(y, x) \mapsto (y, z)$. More generally, the right translations of G are the π_X-preserving diffeomorphisms of $Y \times X$ which commute with π_Y, i.e., are of the form $\mathrm{Id} \times f$. We infer that the right-invariant vector fields ξ on G are of the form $(\eta, 0)$ where η is a vector field on $Y = X$. Since $[(\eta, 0), (\mu, 0)] = ([\eta, \mu], 0)$, it follows that

$$\underline{\pi_{Y*}} \colon \mathfrak{g} \to \underline{TX}$$

is a Lie algebra isomorphism, hence the bracket $[\![\ ,\ ]\!]$ on $\mathfrak{g}$ is simply the usual bracket of germs of vector fields on X. From the nature of the right-invariant vector fields we also infer that

$$\operatorname{Exp} t\Xi \colon \mathbf{x} \mapsto (\exp t\eta(x), x)$$

where $\eta = \pi_{Y_*}\Xi$ and $\mathbf{x} \equiv (x, x) \in X$. A local section $\sigma : U \to G$, U open in X, with respect to α is a map of the form $\sigma\colon x \mapsto (f(x), x)$ where $f = \beta \circ \sigma\colon U \to X$ is differentiable, hence $\sigma = f \times \mathrm{Id}$. The section σ is admissible if and only if f is a diffeomorphism of U onto the open set $f(U)$. If $\tau = g \times \mathrm{Id}$ is composable with σ then $\tau \bullet \sigma = (g \circ f) \times \mathrm{Id}$. We infer that the map

$$j_k\sigma(x) \in G_k \mapsto j_k f(x) = j_k(\beta \circ \sigma)(x) \in \Pi_k X$$

is a Lie groupoid isomorphism which induces a Lie algebra isomorphism

$$\mathfrak{g}_k \to \mathfrak{h}_k$$

where $\mathfrak{h}_k$ is the Lie algebra of $\Pi_k X$, the bracket being induced by the bracket of right-invariant vector fields on $\Pi_k X$. We shall prove that $\mathfrak{h}_k$ is canonically isomorphic to

$$\underline{J_k T} \simeq \mathcal{J}_k T$$

where $T = TX$ and the Lie algebra structure is given by the bracket $[\![\ ,\ ]\!]$ defined in Section 10. Any local diffeomorphism φ of X, defined on the open set U, can be prolonged to a local diffeomorphism φ^k, defined on $\beta_k^{-1}(U) \subset \Pi_k(X)$, by setting

$$\varphi^k(A) = [j_k\varphi(y)] \cdot A, \quad y = \beta_k(A) .$$

If θ is a local vector field, defined on the open set U of X, the local 1-parameter group (φ_t) generated by θ prolongs to a local 1-parameter group (φ_t^k), defined on $\beta_k^{-1}(U)$, and determines the vector field

$$\theta^k = \frac{d}{dt}\varphi_t^k \Big|_{t=0}$$

which is called the k-th prolongation of θ. For any $A \in \beta_k^{-1}(U)$, the vector

$\theta^k_A \in T_A \Pi_k X$ depends only on $j_k\theta(y)$, $y = \beta_k(A)$. If $x = \alpha_k(A)$, then θ^k_A is tangent to $\alpha_k^{-1}(x)$ and the map

$$(35.1) \qquad j_k\theta(y) \in J_k T \mapsto \theta^k_A \in T_A \alpha_k^{-1}(x)$$

is a linear isomorphism. If $B \in \Pi_k X$, $\beta_k(B) = x$, and $C = A \cdot B$ then

$$\theta^k_C = \phi_{B*}\theta^k_A$$

where ϕ_B is the right translation produced by B. We infer that θ^k is a right-invariant local vector field on $\Pi_k X$, a fact that could also be derived by observing that (φ^k_t) is a local 1-parameter group of left translations. Any section $\Xi \in \Gamma(U, J_k T)$ determines, via the isomorphism (35.1), a right-invariant vector field ξ on the open set $\beta_k^{-1}(U)$. Conversely, any right-invariant vector field on $\beta_k^{-1}(U)$ defines a section of $J_k T$ over U. Passing to germs, we obtain the $\mathcal{O}_X$-linear sheaf isomorphism

$$(35.2) \qquad \underline{J_k T} \to \mathfrak{h}_k \ .$$

We claim that (35.2) is also a Lie algebra isomorphism. To prove this, it is sufficient to show (cf. (10.1)) that the Lie algebra structure of $\mathfrak{h}_k$, transported to $\underline{J_k T}$ via (35.2), satisfies the identity

$$(35.3) \qquad [\![f j_k \mu, g j_k \eta]\!] = fg j_k[\mu, \eta] + f[\mathfrak{L}(\mu)g] j_k \eta - g[\mathfrak{L}(\eta)f] j_k \mu$$

where $f, g \in \mathcal{O}_X$ and $\mu, \eta \in \underline{TX}$. In fact, using 1-parameter families instead of 1-parameter groups to define the prolongation of vector fields, it is trivial to check that the prolongation process $\theta \mapsto \theta^k$ is R-linear and preserves the bracket. Moreover, (35.2) is $\mathcal{O}_X$-linear and

$$\mathfrak{L}(\xi)(f \circ \beta_k) = [\mathfrak{L}(\theta)f] \circ \beta_k$$

where f is any local function on X, ξ a right-invariant vector field on $\Pi_k X$ determined by a section Ξ of $J_k T$ and $\theta = \beta_{k*}\xi = \beta_k \circ \Xi$. The relation (35.3) is now evident.

In Section 19 we defined the groupoid $\Gamma_k X$ as the quotient of $\widetilde{\mathfrak{Aut}}\, X^2$ modulo the relation $\sim_k$, or equivalently, as the sheaf of germs of β_k-admissible local sections of $\Pi_k X$ with respect to the fibration α_k. We also defined the group $\Gamma_a(X, \Gamma_k X)$ of admissible global sections which is canonically isomorphic to $\Gamma_a(X, \Pi_k X)$. The uniqueness property of the Exponential maps defined in Sections 19 and 33 implies that these Exponential maps are transported one into the other via the isomorphisms

$$\varepsilon_k : \tilde{\mathcal{J}}_k T \to \mathcal{J}_k T \simeq \underline{J_k T}\ ,$$

or better, the extension

$$\Gamma_c(X, \tilde{\mathcal{J}}_k T) \to \Gamma_c(X, \underline{J_k T})$$

and

$$\Gamma_a(X, \Gamma_k X) \to \Gamma_a(X, \Pi_k X)\ .$$

The ℓ-th prolongation of $\Pi_k X$ is not equal to $\Pi_{k+\ell} X$. In fact, there is a canonical Lie groupoid inclusion $\Pi_{k+\ell} X \to (\Pi_k X)_\ell$ whose image is a regularly imbedded submanifold. Let $\underline{(\Pi_k X)_1}$ be the sheaf of germs of local sections of $\alpha: (\Pi_k X)_1 \to X$. The non-linear operators $\mathfrak{D}$, $\tilde{\mathfrak{D}}$ and $\hat{\mathfrak{D}}$ defined in Chapters IV and V are of order 1. They are the composite of

$$j_1 : \Gamma_{k+1} X \to \underline{(\Pi_{k+1} X)_1}$$

with (the extension to germs of sections of) certain bundle maps whose domains are the total space of the locally trivial bundle $\alpha: (\Pi_{k+1} X)_1 \to X$. The explicit description of these bundle maps is left to the reader.

In [10 (b)], Ngô Van Quê defines a non-linear complex for Lie groupoids in the following way. Let G be a Lie groupoid with units manifold X, $\underline{G}$ the sheaf of germs of admissible α-sections, G_k the k-th prolongation of G and $N(G_k) = N_k$ the bundle of groups defined by the exact sequence

$$1 \longrightarrow N(G_k) \longrightarrow [G_k]_1 \xrightarrow{\beta_1} G_k \longrightarrow 1\ . \tag{35.4}$$

Since $\mathfrak{g}(G_k) \simeq \mathcal{J}_k\mathfrak{g}$ and $\mathfrak{g}([G_k]_1) \simeq \mathcal{J}_1(\mathcal{J}_k\mathfrak{g}) \simeq \mathcal{J}_1 \otimes_{\mathcal{O}} \mathcal{J}_k\mathfrak{g}$, the above sequence yields, on the infinitesimal level, the exact sequence of Lie algebra sheaves

$$(35.5) \qquad 0 \longrightarrow \mathfrak{g}(N_k) \longrightarrow \mathcal{J}_1(\mathcal{J}_k\mathfrak{g}) \xrightarrow{\beta_1} \mathcal{J}_k\mathfrak{g} \longrightarrow 0 .$$

From the exactness of

$$(35.6) \qquad 0 \longrightarrow \underline{T}^* \otimes_{\mathcal{O}} \mathcal{J}_k\mathfrak{g} \xrightarrow{\zeta_1 = -\varepsilon_1} \mathcal{J}_1 \otimes_{\mathcal{O}} \mathcal{J}_k\mathfrak{g} \xrightarrow{\beta_1} \mathcal{J}_k\mathfrak{g} \longrightarrow 0$$

where $T^* = T^*X$, we infer that the Lie algebra sheaf $\mathfrak{g}(N_k)$ identifies with the Lie subalgebra $\underline{T}^* \otimes_{\mathcal{O}} \mathcal{J}_k\mathfrak{g}$ of $\mathfrak{g}([G_k]_1)$.

The following non-linear complex

$$(35.7) \qquad 1 \longrightarrow \underline{G} \xrightarrow{j_{k+1}} \underline{G_{k+1}} \xrightarrow{\mathcal{D}} \underline{N(G_k)}$$

is exact, where

$$\mathcal{D}(\sigma) = \sigma^{-1} \bullet j_1(\rho_k\sigma) ,$$

G_{k+1} is imbedded in $[G_k]_1$ via the natural inclusion $j_{k+1}\tau(x) \mapsto j_1(j_k\tau)(x)$ and the right hand composition is taken with respect to the groupoid structure of $[G_k]_1$ (or better, the extension to germs of admissible sections of the groupoid structure of $[G_k]_1$).

The complex (35.7) is a finite form of

$$0 \longrightarrow \mathfrak{g} \xrightarrow{j_{k+1}} \mathcal{J}_{k+1}\mathfrak{g} \xrightarrow{D} \underline{T}^* \otimes_{\mathcal{O}} \mathcal{J}_k\mathfrak{g}$$

where D is the linear operator defined in Section 6.

Consider, in particular, the trivial groupoid $G = X \times X$ where X is a manifold. Then (35.7) reduces to the exact complex

$$(35.8) \qquad 1 \longrightarrow \mathcal{A}ut\ X \xrightarrow{j_{k+1}} \underline{\Pi_{k+1}X} \xrightarrow{\mathcal{D}} \underline{N(\Pi_k X)} .$$

Moreover, since in this special instance $\mathfrak{g} \simeq \underline{T}$, we infer that $\mathfrak{g}(N_k)$ identifies with the Lie algebra subsheaf $\underline{T}^* \otimes_{\mathcal{O}} \mathcal{J}_k T$ of $\mathcal{J}_1 \otimes_{\mathcal{O}} \mathcal{J}_k T$. A simple computation shows that the following diagram is commutative:

$$\begin{array}{ccccccccc}
1 & \longrightarrow & \mathcal{A}ut\, X & \xrightarrow{\tilde{j}_{k+1}} & \Gamma_{k+1}X & \xrightarrow{\mathfrak{D}} & \underline{T}^* \otimes_{\mathcal{O}} \mathcal{J}_k T \\
 & & \| & & \downarrow & & \downarrow \text{Exp} \\
1 & \longrightarrow & \mathcal{A}ut\, X & \xrightarrow{j_{k+1}} & \underline{\Pi_{k+1}X} & \xrightarrow{\mathfrak{D}} & \underline{N(\Pi_k X)}
\end{array}$$

where the first line is the initial portion of the first non-linear complex $(21.8)_{k+1}$.

It is an interesting problem to try to extend the complex (35.7) by an operator $\mathfrak{D}_1$. A more interesting problem is to try to define, for arbitrary Lie groupoids, a complex analogous to the second non-linear complex $(23.9)_{k+1}$.

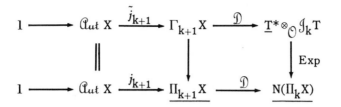

REFERENCES

[1] BKOUCHE R., *Modules involutifs,* Séminaire sur les structures infinitésimales, Faculté des Sciences de Brest (1967).

[2] EHRESMANN C., (a) *Introduction à la théorie des structures infinitésimales et des pseudo-groupes de Lie,* Coll. Géom. Différ. Strasbourg, C.N.R.S. (1953), 97-110.
(b) *Catégories topologiques et catégories différentiables,* Coll. de Géom. Différ. Globale, Bruxelles (1959), 137-150.
(c) *Esquisses d'un folklore de géométrie différentielle,* Secrétariat Mathématique, Paris (1967).

[3] FRÖLICHER A. and NIJENHUIS A., *Theory of vector-valued differential forms, Part I,* Kon. Nederl. Akad. Wetensch. Proc. A.59 = Indag. Math. 18 (1956), 338-359.

[4] FRÖLICHER A., KOBAYASHI E. and NIJENHUIS A., *Deformation theory of complex manifolds,* Technical Report no. 10, Univ. of Washington, Seattle (1959).

[5] GOLDSCHMIDT H., (a) *Existence theorems for analytic linear partial differential equations,* Annals of Math. 86 (1967), 246-270.
(b) *Integrability criteria for systems of non-linear partial differential equations,* J. Differ. Geom. 1 (1967), 269-307.
(c) *Sur la structure des équations de Lie: le troisième théorème fondamental,* J. Differ. Geom. (to appear).
(d) *Prolongements d'équations différentielles linéaires*: III, *La suite exact de cohomologie de Spencer,* Univ. de Grenoble, Inst. de Math. Pures, November 1971 (to be published).

[6] GUILLEMIN V. and STERNBERG S., *Deformation theory of pseudo-group structures*, Mem. A.M.S. 64 (1966).

[7] KUMPERA A., *Invariants différentiels* (to appear).

[8] LEHMANN D., (a) *Introduction aux jets et aux opérateurs différentiels*, Séminaire sur les Structures Infinitésimales, Faculté des Sciences de Brest (1967).
(b) *Le théorème de Quillen sur la complète intégrabilité formelle pour les systèmes différentiels éventuellement singuliers*, Séminaire sur les Structures Infinitésimales, Faculté des Sciences de Brest (1967).

[9] MALGRANGE B., (a) *Cohomologie de Spencer (d'après Quillen)*, Secrétariat Mathématique d'Orsay (1965).
(b) *Théorie analytique des équations différentielles*, Séminaire Bourbaki, exposé 329 (1966/1967).
(c) *Pseudo-groupes de Lie elliptiques*, Séminaire sur les équations aux dérivées partielles, Collège de France (1969-70).
(d) *Equations de Lie*, J. Differ. Geom. (to appear).

[10] NGÔ VAN QUÊ, (a) *Du prolongement des espaces fibrés et des structures infinitésimales*, Ann. Inst. Fourier, 17 (1967), 157-223.
(b) *Non-abelian Spencer cohomology and deformation theory*, J. Differ. Geom. 3 (1969), 165-211.

[11] PRADINES, J., (a) *Théorie de Lie pour les groupoïdes différentiables. Relations entre propriétés locales et globales.* C. R. Acad. Sc. Paris, 263 (1966), A, 907-910.
(b) *Théorie de Lie pour les groupoïdes différentiables. Calcul différentiel dans la catégorie des groupoïdes infinitésimaux.* C. R. Acad. Sc. Paris 264 (1967) A, 245-248.
(c) *Troisième théorème de Lie pour les groupoïdes différentiables.* C. R. Acad. Sc. Paris 267 (1968) A, 21-23.

[12] QUILLEN D. G., *Formal properties of over-determined systems of linear partial differential equations,* Thesis, Harvard (1964).

[13] SPENCER D. C., (a) *Deformation of structures on manifolds defined by transitive, continuous pseudogroups, Parts I and II,* Annals of Math. 76 (1962), 306-445.
(b) *Deformation of structures on manifolds defined by transitive, continuous pseudogroups, Part III,* Annals of Math. 81 (1965), 389-450.
(c) *On deformation of pseudogroup structures,* Global Analysis: Papers in Honor of K. Kodaira, Univ. of Tokio Press and Princeton Univ. Press (1969), 367-395.
(d) *Overdetermined systems of linear partial differential equations,* Bull. A.M.S. 75 (1969), 179-239.

[14] SWEENEY W.J., *The δ-Poincaré estimate,* Pacific J. Math. 20 (1967), 559-570.

ADDITIONAL REFERENCES TO THE INTRODUCTION

[15] ARENS R., *Normal form for a Pfaffian,* Pacific J. Math. 14 (1964), 1-8.

[16] BA B., *Structures presque complexes, structures conformes et dérivations,* Cahiers Topol. et Géom. Differ. VIII (1966).

[17] de BARROS C. M., (a) *Espaces infinitésimaux,* Cahiers Topol. et Géom. Differ. VII (1965).
(b) *Variétés presque hor-complexes,* C. R. Acad. Sc. Paris 260 (1965), 1543-1546.
(c) *Sur la géométrie différentielle des formes différentielles extérieures quadratiques,* Atti Conv. Int. Geom. Differ. Bologna (1967).

[18] BERNARD D., *Sur la géométrie différentielle des G-structures,* Ann. Inst. Fourier 10 (1960), 151-270.

[19] BERS L., *Introduction to Riemann surfaces*, lecture notes, New York Univ. (1951-52).

[20] CHERN S. S., (a) *An elementary proof of the existence of isothermal parameters on a surface*, Proc. Amer. Math. Soc. 6 (1955), 771-782.
(b) *Pseudo-groupes continus infinis*, Coll. Inter. Géom. Differ. Strasbourg (1953), 119-136.

[21] COURANT R. and HILBERT D., *Methods of Mathematical Physics*, Vol. 2, Eng. Transl. Interscience (1953).

[22] ECKMANN B., *Sur les structures complexes et presque complexes*, Coll. Inter. Géom. Differ. Strasbourg (1953), 151-159.

[23] ECKMANN B. and FRÖLICHER A., *Sur l'intégrabilité des structures presque complexes*, C. R. Acad. Sc. Paris 232 (1951), 2284-2286.

[24] EHRESMANN C., (a) *Sur les variétés presque complexes*, Proc. Int. Cong. Math. 2 (1950), 412-419.
(b) *Sur les structures infinitésimales régulières*, Proc. Int. Cong. Math. 1 (1954), 479.
(c) *Sur les pseudogroupes de Lie de type fini*, C. R. Acad. Sc. Paris 246 (1958), 360-362.

[25] ELIOPOULOS H.A., *Structures presque tangentes sur les variétés différentiables*, C. R. Acad. Sc. Paris 255 (1962), 1563-1565.

[26] FRÖLICHER A., *Zur Differentialgeometrie der komplexen Strukturen*, Math. Ann. 129 (1955), 50-95.

[27] GODBILLON C., *Géométrie différentielle et mécanique analytique*, Hermann, Paris (1969).

[28] GOURSAT E., *Leçons sur le problème de Pfaff*, Hermann, Paris (1922).

[29] GRAY J.W., *Some global properties of contact structures*, Annals of Math. 69 (1959), 421-450.

[30] GRIFFITHS P.A., *On the theory of variation of structures defined by transitive, continuous pseudogroups*, Osaka Journ. of Math. 1 (1964), 175-199.

[31] GUGENHEIM V. K. A. M. and SPENCER D. C., *Chain homotopy and the de Rham theory*, Proc. Amer. Math. Soc. 7 (1956), 144-152.

[32] GUILLEMIN V., *The integrability problem for* G-*structures*, Trans. Amer. Math. Soc. 116 (1965), 544-560.

[33] GUILLEMIN V. and STERNBERG S., *Subelliptic estimates for complexes*, Proc. Nat. Acad. Sci. 67 (1970), 271-274.

[34] HAANTJES J., *On the* X_m*-forming sets of eigenvectors*, Proc. Kon. Ned. Ak. Wet. Amsterdam, A58 (1955), 158-162.

[35] HÖRMANDER L., (a) *The Frobenius-Nirenberg Theorem*, Arkiv för Matematik, 5 (1964), 425-432.
(b) *An introduction to complex analysis in several variables*, Van Nostrand (1966).
(c) *Fourier integral operators*. I, Acta Math. 127 (1971), 79-183.

[36] KOBAYASHI E.T., *A remark on the Nijenhuis tensor*, Pacific J. of Math. 12 (1962), 963-977.

[37] KODAIRA K., *On deformations of some complex pseudo-group structures*, Annals of Math. 71 (1960), 224-302.

[38] KODAIRA K., NIRENBERG L. and SPENCER D. C., *On the existence of deformations of complex analytic structures*, Annals of Math. 68 (1958), 450-459.

[39] KODAIRA K. and SPENCER D. C., (a) *On deformations of complex analytic structures, I-II,* Annals of Math. 67 (1958), 328-466.
(b) *On deformations of complex analytic structures, III. Stability theorems for complex structures,* Annals of Math. 71 (1960), 43-76.
(c) *Multifoliate structures,* Annals of Math. 74 (1961), 52-100.

[40] KOHN J. J., *Harmonic integrals on strongly pseudo-convex manifolds I,* Annals of Math. 78 (1963), 112-148.

[41] KORN A., *Zwei Anwendungen der Methode der sukzessiven Annäherungen,* Schwarz Festschrift (1914), 215-229.

[42] KUMPERA A., *Equivalence locale des structures de contact de codimension un,* Can. J. Math. 22 (1970), 1123-1128.

[43] KUMPERA A. and SPENCER D. C., *Systèmes d'équations aux dérivées partielles linéaires et déformation des structures de pseudogroupes,* Presses Univ. de Montréal (to appear).

[44] KURANISHI M., *New proof for the existence of locally complete families of complex structures,* Proc. Conference on Complex Analysis, Minneapolis (1964), Springer-Verlag (1965), 142-154.

[45] LEHMANN-LEJEUNE J., (a) *Etude des formes différentielles liées à certaines G-structures,* C. R. Acad. Sc. Paris 260 (1965), 1838-1841.
(b) *Intégrabilité des G-structures définies par une 1-forme 0-déformable à valeurs dans le fibré tangent,* Ann. Inst. Fourier 16 (1966), 329-387.

[46] LIBERMANN P., (a) *Problèmes d'équivalence relatifs à une structure presque complexe sur une variété à quatre dimensions,* Acad. Roy. Belgique, Bull. Cl. Sci. 36 (1950), 742-755.
(b) *Sur les structures presque complexes et autres structures infinitésimales régulieres,* Bull. Soc. Math. Fr. 83 (1955), 195-224.

(c) *Pseudogroupes infinitésimaux attachés aux pseudogroupes de Lie*, Bull. Soc. Math. Fr. 87 (1959), 409-425.

(d) *Connexions d'ordre supérieur et tenseurs de structure*, Atti Conv. Int. Geom. Differ. Bologna (1967).

(e) *Sur les prolongements des fibrés principaux et des groupoïdes différentiables banachiques*, Analyse Globale, Presses Univ. de Montréal (1971).

[47] LICHNEROWICZ A., *Théorèmes de réductivité sur des algèbres d'automorphismes*, Rendiconti di Mat. (1-2) 22 (1963), 197-244.

[48] LICHTENSTEIN L., *Zur Theorie der konformen Abbildung. Konforme Abbildung nichtanalytischer, singularitätenfreier Flächenstücke auf ebene Gebiete*, Bull. Int. Acad. Sci. Cracovie, A (1916), 192-217.

[49] MALGRANGE B., (a) *Théorème de Frobenius complexe*, Séminaire Bourbaki, exposé 166 (1957/1958).

(b) *Sur l'intégrabilité des structures presque complexes*, Roma, Istit. Naz. di Alta Mat., Symposia Mat. 2 (1968), 289-296.

[50] MATSUSHIMA Y., *Pseudogroupes de Lie transitifs*, Séminaire Bourbaki, exposé 118 (1954/1955).

[51] NEWLANDER A. and NIRENBERG L., *Complex analytic coordinates in almost-complex manifolds*, Annals of Math. 65 (1957), 391-404.

[52] NICKERSON H.K. and SPENCER D. C., *Differentiable manifolds and sheaves*, mimeographed notes, Princeton Univ. (1955).

[53] NIJENHUIS A., *Graded Lie algebras and their applications*, Math. Inst., Amsterdam (1964).

[54] NIJENHUIS A. and WOOLF W. B., *Some integration problems in almost-complex and complex manifolds*, Ann. of Math. 77 (1963), 424-489.

[55] NIRENBERG L., (a) *A complex Frobenius theorem*, Seminar on analytic functions, Inst. for Adv. Study, 1 (1957), 172-179.
(b) *A complex Frobenius theorem*, mimeographed notes, New York Univ., Inst. of Mathem. Sc. (1958).
(c) *Partial differential equations with applications in geometry*, Lectures on modern mathematics, Vol. II, John Wiley (1964), 1-41.

[56] SAMELSON H., *Differential geometry*, lecture notes, Univ. of Michigan (1955).

[57] SCHOUTEN J. A., *Ricci Calculus*, Springer (1954).

[58] SINGER I. M. and STERNBERG S., *The infinite groups of Lie and Cartan, Part I (the transitive groups)*, Jour. d'Anal. Math. XV (1965).

[59] SPENCER D. C., (a) *Potential theory and almost-complex manifolds*, Conf. on Complex Variables, Univ. of Michigan (1953).
(b) *Some remarks on perturbation of structures*, Analytic functions, Princeton Univ. Press (1960), 67-87.
(c) *Some remarks on homological analysis and structures*, Proc. Symposia in Pure Math. 3 (1961), 56-86.
(d) *Overdetermined operators: some remarks on symbols*, Actes Congrès Int. des Mathématiciens, Nice (1970), Tome 2,

[60] STERNBERG S., *Lectures on Differential Geometry*, Prentice Hall (1964).

[61] SWEENEY W. J., *Coerciveness in the Neumann problem*, J. Differ. Geom. (to appear).

[62] YUEN PING CHENG, *Sur les prolongements de G-structures*, Thèse, Paris (1970).

INDEX

An asterisk attached to a page number indicates a reference by number only to the bibliography on pages 278-285.